Principles of
Forest Entomology

McGraw-Hill Series in Forest Resources

Avery: Natural Resource Measurements
Boyce: Forest Pathology
Brockman and Merriam: Recreational Use of Wild Lands
Brown and Davis: Forest Fire: Control and Use
Chapman and Meyer: Forest Mensuration
Dana and Fairfax: Forest and Range Policy
Daniel, Helms, and Baker: Principles of Silviculture
Davis: Forest Management
Davis: Land Use
Duerr: Fundamentals of Forestry Economics
Guise: The Management of Farm Woodlands
Harlow, Harrar, and White: Textbook of Dendrology
Heady: Rangeland Management
Knight and Heikkenen: Principles of Forest Entomology
Panshin and De Zeeuw: Textbook of Wood Technology
Panshin, Harrar, Bethel, and Baker: Forest Products
Rich: Marketing of Forest Products: Text and Cases
Sharpe, Hendee, and Allen: An Introduction to Forestry
Shirley: Forestry and Its Career Opportunities
Stoddart, Smith, and Box: Range Management
Trippensee: Wildlife Management
   Volume I—Upland Game and General Principles
   Volume II—Fur Bearers, Waterfowl, and Fish
Wackerman, Hagenstein, and Michell: Harvesting Timber Crops
Worrell: Principles of Forest Policy

Walter Mulford was Consulting Editor of this series from its inception in 1931 until January 1, 1952.
Henry J. Vaux was Consulting Editor of this series from January 1, 1952, until July 1, 1976.

Principles of Forest Entomology

Fifth Edition

Fred Barrows Knight
Dwight B. Demeritt Professor of Forest Resources
School of Forest Resources
University of Maine at Orono

Herman John Heikkenen
Associate Professor of Forest Entomology
Virginia Polytechnic Institute and State University

McGraw-Hill Book Company

New York St. Louis San Francisco Auckland Bogotá Hamburg
Johannesburg London Madrid Mexico Montreal New Delhi
Panama Paris São Paulo Singapore Sydney Tokyo Toronto

PRINCIPLES OF FOREST ENTOMOLOGY

4 5 6 7 8 9 0 DODO 8 9 8 7 6 5 4 3 2

This book was set in Times Roman by Cobb/Dunlop Publisher Services Incorporated.
The editor was Marian D. Provenzano; the production supervisor was Donna Piligra.
R. R. Donnelley & Sons Company was printer and binder.

Library of Congress Cataloging in Publication Data

Knight, Fred Barrows, date
    Principles of forest entomology.

    (McGraw-Hill series in forest resources)
    Fourth ed. (1965) by S. A. Graham and F. B. Knight.
    Includes index.
    1. Forest insects. 2. Forest insects—Control.
I. Heikkenen, Herman John, joint author. II. Graham,
Samuel Alexander, date   Principles of forest entom-
ology. III. Title.
SB761.G75 1980   634.9'6'7   79-15390
ISBN 0-07-035095-7

Professor Samuel A. Graham at work
in the forest. It was a delight to talk
to Dr. Graham in this environment,
where he could show clearly the real
interrelationships among the insects, the plants, and the other animals. He was an exceptional forest
ecologist.

We dedicate this fifth edition of *Principles of Forest Entomology* to
Samuel Alexander Graham, a friend and adviser to us and to many
others. Professor Graham was born in Salisbury, Maryland, on March
22, 1891, and died in Ann Arbor, Michigan, on September 22, 1967.
He was one of the leading figures in forest entomology and the original
author of this text. He was a forestry graduate (1914) of the University
of Minnesota and completed graduate work in forest entomology (M.S.,
Cornell 1916; Ph.D., Minnesota 1921). He retired from the University
of Michigan in 1961.

Samuel Graham knew the woods as few persons do. He was a fine
teacher to be with in the forests; the deepest memories of his many
students were the personal, friendly conversations in that environment.
He was nationally recognized as a teacher, a scholar, and a scientist.
He was a fellow of the Society of American Foresters and of the
American Association for the Advancement of Science. In 1953, the
University of Minnesota honored him with its Outstanding Achieve-
ment Award, and in 1964, the Society of American Foresters presented
him its Award for Achievement in Science (the Barrington Moore
Award).

Dr. Graham was a true friend, a highly respected counselor, and
a fine family man. We are proud to honor him with this edition of his
text.

Contents

Preface

Principles of Forest Entomology was first published by Professor Samuel A. Graham in 1929; since then, there have been three revisions. This fifth edition begins a new phase. Professor Graham is gone, but the authors both knew him well and have retained much of his philosophy and organization.

By design, this is an introductory text on forest entomology for undergraduate *forestry* students, rather than a work on forest insects for the benefit of entomologists. As in previous editions, ecological matters are emphasized, and we have attempted to show the place of forest entomology in the field of forest management and protection. This is not intended as a manual or an advanced text, and the serious students of forest entomology are advised to obtain additional writings to advance their knowledge.

The authors feel that the length of the text must be limited so that the material can be covered in one term. At the same time, it should serve forest entomology courses of varying design. The text provides a good base for the typical laboratory course in forest entomology, in which insect morphology, anatomy, and identification are taught in the laboratory. Most instructors use a laboratory manual or individual exercises for their laboratory sessions in such courses. It is also suitable for a forest protection course, in which a minimum amount of time is devoted to insect identification, but in which insect control and examples of serious pests are discussed in more detail. Selected species are used to illustrate principles or as examples of the various ecological groups of forest insects. The instructor may wish to use other species that are more appropriate to a local situation.

The Bibliography at the end of each chapter has been selected with the students and the space limitations in mind. Whenever possible, except for a few historic articles and some special reports, we have selected references that adequately cover the subject discussed and are most likely to be available in libraries. These provide an entrée to the literature available on each topic.

We have made some changes in the book that allow for greater flexibility in its use without adding greatly to its length. There are two completely new chapters. The first, Chapter 3, covers insect development and classification. This is a response to expressed need for coverage of this material and should reduce the dependency on other sources. The second, Chapter 20, discusses insects and other forest anthropods of medical importance. It has been added because of the multiple-use aspects of forest management and the many problems associated with nuisance insects in forested recreation areas.

We have eliminated some of the stress on insect control and have reorganized the material in a pest management context. The two chapters on biological control and the two on chemicals have each been combined into one (Chapters 9 and 11, respectively), and the stress on specific chemical materials has largely been eliminated. However, because changes take place rapidly and current registrations of pesticides are short-lived, it is the responsibility of instructors to provide current information. The chapters have been updated, and discussions on pest management and regulation have been added. At the end of the book, we have added a list of scientific and common names of the insects discussed (Appendix A) and a glossary of terms used in the text. The organization of *Principles of Forest Entomology* is such that current developments are readily fitted into the course.

During the preparation of this fifth edition, many foresters and entomologists have rendered valuable advice and assistance. We are acknowledging some of them here, but many more provided help as we progressed. The prereviewers provided McGraw-Hill with information on the usefulness of the text and some of the needed changes and advised that the book should be revised. These people were: C. W. Berisford, T. C. Eiber, R. C. Fox, H. M. Kulman, and J. B. Simeone. McGraw-Hill also used reviewers as the work progressed. We are indebted to Dr. E. A. Cameron, Dr. C. W. Berisford, and Dr. Norman Sloan for the work they did on the text. We utilized their suggestions and appreciate their attention to detail.

We have also obtained reviews on many chapters on an individual basis, so that many chapters have been reviewed by at least one or more persons in addition to the McGraw-Hill reviewers. For example, five persons reviewed the biological control chapter; others provided photographs and figures; and Dr. M. M. Neilson wrote a section on Canadian laboratories. We cannot name every person and the organization he or she represents, but the names of many are listed with our thanks: D. C. Allen; G. D. Amman; R. H. Beal; J. A. Copony; H. C. Coppel; J. B. Dimond; A. T. Drooz; G. R. Esenther; K. E. Gibbs; A. F. Hedlin; T. Ide; D. T. Jennings; L. T. Kok; M. Kosztarab; J. K. Mauldin; E. P. Merkel; W. E. Miller; M. M. Neilson; E. A. Osgood; C. L. Park; R. M. Pelletier; R. C. Reardon; B. R. Richards; R. M. Romancier; F. Schwerdtfeger; T. P. Sibury; G. A. Simmons; R. V. Smythe; R. L. Talerico; H. A. Thomas; C. T. Turner; F. G. Wagner, Jr.; W. E. Wallner; C. B. Williams, Jr.; L. H. Williams: J. A. Witter; K. H. Wright; and H. O. Yates III.

Finally, we thank three very dedicated and supportive women: our wives, Jane Wooster Knight and Gail Foster Heikkenen, for their helpfulness, sympathy, and patience during the preparation of the manuscript, and Mrs. Sybil Fleming Graham, who encouraged us to prepare this fifth edition. We especially wrote Chapter 20 for Mrs. Graham because of *On Your Own,* the small book written by her husband many years ago; when reviewed, it is filled with advice for people who work and play outdoors that is as fresh today as it was then. A number of foresters and one of the prereviewers requested the chapter, and because no other course required of forestry students is likely to include such information, we are pleased to present it here.

Fred Barrows Knight

Herman John Heikkenen

Introduction

The field of forest entomology is concerned with both trees and insects. It deals with the effects of insects on forests and forest products and how to prevent adverse effects from reaching serious proportions economically while using methods that meet the social constraints of citizens. Foresters managing recreational areas will also be directly concerned with annoying insects and others having a more direct impact on the health of forest visitors.

The forest entomologist studies the characteristics, the habits, and the physiological reactions of forest insects because by doing so he or she will be better able to regulate their activities. But also one must understand the forest: the life history and requirements of the individual tree species, their reactions to the habitat and to one another, and the characteristics that make some forest types either susceptible or resistant to insect injury. Thus, the forest entomologist must be both a forester and an entomologist.

The injurious insects discussed in this book have a direct effect on the trees themselves, on the products derived from the trees, or on the people utilizing the forest for various reasons. Some species of insects, however, have an indirect effect in that they are predaceous or parasitic and prey upon the tree pests. In addition to the pest species and their natural enemies, a multitude of insects enter into the forest economy that are neither pests of trees nor parasitic or predatory enemies of pests. Examples of these are insects that live on the plants of the undergrowth, those that aid in the disintegration of waste wood in the forest, and those that feed on the organic matter in the duff layer of forest soils. There can be no doubt that these insects play an important role in the development of the forest ecosystem.

IMPORTANCE OF FOREST ENTOMOLOGY IN FORESTRY

Years ago, a person may have heard the statement that forestry is 90 percent protection. We may not accept this high percentage as a fair estimate of the importance of protection, but it must be admitted that if our forests are not protected from the devastation of forest fire and the ravages of insects and diseases, there will be little opportunity to practice forestry. Diseases, insects, and fire are the greatest agents of destruction in our forests. Any program of protection that ignores any member of this formidable triumvirate endangers our future timber supply and certainly invites disaster.

During every stage in the growth of wood, from the seed to the finished product, important insect problems are continuously presenting themselves. Even before the seeds are collected, they may be attacked and injured by certain insects. These are for the most part the larval stages of moths, beetles, and wasps. Although seeds of forest trees are produced in prolific quantities, often numbering millions per acre in a single season, the damage caused by seed-destroying insects is sometimes serious. These pest problems are of great importance where a large investment is being made in producing seed from superior trees in orchards specially tended for that purpose (Fig. 1–1).

In the nursery, the seedlings or transplants may be injured by such defoliators as climbing cutworms or by root-eating insects, such as white grubs and wireworms. Bark beetles, leaf or bud miners, plant lice, and scale insects all take their toll on the trees in the forest nursery and also on the advance growth of young trees under natural conditions.

Trees in the sapling stage are sometimes attacked and severely injured by defoliators, phloem insects, and sucking insects. The vigorous period between the sapling stage and commercial maturity is, as a rule, the stage most resistant to insect attack. However, even in this period, the trees may occasionally succumb to the attack of defoliators or primary bark beetles. With approaching maturity, vitality is reduced, and the trees become increasingly susceptible to insect injury. Bark beetles that cannot kill vigorous young trees may breed successfully in the trees of the mature forest, and defoliators become much more dangerous than they were when the trees were in the full vigor of youth.

Later, when the trees die or are cut, they promptly become subject to the attack of the many kinds of wood-deteriorating insects. Bark beetles, ambrosia

Figure 1-1 The mist blower may be used in seed orchards for control of cone-infesting insects. *(U.S. Forest Service photo.)*

beetles, roundheaded borers, and flatheaded borers all attack and injure dying or recently killed trees and freshly cut wood. Not only do these insects injure the wood directly by their borings, but they are often responsible for the introduction of wood-staining and wood-rotting organisms. As the wood seasons or decays, it becomes subject to the attack of numerous other insects.

With such a multiplicity of insect species attacking trees and wood products, it is difficult indeed for a forester to find any line of forestry work in which there is not an immediate insect problem to solve. Even the lumber retailer may be called upon to pacify a customer who finds powderpost beetles emerging from a newly laid hardwood floor. Also, in lumber manufacturing, in the pulp and paper industry, in forest by-product industries, and in the more technical phases of forestry work, entomological problems are forever intruding.

Entomology has often been looked upon in the forestry profession as something to be ignored whenever possible in spite of the tremendous losses of forests, trees, and forest products caused by insects. This has been due, in part at least, to the feeling among foresters and people in the lumber business that insects in the forest could not be controlled, and therefore they were given scant attention. This point of view has changed during the past 25 years, so that today, when wood is needed in ever increasing quantities and forests are providing many additional needs of the public, forest entomology is properly an integral and important part of forest protection and forestry. Forest managers are facing their insect problems with much the same attitude that they exhibit toward their problems in silviculture, management, and utilization.

Therefore, every forester should be able to recognize evidence of possibly dangerous insect activities and should know enough about insects and their control to act intelligently. The forester should know how and where to obtain information about insects and should be able to apply the necessary remedies. A person totally ignorant of insects and their ways cannot hope to get the best results, any more than an inadequately trained physician can hope to give as good service as one who is well acquainted with the disease he or she is called upon to cure.

LOSSES CAUSED BY FOREST INSECTS

It is undoubtedly true that in North America more wood has been destroyed by insects, fungi, and fire than has ever been cut and used. Of the various wood destroyers, insects are by no means the least important (Fig. 1–2).

Accurate estimates of losses caused by forest insects are very difficult to obtain. In only a few instances have systematic loss records been made year

Figure 1-2 These lodgepole pines in the Arapahoe National Forest were defoliated by the pandora moth, *Coloradia pandora*. *(U.S. Forest Service photo.)*

after year, and even these are relatively fragmentary. The records of losses in the West because of bark beetles are perhaps the best available (Crafts, 1958; U.S. Forest Service, 1973).

Injury to trees by forest insects may result in their death, reduction of growth rate of injured trees, or degrading of wood products. These effects combined represent the total damage. Crafts (1958) defined the term *growth impact* to include mortality and growth losses in the forest. The term does not include value losses as a result of defects and changed species composition; it is confined to direct and measurable losses to the trees.

In *Timber Resources for America's Future* (Crafts, 1958), nationwide estimates of growth impact were made for the first time for insects, diseases, fire, and other decimating agencies. According to this review, insects kill twice as much timber as disease-causing organisms do, and seven times the amount killed by fire. However, where decay and growth losses are added, the prevalence of stem rots changes the ratio, so that greater damage is said to be caused by diseases when expressed in terms of the broader definition of growth impact.

The total yearly losses caused by insects in the United States, including coastal Alaska, were estimated to be 8.6 billion bd ft. Table 1-1 shows the estimated distribution of these losses in sawtimber for the year 1952.

"The Outlook for Timber in the United States" (U.S. Forest Service, 1973) does not break out the specific losses in detail comparable to that found in the 1958 Crafts report, but it does present the significant data on losses caused by natural agencies. The total mortality in terms of sawtimber losses in 1970 was similar to the figures estimated for 1952 (Table 1-2).

The Douglas-fir tussock moth, *Orgyia pseudotsugata,* has been a periodic problem in the western United States and Canada. A recent outbreak (1971

Table 1–1 Loss in Sawtimber in the United States in 1952

Cause	Mortality, million bd ft	Growth impact, million bd ft
Bark beetles	4530	5410
Defoliators	30	1310
Other insects	480	1900
All insects	5040	8620

Table 1-2 Sawtimber Mortality on Commercial Forest Lands, by Section of Country, 1952 and 1970 (in million bd ft)

Year of estimate	Section of country				
	North	South	Rocky Mountains	Pacific Coast	United States
1952	1.5	2.6	2.5	8.4	15.0
1970	2.3	3.2	2.6	7.1	15.2

to 1975) in Idaho, Oregon, and Washington covered 800,000 acres (323, 887 ha) and resulted in 1 billion bd ft of timber mortality (Furniss and Carolin, 1977). One can use any period as a bench mark for citing such damage, but continuing with the 1970s, we can cite the losses caused by the southern pine beetle, *Dendroctonus frontalis,* as an example in our southern region. Price and Slentz (1973) estimated that in July and August 1973, there were 1,139,000 infested pine trees plus 552,000 recently killed trees in central and northern Georgia. This was only a part of the southern pine beetle outbreak that in 1973 was destroying large numbers of trees in a 10-state area (Price and Doggett, 1978). Both the Douglas-fir tussock moth and the southern pine beetle have been given added research attention through funds provided in combined forest-pest research and development programs initiated in 1974.

Canadian losses have not been confined to the spruce budworm, *Choristoneura fumiferana.* In Canada, as in the United States, there are a variety of problems. The spruce beetle, *D. rufipennis,* has killed more than 3 billion bd ft of mature spruce in British Columbia (Schmid and Frye, 1977), and the hemlock looper, *Lambdina fiscellaria fiscellaria,* caused severe losses in Newfoundland between 1966 and 1971 (Otvos and Warren, 1975).

The losses during the beginning of the century were not nearly so serious from an economic viewpoint as similar outbreaks are today. Today, formerly inaccessible areas are needed to meet the multiple use requirements of our citizens. It becomes evident that the economic loss from outbreaks cannot always be expressed fairly in terms of volume of timber lost. There are many other factors to consider, including the availability of markets, the quality of the productive site, the social needs of the citizens, and the multiple uses of the forest. The consideration of multiple uses of the forest should include the additional benefits gained from higher levels of manufacture. The quality wood in beautiful furniture represents a far greater value than would be obtained if the same wood fiber were burned for home heating. Then there are the values associated with recreational use of the forest, the necessity of watershed protection, and the need for maintenance of productive wildlife habitats.

The losses resulting from outbreaks of insects such as those cited here represent only a part of the total damage for which forest insects are responsible. There must be added the less conspicuous but nonetheless real damage caused by insects present in normal numbers. No satisfactory estimates of these losses have ever been made, but in the aggregate, they undoubtedly amount to millions of board feet annually. These unestimated losses should be added to the impact on growth if loss estimates are to be realistic. But even without them, the damage caused annually by forest insects probably exceeds 5 billion bd ft.

The destruction of manufactured wood products by insects amounts to a very high total, but no definite data are available at present on which to base an estimate. Termites, especially in tropical and subtropical regions, are particularly injurious to unprotected wooden structures. Even in the temperate regions of this country, particularly along the Atlantic and Pacific Coasts, termites are sufficiently numerous to cause injury to wooden structures. Other insects, such as the powderpost beetles and the pole borer, *Parandra brunnea,* attack and destroy seasoned wood and finished products, but the data available are insufficient to form the basis for satisfactory damage estimates.

Incomplete and unsatisfactory as our statistics may be, they are at least sufficient to indicate that insects are an important economic factor in our forest industry and should receive a prominent place in our plans for the protection of forests and forest products.

SCOPE AND SUBDIVISIONS OF FOREST ENTOMOLOGY

The scope of forest entomology is surprisingly wide. It includes a great variety of subject matter leading to the better understanding of the biological phenomena in the forest ecosystem. The ultimate aim of forest entomology is to make possible the regulation, in the interest of man and within realistic economic constraints, of insect activities in forests and forest products. In the control of forest insects, the possibilities of using direct protective methods are limited because of the relatively high cost of such operations. Preventive rather than curative methods should be favored. This preventive entomology calls for a much more profound knowledge of both the insects and the forest environment than would be required if one could depend largely on direct control. One of the first requisites for forest-insect work, therefore, is a sound basic knowledge of silvics and silviculture. The forest entomologist who does not know the trees and how they grow cannot effectively protect the forest from the serious pest problems.

Anyone engaged in forest-insect work must also know the insects. For this, the professional must be able not only to recognize the genus and species to which a specimen belongs but also to understand its functions, its reactions to its environment, and its physical limitations. Thus, all the major divisions of the science of entomology are needed in the solution of forest entomological problems. Taxonomy[1] is needed in classifying insects and in indicating their relationships and origins. Of course, it aids the forester by providing names. But in addition, whenever a new forest entomological problem presents itself, taxonomy may actually furnish the key to satisfactory control measures because closely related insects can often be controlled by similar means. Thus, the service of taxonomy may often prove invaluable in showing relationships of new to old pests. Studies in morphology, histology, and physiology of forest insects leads to a more complete knowledge of the insects studied and aid

---

[1]Taxonomic services are available from specialists in the United States National Museum, Washington, D.C.; the Biosystematics Research Institute, Ottawa, Canada; and from specialists at various universities, laboratories, and institutes in both countries.

directly or indirectly in the solution of forest entomological problems. Chemical control, which includes spraying, dusting, and fumigating, has its place in the regulation of insect pests, especially in emergencies. Ecological studies, including life history investigations, the effect of climatic and other environmental factors on forest insects, and the interrelation of parasites and predatory species with their hosts, are all of fundamental importance.

Obviously, no forest entomologist can be expected to have the detailed knowledge of taxonomy that is expected of a specialist in the subject and at the same time be a specialist in morphology, histology, physiology, ecology, and toxicology (Fig. 1-3). The field is too large to permit such a wide scope of endeavor. However, the forest entomologist is expected to have a general knowledge of each of these fields and to have a detailed knowledge of some of them.

The development of the many remarkable insecticides during the past 30 to 40 years has resulted in a greatly increased use of direct control for certain forest insects. Therefore, until recently, many forest entomologists in the United States were heavily involved in the supervision of control projects and

Figure 1-3 The insect causing this damage to a young aspen sucker was first identified as *Agrilus anxius*. The damage suggested another species, and further investigation revealed that *Agrilus horni* was the species responsible. The correct identification was the result of work by both a taxonomist and a forest entomologist. *(Photo by J. C. Nord.)*

in development of methods for direct control of insects. In Canada, a much greater emphasis was given to development of biological and ecological methods of preventive control, though there were many exceptions in both countries. At present, there are many scientists and practitioners involved in integrated pest management systems in which all approaches to solving the pest problem are taken into consideration. Direct control will continue to be a part of the protection of our valuable forest resource, but it is no longer considered the way of solving all pest problems.

The problems of forest-insect control have become more and more complex and require the services of scientists who have specialized in chemical control, insect pathology, ecology, physiology, computer science, and many other fields. The major problem associated with specialization is the chance that the individual will lose the ability to grasp the breadth of the problem and the relationship of the work to the objectives in forest entomology and especially in forest management. It is essential that specialization occur, but the accomplishment of meaningful research requires that specialists work together in teams to solve critical problems in the field.

FOREST ENTOMOLOGICAL LITERATURE

Every worker must rely on the writings of others for a foundation of information in any given line. The ability to locate all available literature on any subject and to use that literature efficiently is essential for best results. The library is one of the most important tools. In the field of forest entomology, there is particular need for training in the ways and means of locating information because the writings in this field are so scattered that they are sometimes difficult to find and may easily be overlooked.

There are many valuable literature sources of forest-insect data that would be highly significant to the scientist or specialist. The bibliographies and indexes provide the access needed to much of the entomological and forestry literature of the United States and Canada. The forestry student is expected to learn how to use publications such as Forestry Abstracts and the other indexes.

The following are some of the sources in which one ordinarily expects to find the original publications dealing with American and Canadian forest entomology:

1   **U.S. Department of Agriculture**
     Department bulletins and circulars
     Forest-insect and disease leaflets
     Cooperative plant-pest reports
     Farmers' bulletins
     Reports

U.S. Forest Experiment Station notes, papers, and reports
U.S. Forest Service separates
2 **State Publications**
Experiment station bulletins, circulars, and reports
University bulletins and memoirs
State entomology and forestry publications
3 **Canadian Publications**
Department of Fisheries and Environment annual reports; bulletins,
  circulars, and reports; bimonthly research notes
Provincial forestry and entomology organizations' bulletins, circulars,
  and reports
University bulletins, circulars, and reports
4 **Periodicals**
Annals of the Entomological Society of America
Canadian Entomologist
Canadian Journal of Forestry Research
Canadian Journal of Research
Canadian Journal of Zoology
Ecology
Environmental Entomology
Forestry Chronicle
Forest Science
Journal of Economic Entomology
Journal of Forestry
Southern Journal of Applied Forestry

These lists are intended to give the reader a sense of the variety of sources
available; references to many of these sources and to others will be found
throughout the literature cited in this text. One could go on with similar lists
of many books and publications in the overseas literature, but the student may
develop that information by contacting the officials in the country of interest.
The North American books available to students are not numerous; the following are available:

1 *Aerial Control of Forest Insects in Canada* (M. L. Prebble, ed., 1975)
2 *Concepts of Forest Entomology* (K. Graham, 1963)
3 *Eastern Forest Insects* (W. L. Baker, 1972)
4 *Forest and Shade Tree Entomology* (R. F. Anderson, 1960)
5 *Insects that Feed on Trees and Shrubs* (W. T. Johnson and H. H.
Lyon, 1976)
6 *Perspectives in Forest Entomology* (J. F. Anderson and K. Kaya, 1976)
7 *Western Forest Insects* (R. L. Furniss and V. M. Carolin, 1977)

Because forest entomological literature is so widely scattered in so many
different publications, no local library can have copies of all the publications
on its shelves. Even in some of our best entomological libraries, we cannot find
every article that may be pertinent to any one subject. Fortunately, however,

most of the local libraries have the privilege of borrowing from other libraries and if the publications themselves cannot be loaned, photostatic copies of parts of articles or books can usually be obtained. Many foreign publications and doctoral theses from various universities and colleges are now being made available on microfilm. References to microfilms can be found in most university and many public libraries. Consequently, if we have a definite reference, it is usually possible to see a certain publication even though the facilities of a local library are limited. The first and most important bibliographic aids are the indexes and the abstracting organs such as *Forestry Abstracts, BIOSIS,* and the *Review of Applied Entomology.* With these at our command, we can search the literature effectively; without them, we are helpless. Therefore, every library should be equipped with at least these very necessary tools.

BIBLIOGRAPHY

Anderson, J. F., and Kaya, H. K. 1976. *Perspectives in forest entomology.* New York: Academic Press.

Anderson, R. F. 1960. *Forest and shade tree entomology.* New York: John Wiley & Sons.

Baker, W. L. 1972. Eastern forest insects. *USDA Misc. Pub.* no. 1175.

Crafts, E. C. 1958. Timber resources for America's future. Forest Resources Report *USDA Forest Service* separate no. 1.

Furniss, R. L., and Carolin, V. M. 1977. Western forest insects. *USDA Misc. Pub.* no. 1339.

Graham, K. 1963. *Concepts of forest entomology.* New York: Reinhold Publishing Company.

Johnson, W. T., and Lyon, H. H. 1976. Insects that feed on trees and shrubs. *Comstock Publishing Associates.* Ithaca, N.Y.: Cornell University Press.

Keen, F. P. 1952. Insect enemies of western forests. *USDA Misc. Pub.* no. 273.

Otvos, I. S., and Warren, G. L. 1975. Eastern hemlock looper: The Newfoundland Project 1968, 1969. In *Aerial control of forest insects in Canada,* ed., M. L. Prebble. Ottawa: Dept. of the Environment, pp. 170–173.

Prebble, M. L., ed. 1975. Aerial control of forest insects in Canada. Ottawa: Dept. of the Environment.

Price, T. S., and Doggett, C. 1978. A history of southern pine beetle outbreaks in the southeastern United States. Southern Forest Insect Working Group, Georgia Forestry Commission, Macon, Ga.

Price, T. S., and Slentz, H. 1973. Evaluation of southern pine beetle infestations in Georgia. *Georgia Forest Commission Annual Aerial Surveys.*

Schmid, J. M., and Frye, R. H. 1977. Spruce beetle in the Rockies. *USDA For. Serv. General Tech. Report* RM-49.

U.S. Forest Service. 1973. The outlook for timber in the United States. *USDA Forest Resource Rept.* no. 20.

Chapter 2

Development of the Science and Practice

The science of forest entomology, like the science of forestry of which it is a part, is relatively young. It had its beginning during the first part of the nineteenth century.

The first forest entomologist, J. T. C. Ratzeburg, stated that "by forest insects we do not mean all insects living in the forest but only those which have influence on the thriving and the utility of those wood plants with which the forester is concerned" (Schwerdtfeger, 1973). That clear statement expresses the basis for the science. The brief historical review in the next few pages will provide a perspective for the remainder of the text. Those who wish to study the early developments should read the chapter by Schwerdtfeger in *History of Entomology* (1973).

DEVELOPMENT IN EUROPE

Inasmuch as Germany was the first country to develop forestry, it is natural that forest entomology should have had its inception there. As the value of trees became more and more appreciated and as methods of silvicultural practice were developed, the necessity of protecting the trees from the ravages of insects became increasingly important.

Early Period

The outbreak of a forest pest was something with which the early foresters were unable to cope, and so they were forced to appeal to zoologists for help. As a result, a number of publications appeared that dealt with various individual forest-insect problems. For instance, one finds such treatises as that of Gmelin, "Abhandlung über die Wurmtroknis," published in Leipzig in 1787, and that of C. W. Hennert, "Ueber den Raupenfrass und Windbruch in den Königl. Preuss, Forsten in den Jahren 1791 bis 1794," printed in 1797 (Schwerdtfeger 1973). The first relates to an outbreak of the bark beetle, *Ips typographus;* the second concerns a defoliator of pines, *Dendrolimus pini.*

The purpose behind most of these early studies of forest insects was the development of methods by which a certain pest or the pests of a certain tree might be controlled. It is true that this end was seldom accomplished and that the chief contribution of these studies to science was either taxonomic or biologic in character. Sometimes, however, effective measures were suggested and applied. For instance, Linnaeus is said to have recommended that freshly cut logs be floated in water to prevent injury by borers, a very effective method that is in use today.

Prior to 1800, forest entomology as such did not exist. There were no specialists in this subject, and the studies of tree insects were conducted by men whose primary interest was in other lines. The first attempt to gather all available information concerning forest insects was that of Bechstein and Scharfenberg (1804–1805), who published three volumes entitled "Vollständige Naturgeschicte der schädlichen Forstinsekten." This work was a pretentious compendium of forest-insect information and for a period of 30 years was the only general work available on the subject.

Natural History Period

Then there appeared a monumental work that even today has never been surpassed in excellence and scope. This was Ratzeburg's "Die Forstinsekten" (1837, 1840, 1844). Ratzeburg, frequently called the "father of forest entomology," was the first man to devote all his energies to the field. He lived in an age when specialization was the exception rather than the rule. Even he attempted at first to cover forest pathology as well as forest entomology, but he soon found that these two subjects were too extensive to be handled effectively by any one person. His later work, therefore, was confined to the field of entomology (Fig. 2-1).

The first volume of "Die Forstinsekten" appeared in 1837, the second in 1840, and the third in 1844. He also published a handbook, "Die Waldverder-

Figure 2-1 J. T. C. Ratzeburg (1801–1871) has often been cited as the father of forest entomology. *(Photo supplied by F. Schwerdtfeger.)*

ber und ihre Feinde," that summarized in more condensed form the material contained in "Die Forstinsekten." The demand for this handbook was so great that by 1869 it appeared in a sixth edition. After Ratzeburg's death in 1871, this book continued to appear in new editions edited by his successors. In 1885 Judeich and Nitsche published a two-volume book that purported to be a revision of Ratzeburg's work under the title "Lehrbuch der mitteleuropäischen Forstinsektenkunde." Between 1914 and 1942, Escherich published a new four-volume edition, which is a masterpiece of its kind, under the title "Die Forstinsekten Mitteleuropas." Escherich added much new material and rewrote the older portion of the book, thus making a thoroughly modern presentation of the subject.

Ratzeburg published numerous other articles and books throughout his life. One of his best-known works is his "Ichneumoniden der Forstinsekten," published in three parts in 1844, 1848, and 1852. Ratzeburg dominated the period in which he lived. But although he stood head and shoulders above his contemporaries, there were other workers who made very valuable contributions to the science of forest entomology. Among them we find Köllar, Hartig, and Nordlinger in Germany and Perris in France. An important contribution by Köllar (1840) has been translated into English and is available under the title "Treatise on Insects Injurious to Gardens, Forests, and Farms." Perris is credited with the first experimental studies in forest entomology. He cut trees at different seasons and studied the life history and habits of the various insects that attacked them. His greatest contribution was the "Histoire des insectes du pin maritime," which appeared in ten parts between 1851 and 1870 in the *Annals of the Entomological Society of France.*

Taxonomic and Biological Period

Up to the time of Ratzeburg's death, the chief emphasis in forest entomology was biological. These investigations were often in the nature of general natural history studies and were usually not based on controlled experimental evidence. The work of Eichhoff ushered in a new era that made forest entomology a more exact science than it had previously been. By combining careful biological experiments and detailed taxonomic studies, he cleared up many misconceptions concerning the biology of bark beetles and set up a model for other investigators to follow. His outstanding contribution, entitled "Europaischen Borkenkäfer," was published in 1881.

During the later part of the nineteenth century, there were, particularly in Germany, many workers who devoted a part or all of their time to the study of forest insects. Among them was Altum, at Eberswalde, who did much to stimulate investigation and discussion as a result of his proposed theories and hypotheses. Nitsche of the Forstakademie in Tharand and Henschel at the Agricultural College at Wien were both outstanding teachers and investigators.

Modern Period

From the very beginning of the nineteenth century until its end, European forest entomologists looked to Germany for leadership. This was the natural result of the tremendous amount of valuable pioneering work on forest insects produced by the entomologists of that country. Since the beginning of the twentieth century, however, a decided change has taken place. Now, in the modern period, leaders in forest entomology, instead of being centered in a single country, are to be found throughout those parts of the Old World where forests are economically important. In Europe, Escherich has perhaps exerted the greatest influence, and associated with him have been other capable forest entomologists too numerous to mention individually. Escherich's influence has been felt strongly through his editorship of the periodical *Zeitschrift für angewandte Entomologie.*

In Europe, the science of forest entomology has passed through several more or less definite periods. The first was the natural history period, characterized and dominated by the work of Ratzeburg. The second was the period of great taxonomic activity supplemented by experimental life history studies. Eichhoff characterizes this stage, which was really the connecting link between the first period and the third. The third, or modern, period places the great emphasis on experimental biology.

DEVELOPMENT IN NORTH AMERICA

During Ratzeburg's period of domination in Europe, an interest in tree insects was developing in America. Much of this work was prompted by an interest in ornamental trees.

Natural History Period

The early contributions in America, as in Europe, dealt largely with forest-insect biology from the natural history viewpoint. These early writings are both useful and interesting to us today, but unfortunately they are so scattered in various publications that they are sometimes difficult to obtain. Most of this material, however, is referred to in the indexes already discussed. Many articles on tree insects are included in Harris's "Treatise on Some of the Insects Injurious to Vegetation" (1886) and in Fitch's reports entitled "Noxious, Beneficial and Other Insects of the State of New York" (1856 to 1870). The reports and articles of Walsh, Riley, Lintner, Comstock, and Forbes also contain much information concerning forest insects.

It was not until 1890 that the first compendium on American forest insects was published. At that time, Packard brought together all the available material in the fifth report of the Entomological Commission of the U.S. Department of Agriculture, entitled "Insects Injurious to Forest and Shade Trees." This report contains a mass of valuable information concerning tree insects and is well illustrated by numerous plates and figures. Packard included verbatim many of the important articles by other authors that were not easily accessible, thus adding materially to the reference value of his book. For years, his report was the only comprehensive work on American forest insects and is still an invaluable reference book. In 1905, Memoir no. 8 of the New York State Museum, a memoir by Felt entitled "Insects Affecting Park and Woodland Trees," added some new information not included in Packard's report, but its chief value was in its remarkably fine colored illustrations.

Taxonomic and Biological Period

There followed a period when taxonomic-biological studies predominated in forest entomological work. This period, though coming later, was similar in character to the Eichhoff period in Europe, the influence of which was evident on American work of the early part of the present century. Hopkins in his bark beetle studies added much to our knowledge of that group of insects both biologically and taxonomically (Fig. 2-2). He was able, as Eichhoff was, to dispel many incorrect views previously held concerning both the bark beetles and the bark weevils and to lay the foundation for modern control practices. He also made especially notable contributions to the subject of bioclimatology. Hopkins died in 1948.

Much of the work of both Swaine and Blackman belongs in the taxonomic-biological category, and some of the earlier work of Craighead was along similar lines; later, all three of these men worked along the lines of experimental biology. Swaine shifted his activities from forest entomology and became director of research and then head of Science Service in the Department of Agriculture of Canada. After his retirement and until his death, he returned to the study of bark beetles.

Figure 2-2 A. D. Hopkins was an early leader in forest entomology in the United States. *(U.S. Forest Service photo.)*

During the early periods in America, the emphasis was placed on shade-tree insects. Virgin forests were still supplying an abundance of wood products, and the practice of forestry was practically unknown because it was unnecessary. The few methods of forest-insect control that were mentioned in the publications of that time were for the most part borrowed from continental Europe.

Modern Period

Forest entomology in Europe has from early times emphasized the practical aspects. This has been the direct result of economic conditions in the densely populated Old World, where scarcity of wood has made the practice of forestry mandatory. Furthermore, in Europe, forest entomological investigations have been centered in schools of forestry. In America, on the other hand, wood has been abundant, and prior to 1915, owners of timberland paid little attention to insect damage. As a result, the insects have in the past been studied without considering seriously their ecological relations with the forest. Often these investigations have been conducted by entomologists whose training has in no way been connected with forests or forestry. As a result, the development of practical control measures has suffered.

The viewpoint has changed. Today, there are many workers devoting their entire time to practical forest entomology. This change has paralleled the development of forestry in America. The more intensive treatment of the forest has spread gradually over the continent, with a greater attention to the insect problem as a parallel development. American forest entomology is developing independently of European, and methods of control are fitted to specific existing conditions. Modern forest entomologists are interested both in the insects themselves and in the forest, but their primary concern is the influence of the insects on the forest.

This point of view has resulted in the application of experimental biological methods to forest entomological problems. Purely observational methods of study have been largely abandoned, and in forest entomology, taxonomy is no longer thought to be an end in itself but nevertheless is one of the useful tools of the science. Life history studies, too, are regarded as means to an end rather than ends in themselves. The interrelation of insects with one another and with the various other elements of the forest environment is coming to be regarded as more and more important. This ecological viewpoint was slowed during the insecticide period, about 1945 to 1965, but since that time, the emphasis on integrated methods has increased rapidly as managers have become convinced that insecticides will not solve the basic problems related to insect damage.

Thus, the development of forest entomology in America has passed through stages that are comparable to those in Europe. The first, or natural history, period was contemporaneous with a similar period in Europe. The taxonomic-biological period was also contemporaneous with similar efforts in the Old World. Similarly, the application of the methods of experimental biology appeared in Europe and America at about the same time. These methods have led to the recent rapid development in the fields of ecology, physiology, genetics, biometry, and pest management. Forest entomology, like the other biological sciences, is now being developed on the basis of experimental work.

CONTEMPORARY WORK IN AMERICA

Students frequently ask where and by whom forest entomological work is being conducted. Because this information is not readily obtainable from current literature, a brief statement is introduced here, recognizing, of course, that the picture as it exists at present will soon change.

Federal Forest-insect Work

The greater part of the forest entomological work in the United States is carried on in the U.S. Department of Agriculture Forest Service. The leaders in forest entomology in the Department of Agriculture from the beginning of the century to the 1960s included A. D. Hopkins, F. C. Craighead, and J. A. Beal (Fig. 2-3).

Figure 2-3 F. C. Craighead and J. A. Beal were responsible for directing forest-insect investigations in the U.S. Department of Agriculture. This photograph of these two pioneers of U.S. entomology was recorded in 1926. *(U.S. Forest Service photo.)*

In 1953, as the result of a reorganization of the U.S. Department of Agriculture, all federal forest entomological research was transferred to the U.S. Department of Agriculture Forest Service and located in the regional forest experiment stations. Survey activities continued to be the responsibility of these stations until 1961, when this function was transferred to regional administrative offices.

Attached to each regional forest experiment station are subsidiary laboratories in which special types of research are conducted. In addition, laboratories for basic research in forestry in several regions have been constructed or are proposed. In each of these, forest entomological research constitutes an important part of the program. Thus, since 1950, the number of federal laboratories in which forest entomological research is conducted has greatly increased. Also, the diversity and complexity of the federally supported research program have grown correspondingly.

Forest-insect Laboratories

The oldest forest-insect laboratories in the East were located at Melrose Highlands, Massachusetts; New Haven, Connecticut; and Asheville, North Carolina. The laboratory at Melrose Highlands was known as the Gypsy Moth Laboratory. There pioneer studies of that introduced pest, its parasites, and its predators were conducted for many years. Since its abandonment, the New Haven laboratory (now located in Hamden, Conn.) has been the center of all forest-insect research in New England and the Middle Atlantic states. For many years, a special laboratory was maintained at Morristown, New Jersey. Intensive studies of the Japanese beetle, *Popillia japonica,* were conducted there. This work was transferred to Columbus, and then to the town of Delaware, Ohio, just outside of Columbus, where the lab is located at present. There a variety of insect problems are being studied. Additional federal insect work of the Northeastern Forest Experiment Station is being done at Orono, Maine.

The headquarters of the Southeastern Station is at Asheville, North Carolina, where Craighead and his associates pioneered in the study of the southern pine beetle, *Dendroctonus frontalis.* There they determined the important relationship existing between the beetle and the fungi that cause blue staining of the wood. Currently, insect work in the Southeast is centered at Athens, Georgia; Olustee, Florida; and Research Triangle Park, North Carolina.

The Southern Forest Experiment Station is located in New Orleans, Louisiana. Work at the laboratory in New Orleans has been concerned chiefly with wood-destroying insects. The scope of the work has broadened to include the study of bark beetles, insects injurious to hardwood trees, pests of forest reproduction both natural and planted, and other pests of the region. Field laboratories are located at Alexandria, Louisiana, and Stoneville and Gulfport, Mississippi.

The North Central Forest Experiment Station is located in St. Paul, Minnesota, where a variety of forest-insect problems common to the Minnesota, Wisconsin, and Michigan region have been studied. Among these are several defoliators, root-eating scarabaeids, the hemlock borer, *Melanophila fulvoguttata,* and insects injurious to coniferous forest plantations. A field laboratory is maintained at East Lansing, Michigan.

A forest-insect laboratory is located in Fort Collins, Colorado, at the Rocky Mountain Forest and Range Experiment Station, where work is conducted in cooperation with Colorado State University. That laboratory serves the southern and central Rocky Mountains and the central portion of the Great Plains. The study of bark beetles, especially the mountain pine beetle, *D. ponderosae,* and the spruce beetle, *D. rufipennis,* has required increasing attention. Studies have also been conducted on some defoliators of pine and Douglas-fir and on pine tip moths. Insects of the shelterbelts and nurseries are being investigated at Bottineau, North Dakota.

The Intermountain Forest and Range Experiment Station has headquarters at Ogden, Utah. Forest-insect research is concentrated on bark beetles at Ogden and on forest insects of the northern Rocky Mountains at Moscow, Idaho. Some of the basic work on the western spruce budworm, *Choristoneura occidentalis,* the Douglas-fir tussock moth, *Orgyia pseudotsugata,* and the larch casebearer, *Coleophora laricella,* has been done at Moscow.

At the Pacific Northwest Forest and Range Experiment Station at Portland, Oregon, important studies are being conducted on bark beetles in ponderosa pine, sugar pine, and lodgepole pine. While in charge of the Portland laboratory, F. P. Keen developed his ponderosa pine tree classification used for the identification of trees that are susceptible or relatively resistant to bark beetle attack. The Pacific Northwest Station has been active in the research programs on the Douglas-fir tussock moth, *O. pseudotsugata,* and the western spruce budworm, *C. occidentalis.* Much of the research formerly done in Portland is now concentrated in the laboratory at Corvallis, Oregon. The work in Alaska is being done at the laboratory in Juneau.

The oldest forest-insect laboratory in the West is at Berkeley, California, now a division of the Pacific Southwest Forest and Range Experiment Station. Numerous important insects have been the subject of investigation, including the lodgepole needleminer, *Coleotechnites milleri,* and the fir engraver, *Scolytus ventralis.* Since about 1920, however, investigations have emphasized *Dendroctonus* beetles, especially the western pine beetle, *D. brevicomis,* because of the tremendous losses they cause in mature stands of timber. A special laboratory concentrating on chemical insecticide evaluation is located at Davis, California.

Forest-pest Management

The federal government also has responsibility for work on pest control and technical assistance on control projects. The U.S. Forest Service has the responsibility for this work, which has been assigned since 1961 to the State and Private Forestry branch. The service has brought together in the same unit all activities concerned with evaluation of insect outbreaks and the application of control practices. Such an arrangement assumes that standard survey methods have been developed that can be used in a more or less routine manner. It also assumes that control practices have become standardized. Thus, the administrative officers would be in a position to evaluate conditions and implement control procedures against forest insects just as a ranger might make a timber sale. Unfortunately, neither survey nor control practices have been standardized, though more effective and efficient procedures are evolving.

The U.S. Forest Service has a vital role in helping the states in all aspects of control projects involving federal cooperation. Of special significance is the assistance in preparation of environmental-impact statements as required by federal and state laws.

The workers in this effort are located in all regions of the country. In the East, they are assigned to the Resource Protection sections of the State and Private Forestry organizations of the Northeast and Southeast, with headquarters in Broomall, Pennsylvania, and Atlanta, Georgia. In the West, the forest entomologists are located in the National Forest Regional offices at Missoula, Montana; Lakewood, Colorado; Albuquerque, New Mexico; Ogden, Utah; San Francisco, California; Portland, Oregon; and Juneau, Alaska.

Forest Entomology Outside the Federal Government

The amount of research and control outside the federal offices has grown steadily in recent years. The McIntire-Stennis Act of 1962 has had a significant impact on the forestry research work at the universities. A portion of this research has been in entomology. The states and private industry have also increased their efforts, so that the total research effort has significantly increased.

Since early in the century, the land-grant colleges and universities have supported forest-insect work through funds received for research under the Hatch Act of 1887. Other universities were also active in forest entomology research, notably the University of Michigan, where Samuel A. Graham was located. Since the passage of the McIntire-Stennis Act supporting forestry research and the increased emphasis on cooperative research, there has been a very large expansion in university work.

With the increasing interest in forest management by wood-using industries, the demand by industry for entomological services has increased. Some companies now employ full-time forest entomologists who are also professional foresters. Some of these are engaged in research; others apply the research results to forest management and help the land managers in solving their field problems in forest protection.

State forestry organizations employ many forest entomologists to work with landowners in reducing the impacts of forest insects. These individuals generally work very closely with their counterparts in the U.S. Forest Service and with the land managers in their own organizations. The Cooperative Extension Service is also involved in this coordinated effort.

CANADIAN FOREST-INSECT ACTIVITIES

In Canada, the organization of forest-insect activities is very different from that in the United States. The vast majority of research and survey activity is carried out by the federal government. Occasionally, provinces and universities become involved in forest-insect investigations, but these are usually undertaken in cooperation with the federal government and in many cases are part of ongoing federal projects. However, forest-insect control operations are carried out by the provinces, with technical advice and assistance on insect development, insect and tree condition surveys, and biological evaluations of success of programs provided by the federal government. The situation in the Province of Quebec is somewhat different in that it conducts its own surveys and evaluations of control programs and occasionally undertakes independent research programs.

Forest-insect investigations by the federal government are carried out by the Canadian Forestry Service within the Department of Fisheries and Environment. There are six regional forestry research centers, one institute, and two forest products laboratories in which forest entomological studies have been a significant part of the overall forestry research program.

The Newfoundland Forest Research Centre, located in St. John's, Newfoundland, is best known for research on the balsam woolly adelgid, *Adelges piceae,* and the development of survey and damage appraisal techniques for that insect.

The Maritimes Forest Research Centre in Fredericton, New Brunswick, serves the three Maritime Provinces (Prince Edward Island, Nova Scotia, and New Brunswick) and is the oldest forest entomological research establishment in Canada (Fig. 2-4). This is where Tothill conducted his classic studies on the parasites of the spruce budworm, *C. fumiferana,* and the fall webworm, *Hyphantria cunea.* In his report on the fall webworm, he presented a mortality table that was similar to the more refined survival tabulations called by Morris and Miller (1954) *life tables.* To call both survival and mortality tables *life tables* would seem appropriate. Tothill's successors have continued work on the budworm at the New Brunswick station and in addition have contributed much to our knowledge of many other important species, among them the European spruce sawfly, *Gilpinia hercyniae,* and the balsam woolly adelgid, *A. piceae.* More recent work includes studies, using rather novel approaches, on the dispersal of the adult spruce budworm and the development of a forest simulation model for the New Brunswick situation.

(a)   *(b)*   *(c)*

Figure 2-4   Three twentieth-century leaders in Canadian forest entomology have been *(a)* J. D. Tothill, *(b)* J. J. deGryse, and *(c)* M. L. Prebble. *(Canadian Forestry Service.)*

The Province of Quebec is served by the Laurentian Forest Research Centre. The establishment is best known for research on the spruce budworm, *C. fumiferana,* development of formulations, and extensive field testing of bacterial insecticides as well as population studies of the Swaine jack pine sawfly, *Neodiprion swainei.*

The Great Lakes Forest Research Centre in Sault Ste. Marie, Ontario, is probably best known for pioneering work on interrelationships between climate and insect population dynamics and behavior, as well as for innovative research on predator-prey relationships. Most of the earlier work, including identification, synthesis, and first field testing of spruce budworm, *C. fumiferana,* pheromone, was also done at this center.

The three Prairie Provinces (Manitoba, Saskatchewan, and Alberta) as well as the Northwest Territories are served by the Northern Forest Research Centre in Edmonton, Alberta. Population dynamics of the larch sawfly, *Pristiphora erichsonii,* and studies on bark beetles and insect problems of shelterbelts are important projects of this center.

The Pacific Forest Research Centre in Victoria, British Columbia, has done excellent work on a wide range of important forest insects, including the western spruce budworm, *C. occidentalis,* the balsam woolly adelgid, *A. piceae,* the satin moth, *Leucoma salicis,* and bark beetles. Research has also been accomplished on host reaction to insect attack.

The Forest Pest Management Institute in Sault Ste. Marie, Ontario, was formed in 1977 by combining the Chemical Control Research Institute and the Insect Pathology Research Institute. The Chemical Control Research Institute had an impressive record in the development of chemical control techniques, including the testing and evaluation of new chemicals for forest applications. The Insect Pathology Research Institute was well known for research on insect pathogens and other noninsecticidal forms of insect control, including insect-growth regulators and pheromones.

The two forest products laboratories in Vancouver and Ottawa have done work on insect pests of wood and wood products.

FEDERAL AID FOR FOREST-INSECT CONTROL

The enactment of the Forest Pest Control Act by the U.S. Congress on June 25, 1947, marked an important milestone in the development of a nationwide forest-insect control program. This legislation was the culmination of a series of acts beginning with the establishment of the U.S. Forest Service (Anon. 1974, 1976a). To the Forest Service was delegated the responsibility for protecting federally controlled lands from fire, injurious insects, and disease. Since then, the federal responsibility for pest control has been broadened by each successive act of Congress.

The need for federal leadership and aid in forest-insect control has been due in part to the widespread character of many insect outbreaks and because almost every important outbreak of forest insects in the United States has involved mixtures of landownership: federal, state, and private. Attempts at control by individual owners seemed futile unless control was also applied to other adjacent infested lands. The establishment of one coordinating agency to assume leadership seemed essential.

1921 Deficiency Act   The first act to provide specifically for the control of a native forest pest was the Deficiency Act of December 15, 1921. It was aimed at the control of a bark beetle outbreak in northern California and southern Oregon. It provided funds for the Secretary of Agriculture to "prevent further loss from beetle infestations on federally owned lands" in the designated area. Furthermore, it provided for cooperation of the Department of Agriculture with the Bureau of Indian Affairs and with state and private owners.

Federal Cooperation Acts

Even though the Deficiency Act was an appropriation act, rather than a basic law, it set the pattern for federal aid to cooperative insect control projects in which state, federal, and private owners participated. This federal aid principle was established legally by the Clark-McNary Act of June 7, 1924, which was aimed at fire protection, rather than at pest control. It provided for the protection of lands in state and private ownership with federal and state funds in equal amounts. Appropriation for research on forest insects and diseases was specifically authorized by the McNary-McSweeney Act of May 22, 1928. At that time, however, there was no recognition of federal obligation to aid in the control of insect pests when federally owned lands were not involved. It was not until 10 years later that this responsibility was recognized.

The Joint Resolution of 1937, usually called the Incipient and Emergency Pest Control Act, authorized the Secretary of Agriculture to control incipient and emergency outbreaks of insects and diseases. The Agriculture Organic Act of 1944 further defined the policy of cooperation and included provisions for cooperation with foreign governments. Only three forest pests were mentioned: the gypsy moth, *Lymantria dispar,* the browntail moth, *Euproctis chrysorrhoea,* and the Dutch elm disease.

Forest Pest Control Act   On June 25, 1947, legislation was enacted that established a definite pest control policy. This was the Forest Pest Control Act. Appropriations for a number of continuing projects were immediately made. The act provided that

... it shall be the policy of the Government of the United States, independently and through cooperation with governments of states, territories, and possessions, and private timber owners to prevent, retard, control, suppress, or eradicate incipient, potential, or emergency outbreaks of destructive insects and diseases on, or threatening all forest land irrespective of ownership.

The Secretary of Agriculture was authorized either directly or in cooperation with other departments of the federal government or with states or private interests (1) to conduct surveys of any forest lands to detect and appraise forest insects or diseases, (2) to determine the measures to be applied on such lands for control, and (3) to plan, organize, direct, and carry out control measures on state or private lands.

Provision was made for cooperation with other federal agencies and with states to apply control on all lands, public or private. However, none of the funds appropriated under the act could be expended on state or private lands "until such contributions toward the work as the Secretary may require have been made or agreed upon in the form of funds, services, materials, or otherwise." Thus, the act provided for cooperative action but left to the discretion of the Secretary of Agriculture the kind and amount of any contribution that would be required of state and private owners. The trend toward federal leadership and aid in forest-insect control was clear. As a matter of fact, under the Forest Pest Control Act, if the Secretary of Agriculture considered it to be in the public interest, he or she might require little or no contribution from the state or from owners of private lands.

The act was especially significant because it applied to all forest lands and to all insects and diseases and because it recognized the necessity for early detection and appraisal of dangerous insect conditions and provided for the control of outbreaks and the suppression of incipient and potentially dangerous situations.

In every state in which forests were an important resource, the effective administration of this act included the establishment of (1) a detection service that would (a) continually collect information on insect population trends, (b) be ever alert for increases of potential pests, and (c) be prompt in reporting findings and (2) a control organization prepared to meet emergency control requirements and to aid forest operators in protective practices.

The application of the provisions of the Forest Pest Control Act was limited mostly to control projects involving the use of chemicals and generally to situations that have been considered emergencies. Under the wording of the act, other forest-insect control practices appeared to be included. Logically, thinnings, noncommerical cuttings designed to change stand composition, and even the regulated use of fires for insect control should have been eligible for federal aid under the Forest Pest Control Act.

The restriction of the law to chemical control was unfortunate in that it resulted in overemphasis of the importance of that type of control in the minds of foresters. As a result, forest entomologists were sometimes placed under pressure to use chemicals when other control practices would have been more appropriate.

No provision for research was included in the Forest Pest Control Act, but as a result of its implementation, research was stimulated. As the control program developed, the need for accurate information—about the pests themselves, their reactions to the trees on which they feed, and their reactions to one another—has become increasingly obvious. Only through knowledge of the facts can better control practices be devised. The need for such knowledge demonstrated the need for research.

Environmental Policy

The Forest Pest Control Act was followed by other congressional actions requiring specific actions in regard to activities that may affect the environment. The National Environmental Policy Act of 1969 has had a very important and continuing effect on all insect control activities. This act requires that environmental-impact statements be prepared before federal dollars are spent. Section 102 states that all agencies of the Federal Government shall include in every recommendation or report on proposals for legislation and other major Federal actions significantly affecting the quality of the human environment a detailed statement by the responsible official on—

(i)   the environmental impact of the proposed action,

(ii)   any adverse environmental effects which cannot be avoided should the proposal be implemented,

(iii)   alternatives to the proposed action,

(iv)   the relationship between local short-term uses of man's environment and the maintenance and enhancement of long-term productivity, and

(v)   any irreversible and irretrievable commitments of resources which would be involved in the proposed action should it be implemented.

Cooperative Forestry Assistance Act of 1978   Congress passed a Cooperative Forestry Assistance Act in 1978, bringing together several cooperative forestry programs, as well as broadening the scope of cooperation. The act includes insect and disease control as a section and repeals the 1947 Forest Pest Control Act, which it replaces. Insect and disease control is broadened to include the protection of wood products, stored wood, and wood in use. The 1978 act retains the provisions of the 1947 legislation as described on pages 25 and 26. The act effectively brings together all the cooperative programs in one piece of legislation that eliminates much confusion and ambiguity.

STATE RESPONSIBILITY FOR FOREST-INSECT CONTROL

In most states, there is some legal provision for the control by the state of insect pests that constitute a threat to the best interest of the public. Frequently, statutes providing for this work supplement those providing for quarantines, embargoes, inspection, and certification. These laws and the control projects conducted under them are often administered by the state entomologist or a comparable official. In some states, however, special legislation provides for the control of forest insects by the state under the administrative direction of the state forester. Under such an arrangement, the control of forest insects is separated from the administration of regulations designed to check the spread of pests from one locality to another.

Oregon Forest Insect and Disease Act   Oregon is one of the states in which forest-insect control is set up under the state forester. Because the Forest Insect and Disease Act of Oregon is considered a model law, it is used here as an illustration. This law declares that forest-insect pests and forest-tree diseases constitute a public nuisance and orders every owner of timberlands to control, destroy, and eradicate them. It provides, further, that in case of failure, neglect, or inability to do this, the state forester shall, with the approval of the State Board of Forestry, declare an infestation control district and fix the boundaries within which the infested or threatened lands are situated.

The state forester is then instructed, under the act, to institute control measures within the control district. All monies appropriated by the state legislature, contributed by the federal government, or received from any agency, corporation, or individual are deposited in the Forest Insect and Disease Control Fund and so made available for control work.

Under this act, all control activities except those conducted independently by agencies, corporations, or individuals on their own lands must be administered by the state forester. Even though the federal government is contributing to the project, its representatives act only in an advisory capacity. Therefore, the final decision to apply or not to apply control measures rests with the state. This is a logical responsibility for a state to assume, provided, of course, that the state forester is guided by the best advice available from state and federal forest entomologists.

Problem of Mixed Ownership

Mixed landownership, which usually characterizes areas involved in forest-insect outbreaks, presents some difficult problems. With the adoption of the Forest Pest Control Act (now the Cooperative Forestry Assistance Act of 1978) and state forest-pest control legislation, the problems are simplified. However, under the laws of most states, although the state authorities may enter onto private property to abate a nuisance, legislation fails to provide for any assessment against private property for insect control. As a result, some owners who refuse to participate voluntarily receive free benefits. The trend now seems to be in the direction of levying assessments for control against all property benefited.

The organization for control must be clearly spelled out in order to avoid misunderstandings. Each agency, corporation, or individual entering into a control project must know its obligations and responsibilities in advance. To that end, cooperative agreements among the various parties concerned are absolutely necessary. Without them, confusion and misunderstandings are inevitable. From this we can see how important it is that the legislative provisions for forest-insect control be as simple and flexible as possible.

The forester should have an understanding of the development of federal, state, and private responsibility related to pest management activities. This summary may be of some help in understanding the development of the present situation. The contemporary condition of the forest in relation to insects is described each year in reports published in both Canada and the United States (Anon. 1976b, 1977).

BIBLIOGRAPHY

Anon. 1974. The principal laws relating to Forest Service activities. *U.S. Dept. Agr. Forest Serv., Agr. Handbook* no. 453.

———. 1976a. Supplement. *U.S. Dept. Agr. Forest Serv., Agr. Handbook* no. 453.

———. 1976b. Forest insect and disease survey annual report 1975. Ottawa: Canadian Forestry Service.

———. 1977. Forest insect and disease conditions in the United States 1974. Washington, D.C.: U.S. Department of Agriculture, Forest Service.

Bechstein, J. M., and Scharfenberg, G. L. 1804–1805. Vollständige Naturgeschichte der Schädlichen Forstinsekten.

Escherich, K. 1914–1942. Die Forstinsekten Mitteleuropas. Berlin, Germany.

Felt, E. P. 1905. Insects affecting park and woodland trees. New York State Museum Memoir no. 8. Albany, N.Y.

Köllar, V. 1840. A treatise on insects injurious to gardens, forests and farms, London.

Morris, R. F., and Miller, C. A. 1954. The development of life tables for the spruce budworm. *Can. J. Zool.* 32:283–301.

Packard, A. S. 1890. Insects injurious to forest and shade trees. *U.S. Ent. Com.,* 5th report.

Ratzeburg, J. T. C. 1837–1840–1844. *Die Forstinsekten.* Germany.

Schwerdtfeger, F. 1973. Forest entomology. In *History of entomology,* eds. Smith, Mittler, and Smith, pp. 361–386. Palo Alto, Calif.: Annual Reviews.

Chapter 3

The Insects

The forester will be concerned with many destructive organisms, but none will be as varied and as common as the insects. This chapter is included not only because of the great variety of destructive insects that the forester should recognize but also to emphasize those that are of benefit to the forest ecosystem. It is very easy to lose sight of the importance of beneficial insects when studying those that from time to time become destructive. This brief introduction to insect development and classification is not intended as more than a beginning guide and a means of placing those insects discussed later in their proper relationships to each other. There are many publications that deal directly with the study of insects. The student who wishes to investigate in more depth will find the following useful:

1 *A Field Guide to the Insects of America north of Mexico* (Borror and White, 1970)
2 *Introduction to Insect Biology and Diversity* (Daly et al., 1978)
3 *An Introduction to the Study of Insects* (Borror et al., 1976)
4 *A Textbook of Entomology* (Ross, 1965)

Insects are found virtually everywhere on the earth; they are the dominant group of animals and include nearly 75 percent of all species. The varieties of adaptations to environmental conditions are so great that one cannot do justice in describing them. Many are of great significance to humanity; humans could not exist in their present manner without them. Those that compete with us for food and fiber are a tiny minority among the many species. Insects have existed on the earth for well over 300 million years and thus have evolved in almost every habitat available except most of the marine environment. Their numbers relative to other plants and animals in the world are represented in Figure 3-1.

The forester is concerned with any biological or physical factor that may be destructive to wood products or that may affect the health of trees. Thus, not only insects are of concern. Some of the pests other than insects are considered in forest entomology courses and will be discussed in later chapters of this book.

DEVELOPMENT

A typical adult insect has three body regions: the head, which bears eyes, antennae, and mouthparts; the thorax, which has three segments, each usually bearing a pair of legs (in many insects, the second and third segments each bear a pair of wings); and the posterior portion of the body, the abdomen, which has as many as eleven segments and no legs. Insects, like other arthropods, have an exoskeleton, which provides them with certain advantages over their competitors. The exoskeleton provides a large area for internal muscle attachment, good control of evaporation, and protection for the internal organs.

Perhaps the most significant of the factors related to the great success of insects is the development of functional wings, which provide these animals great advantages in feeding, dispersion, and protection from enemies. This, along with the great adaptability of structures, has added much to the evolution of these very successful animals. The development of very complex systems of metamorphosis was a highly significant evolutionary development that, combined with the others mentioned, has contributed to the vast numbers of successful species.

Each insect passes through a process of change from the time of hatching to the adult stage. The insect must periodically shed its skin and replace it with a new one in order to grow. The process of molting is called *ecdysis,* and the old skins shed in the process are called *exuviae.* The total period of time between two molts is a *stadium,* and the insect during that time period is said to be a specific *instar.* We might say of a species that the third larval stadium averages 8 days in length and that the third instar is pale green and about 9 cm in length.

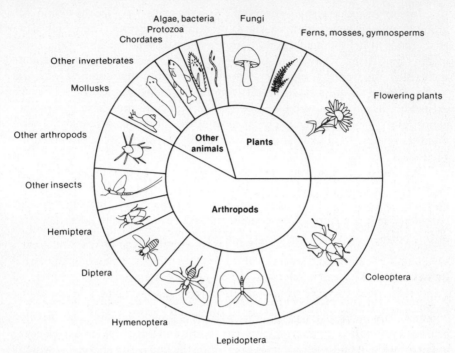

Figure 3-1 Relative numbers of species of arthropods, other animals, and plants. Each 1° equals 4200 species. *(From Daly et al., Introduction to Insect Biology and Diversity, 1978, by permission of McGraw-Hill Book Company.)*

The adult, or *imago,* is the stage having fully developed and functional reproductive organs and in winged species is the stage bearing functional wings.

Metamorphosis

The primitive wingless orders are the Apterygota. These are the orders that are defined as being *ametabolous,* which means without metamorphosis. These animals do grow and molt, as all insects do, but they change very little other than in size from the time of hatch to the adult stage.

The remaining orders of insects are in the subclass Pterygota, which means these are the winged insects. The development of wingless species in these orders has happened secondarily. The Pterygota are divided into two large divisions, the Exopterygota and the Endopterygota. These terms relate to the development of wings. In the Exopterygota, the wings develop externally and are visible on the young insect as small wing pads; in the Endopterygota, the wing rudiments develop internally during early life.

The insects in division Exopterygota have a simple or incomplete metamorphosis. The young of these insects are called *nymphs* and are similar in appearance to the adults. Those with a simple metamorphosis such as grass-hoppers, spittlebugs, and walkingsticks are referred to as being *paurometabolous,* which means a gradual or simple metamorphosis; young and adults live in the same habitats and look alike except for size and wing development. A few of the orders in the Exopterygoptera are described as *hemimetabolous,* relating to more differences between young and adults. Most of these are the orders in which the young live in aquatic habitats and the adults live in terrestrial habitats. The young are often referred to as *naiads* and are aquatic, gill-breathing nymphs. Common examples are dragonflies, mayflies, and stone flies.

The division Endopterygota includes the very large and successful orders of beetles, moths and butterflies, and flies. These insects all have a complete, or *holometabolous,* metamorphosis. There are three distinct postembryonic stages: The early form without wing pads is the *larva;* the quiescent form with wing pads is the *pupa;* and with fully developed wings, the *adult.* The great advantages of this type of metamorphosis relate to the high degree of specialization in the larval and adult stages. Larvae and adults may evolve in entirely different directions. Larvae are specialized for better food gathering; adults, for better dispersal and reproduction.

CLASSIFICATION

The animal kingdom is divided into major groups called *phyla.* Among them is the phylum Arthropoda, which includes the jointed-legged animals such as crayfish, millipedes, centipedes, spiders, and insects. The hierarchical system for classifying animals must include at least six categories: phylum, class, order, family, genus, and species. Commonly among insects we also encounter named superfamilies, subfamilies, and subspecies. The forester should know the classifcation system and the relationship of insects to other organisms. No one is expected to know all the orders, families, or species of destructive insects; but if the classification system is understood, the process of determining the cause of a problem is greatly enhanced.

Orders of Insects

The class Insecta is divided into orders on the basis of structure and form of metamorphosis. There are differences of opinion regarding the specific number of orders and some of the names applied. Daly et al. (1978) place the six-legged animals into two classes, the Entognatha and the Insecta. The classification in this book will list them all as Insecta (Table 3-1). Keys to the various orders and families are available in Borror et al. (1976) and Daly et al. (1978).

The 27 orders listed in Table 3-1 are not all of concern to the forester. Representatives of all the orders may be found within forested ecosystems, but most of the more destructive forest insects are found in only a few orders. The

Table 3-1 Outline of Insect Orders (Class Insecta)

Hierarchical name	Common names	Description
Subclass Apterygota		Primitive wingless insects
Order Protura	proturans	*prot,* first; *ura,* tail
Order Thysanura	bristletails	*thysan,* bristle; *ura,* tail
Order Diplura	diplurans	*dipl,* two; *ura,* tail
Order Collembola	springtails	
Subclass Pterygota		Winged and secondarily wingless insects
Division Exopterygota		With a simple metamorphosis
Order Ephemeroptera	mayflies	*ephemero,* short lived; *ptera,* wings
Order Odonata	dragonflies and damselflies	
Order Orthoptera	grasshoppers, crickets, walkingsticks, mantids	*ortho,* straight; *ptera,* wings
Order Isoptera	termites	*iso,* equal; *ptera,* wings
Order Dermaptera	earwigs	*derma,* skin; *ptera,* wings
Order Embioptera	web spinners	*embio,* lively; *ptera,* wings
Order Plecoptera	stone flies	*pleco,* folded; *ptera,* wings
Order Psocoptera	psocids	
Order Zoraptera	zorapterans	
Order Mallophaga	chewing lice	
Order Anoplura	sucking lice	
Order Thysanoptera	thrips	*thysano,* fringed; *ptera,* wings
Order Hemiptera	bugs	*hemi,* half; *ptera,* wings
Order Homoptera	cicadas, aphids, scales, hoppers	*homo,* alike; *ptera,* wings
Division Endoterygota		With a complete metamorphosis
Order Coleoptera	beetles	*coleo,* sheath; *ptera,* wings
Order Strepsiptera	twisted-winged parasites	*strepsi,* twisted; *ptera,* wings
Order Mecoptera	scorpion flies	*meco,* long; *ptera,* wings
Order Neuroptera	lacewings, ant lions, dobsonflies	*neuro,* nerve; *ptera,* wings
Order Trichoptera	caddis flies	*tricho,* hair; *ptera,* wings
Order Lepidoptera	butterflies and moths	*lepido,* scale; *ptera,* wings
Order Diptera	flies	*di,* two; *ptera,* wings
Order Siphonaptera	fleas	*siphon,* tube; *aptera,* wingless
Order Hymenoptera	sawflies, bees, wasps, ants	*hymeno,* membrane; *ptera,* wings

insects that spend a portion of their life cycles in an aquatic environment are also of significance, as are those that are found in the humus layers of the soil. Some of the aquatic insects in the adult stages are serious nuisances to people using forest recreation areas. Their importance in the aquatic ecosystem may outweigh the problem presented to the manager of the forest recreation site. An understanding of the ecological situations involved can only be gained by first knowing the organisms and their relationships to others in the ecosystem.

The orders of special significance to those studying the aquatic systems within the forest are Ephemeroptera, Odonata, Plecoptera, Hemiptera, Coleoptera, Neuroptera, Trichoptera, and Diptera.

The majority of the more destructive insects affecting forests and forest products may be identified as members of eight orders.

Orthoptera   Grasshoppers, crickets, and walkingsticks are well-known insects in this order; these and most other species are plant feeders, and a few are predaceous. They may be winged or wingless. The fore wings of winged forms are generally long and narrow and are called *tegmina*. The hind wings are broad, membranous, and when the insect is not in flight are folded fanlike beneath the fore wings. The mouthparts are of the chewing type, and the metamorphosis is simple.

Isoptera   Termites are colonial insects with a highly developed caste system. Both winged and wingless individuals are found within a given colony. The front and hind wings of winged forms are approximately equal in size and are held flat over the body when the insect is at rest. The mouthparts are of the chewing type, and the metamorphosis is simple. Termites may be very destructive because they may destroy structures or materials utilized by people. They may also be very beneficial because they assist in the conversion of dead trees and other products to basic nutrients needed for the growth of plants.

Hemiptera   We have all used the term *bug* at some time to refer to insects or other small animals. The insects in the order Hemiptera are the only "true" bugs from the entomologist's point of view. The front wings of the Hemiptera are distinctive in that the basal portion is thickened and leathery; whereas the outer tips are membranous. This type of wing is designated a *hemelytron*. The hind wings are entirely membranous and are usually shorter than the fore wings. The mouthparts of Hemiptera are of the piercing-sucking type, and the metamorphosis is simple (Fig. 3-2).

The majority of the species are terrestrial, but there are many aquatics. Numerous species feed on plant juices, and some may become serious pests; there are also predaceous species, some of which attack humans and other animals.

Homoptera   Adelgids, aphids, scales, spittlebugs, and cicadas are examples from this diverse group of species, which is closely related to the Hemiptera. They also have piercing-sucking mouthparts and a simple metamorphosis. All species of this order are plant feeders, and many are extremely damaging. There are a very few beneficial species, such as the lac insect, which provides shellac.

Winged forms generally have four wings. The fore wings are of uniform structure and either slightly thickened or membranous; the hind wings are membranous. In some groups, one or both sexes may be wingless; and in the

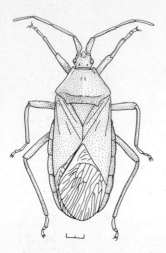

Figure 3-2 *Anasta tristis,* family Coreidae, illustrates clearly the characteristic
hemelytron of the Hemiptera. *(From Daly
et al., Introduction to Insect Biology and
Diversity, 1978, by permission of
McGraw-Hill Book Company.)*

scale insects, the male has one pair of wings on the mesothorax. The life history
of some species is very complex and may include in different generations
bisexual and parthenogenetic forms, winged and wingless specimens, and alternation of food plants.

Coleoptera   The beetles are the most numerous species of animals and
include about 40 percent of the named species of Insecta. They are readily
recognized by their characteristic thickened, often hard or brittle fore wings,
which meet in a straight line down the center of the back. The membranous
hind wings are fully covered by these protective sheaths called *elytra.* The
mouthparts are of the chewing type, and the metamorphosis is complete.
Larvae vary greatly in form and, as is typical of insects with a complete
metamorphosis, do not resemble the adults; they often live in a different habitat
and may be dependent on a totally different source of food (Fig. 3–3).

Beetles are found in every habitat where insects live, and they feed on all
kinds of plant and animal materials. They may be phytophagous, predaceous,
or scavengers. Some are extremely destructive; others are highly beneficial.

Lepidoptera   The moths and butterflies are the common insects most
readily recognized and perhaps liked by people. Many of the adults are very
beautiful in form and coloration, although the caterpillars from which they
develop may be distasteful in appearance. These are often serious pests; most
are phytophagous, and a few feed on various fabrics and stored products.

The mouthparts are generally developed for sucking in the adults, and for
chewing in the larvae. The metamorphosis is complete. The adult wings are
covered with overlapping scales characteristic of the order. There are some
wingless species and others in which one sex is wingless or flightless.

Diptera   The flies are abundant everywhere and are perhaps best known
for their nuisance qualities. Some are very serious pests on humans and other
animals, some are also destructive to plants, and others are extremely benefi-

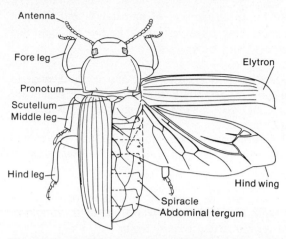

Antenna

Fore leg

Pronotum

Scutellum

Middle leg

Hind leg

Elytron

Hind wing

Spiracle

Abdominal tergum

Figure 3-3 *Tenebrio molitor* is a member of the beetle family Tenebrionidae. The diagram shows the typical elytron of the Coleoptera as well as other characteristics. *(From Daly et al., Introduction to Insect Biology and Diversity, 1978, by permission of McGraw-Hill Book Company.)*

cial. The flies have only two wings that are fully developed; these are the membranous fore wings. The hind wings are reduced to small, knobbed structures called *halteres.*

There is a considerable variation in the mouthparts, but adults in general have some form of sucking structure. The metamorphosis is complete, and the larvae of many species are called *maggots;* most are legless and wormlike. In many species, the larvae are aquatic.

Hymenoptera Ants, bees, wasps, sawflies, and horntails are examples of this order. There are probably more beneficial insects in the Hymenoptera than in all of the rest of the insect orders. Many of the parasites and predators and most of the insects involved in pollination are included in this order. They are an interesting group because of the great variety of specializations that have developed, including the social organization of wasps, bees, and ants.

The winged forms have four membranous wings, with the hind wings smaller than the front ones. The mouthparts are of the chewing type but include a considerable variation. The metamorphosis is complete, and in most of the order, the larvae are grublike or maggotlike. Many of the exceptions are in the suborder Symphyta, which includes most of the destructive forest insects. Many of the larvae in this suborder are caterpillars and commonly feed on living plants.

Families of Forest Insects

The family names of insects may be readily recognized by the suffix *idae,* as in Scolyt*idae.* The coverage of families will be confined to the eight orders already described and, except for three families of parasitoids and three of nuisance insects, to families containing forest insects. Families are identified very briefly by a general comment and by species.

1 Order Othoptera
   a Acrididae: The short-horned grasshoppers are of significance to the forester managing rangelands because they cause significant damage to such habitats. Example: migratory grasshopper, *Melanoplus sanguinipes.*
   b Phasmatidae: The walkingsticks resemble twigs and leaves and feed on trees and shrubs. Example: walkingstick, *Diapheromera femorata.*

2 Order Isoptera
   a Kalotermitidae: These are the drywood or damp wood termites that may infest buildings and other wood products. Example: western drywood termite, *Incisitermes minor.*
   b Rhinotermitidae: The subterranean termites always maintain contact with the soil and may construct tubes to wood not in the soil. Example: eastern subterranean termite, *Reticulitermes flavipes.*

3 Order Hemiptera
   a Miridae: This is the largest family of bugs and may be especially damaging to ornamentals. Example: tarnished plant bug, *Lygus lineolaris.*
   b Rhopalidae: These plant feeders may feed on trees and become a nuisance around homes. Example: boxelder bug, *Leptocoris trivittatus.*
   c Tingidae: The lace bugs are readily distinguished by the sculptured patterns on the upper surface of the wings and body. Example: sycamore lace bug, *Corythucha ciliata.*

4 Order Homoptera
   a Cicadidae: The cicadas may cause damage to trees and shrubs by laying eggs in the terminal portions. Example: periodical cicada, *Magicicada septendecim.*
   b Cercopidae: The spittlebug nymphs are enclosed in a mass of secreted fluid with included air bubbles. Example: Saratoga spittlebug, *Aphrophora saratogensis.*
   c Aphididae: The aphids, or plant lice, are a large group of small, soft-bodied, pear-shaped insects generally found in groups on a variety of plants. A pair of cornicles at the end of the abdomen is characteristic. Example: balsam twig aphid, *Mindarus abietinus.*
   d Phylloxeridae: These are sometimes described as the *woolly adelgids* or the *gall adelgids,* depending on the appearance of waxy secretions or the development of galls. Example: balsam woolly adelgid, *Adelges piceae.*
   e Diaspididae: The *armored scales* are so named because of the secretion of a hard scale over the delicate body. Example: oystershell scale, *Lepidosaphes ulmi.*

f Coccidae: These scales have a softer protective coating and are capable of greater movement. Example: pine tortoise scale, *Toumeyella parvicornis.*

g Eriococcidae: These scales are often called *mealybugs* because their secretions are less evident and movement is much more evident. Example: beech scale, *Cryptococcus fagisuga.*

h Pseudococcidae: This second family of mealybugs includes many additional important species. Example: spruce mealybug, *Puto sandini.*

5 Order Coleoptera

   a Cleridae: The checkered beetles are predaceous insects found on or in trees and logs. Example: blackbellied clerid, *Enoclerus lecontei.*

   b Buprestidae: The wood-boring larvae of the metallic beetles are characterized by the flattened form and wide prothorax. Example: California flatheaded borer, *Melanophila californica.*

   c Anobiidae: These beetles live in and consume vegetable matter, including logs and other woody materials. Example: death watch beetle, *Xestobium rufovillosum.*

   d Bostrichidae: This family of wood-feeding powderpost beetles attacks dead twigs and branches of standing trees as well as logs. Example: leadcable borer, *Scobicia declivis.*

   e Lyctidae: The lyctids are often called the *powderpost beetles* because of their habit of reducing dry wood to powder. Example: southern lyctus beetle, *Lyctus planicollis.*

   f Scarabaeidae: The larvae are well known as the white grubs that may cause severe damage to plant roots. Example: pine chafer, *Anomala oblivia.*

   g Cerambycidae: The long-horned beetles are characterized by their long antennae and elongate form. Larvae are roundheaded borers. Example: southern pine sawyer, *Monochamus titillator.*

   h Chrysomelidae: The leaf beetles are the major family of tree-defoliating beetles. Example: elm leaf beetle, *Pyrrhalta luteola.*

   i Curculionidae: This is the largest family of insects and is characterized by the well-developed snout where the head is prolonged anteriorly. Example: white pine weevil, *Pissodes strobi.*

   j Scolytidae: The bark beetles and ambrosia beetles (Fig. 3-4) are severe pests in the forest environment, causing greater losses in the United States than any other family of insects. Example: mountain pine beetle, *Dendroctonus ponderosae.*

6 Order Lepidoptera

   a Pieridae: These medium-sized butterflies feed on a variety of plants and are generally mostly white or yellow. Example: pine butterfly, *Neophasia menapia.*

   b Nymphalidae: This is a large and varied group of butterflies with much reduced front legs. Example: mourningcloak butterfly, *Nymphalis antiopa.*

Figure 3-4 The Columbian timber beetle, *Corthylus columbianus,* actual size approximately 4 mm in length. *(U.S. Forest Service photo.)*

c   Sphingidae: The sphinx or hawk moths have long, narrow front wings and move very rapidly. Example: catalpa sphinx, *Ceratomia catalpae.*

d   Saturniidae: The giant silkworm moths are some of our largest and showiest of insects. Example: pandora moth, *Coloradia pandora.*

e   Arctiidae: The tiger moths are usually conspicuously spotted or banded, especially on the fore wings, and the larvae are often very hairy. Example: silverspotted tussock moth, *Halisidota argentata.*

f   Noctuidae: This large family varies greatly, but most are medium-sized, dull-colored, and heavy-bodied. Example: darksided cutworm, *Euxoa messoria.*

g   Notodontidae: The larvae of the prominents (Fig. 3-5) generally feed in groups on the foliage of trees and shrubs. Example: saddled prominent, *Heterocampa guttivitta.*

h   Lymantriidae: The larvae of the tussock moths are hairy caterpillars and commonly infest forests. Example: gypsy moth, *Lymantria dispar.*

i   Lasiocampidae: The *tent caterpillars* have received their common name because of the tentlike nests built by the larvae of many species. Example: forest tent caterpillar, *Malacosoma disstria.*

j   Geometridae: The larvae are well known as the inchworms or measuring worms that are commonly seen feeding on the foliage of trees. Example: western hemlock looper, *Lambdina fiscellaria lugubrosa.*

k   Pyralidae: Most of these moths are small and rather delicate; among them are some of our serious cone moths. Example: Zimmerman pine moth, *Dioryctria zimmermani.*

l   Olethreutidae: These are small brownish or gray moths, many with a fringe of hairs along the hind wings. Example: European pine shoot moth, *Rhyacionia buoliana.*

Figure 3-5 The saddled prominent, *Heterocampa guttivitta,* larva feeding. *(Photo by D. C. Allen.)*

m   Tortricidae: This is another group of small moths; many species are leafrollers or leaftiers. Example: spruce budworm, *Choristoneura fumiferana.*

n   Cossidae: The larvae of this family are wood borers, and the adults are large moths with spotted or mottled wings. Example: carpenterworm, *Prionoxystus robiniae.*

o   Sesiidae: The clearwing moths are readily distinguished by the reduced quantity of scales on the wings. Example: ash borer, *Podosesia syringae.*

p   Yponomeutidae: The ermine moths are small, brightly patterned moths with broad wings. Example: arborvitae leafminer, *Argyresthia thuiella.*

q   Coleophoridae: The casebearers are very small moths with narrow, pointed wings. Larvae characteristically work from within a case. Example: larch casebearer, *Coleophora laricella.*

r   Gracillariidae: The leaf blotchminers feed in the larval stages as miners in the leaves of plants. Example: aspen blotchminer, *Phyllonorycter tremuloidiella.*

s   Gelechiidae: This is one of the larger families of small moths. Some are leafminers, others form galls, and many are leafrollers. Example: lodgepole needleminer, *Coleotechnites milleri.*

7   Order Diptera

a   Cecidomyiidae: The gall midges attack plants and cause the growth of galls where they feed. Example: balsam gall midge, *Paradiplosis tumifex.*

b Chironomidae: The midges are very small and numerous flies that are often bothersome to recreationists during evening hours. Example: *Chironomus attenuatus.*

c Culicidae: Mosquitoes are pests of humans and are serious health hazards as carriers of disease. Example: common malaria mosquito, *Anopheles quadrimaculatus.*

d Simuliidae: Blackflies are small, dark-colored insects capable of inflicting severe bites on humans and other animals. Larvae are present in flowing water. Example: *Simulium latipes.*

e Agromyzidae: The leafminer flies are small and seldom recognized; their mines in leaves are much more distinctive. Example: holly leafminer, *Phytomyza ilicis.*

f Tachinidae: This large family of flies is made up of parasitoids of other insects, many of great importance in forest entomology. Example: *Compsilura concinnata.*

8 Order Hymenoptera

a Diprionidae: The conifer sawflies include many serious pests in North America. Example: redheaded pine sawfly, *Neodiprion lecontei.*

b Tenthredinidae: The common sawflies include species that defoliate both hardwoods and conifers. Example: birch leafminer, *Fenusa pusilla.*

c Siricidae: The horntails are wood borers in the larval stage. Adults have a hornlike plate on the abdomen. Example: pigeon tremex, *Tremex columba.*

d Braconidae: This is a large, beneficial group of parasitoids of other insects. Example: *Coeloides dendroctoni.*

e Ichneumonidae: This is one of the largest families of insects and is made up almost entirely of beneficial species. Example: *Megarhyssa lunator.*

f Torymidae: These insects are a part of the superfamily Chalcidoidea. These small insects include species that destroy seeds of trees. Example: Douglas-fir seed chalcid, *Megastigmus spermotrophus.*

g Cynipidae: The gallwasps cause galls to develop on many plants, including trees and shrubs. Example: mossyrose gallwasp, *Diplolepis rosae.*

h Formicidae: Ants live in colonies with a highly developed social system. Example: Texas leafcutting ant, *Atta texana.*

Species of Insects

This discussion of classification has covered the phylum, the orders, and the families of insects, but most important is identification to species. The scientific naming follows the rules outlined in the International Code of Zoological

Nomenclature. The scientific name is *binomial;* that is, it consists of two words: the genus name and the species name, as in *Dendroctonus frontalis*. Notice that this is always written in italics and that the genus name is capitalized.

The trinomial such as *Choristoneura pinus pinus* is used when a subspecies is identified. You will often find the name of the person who first described the species immediately following, as in *Agrilux anxius* Gory. This means that Gory described this species *anxius* in the genus *Agrilus*. The scientific name *Agrilus bilineatus* (Weber) is slightly different. When Weber described the species *bilineatus,* it was placed in a genus other than *Agrilus* and was later changed to *Agrilus;* thus, Weber's name is shown in parentheses.

When you see a species referred to as *Ips* sp., this means that the specimen is of the genus *Ips* but that it has not been identified to species. If you should find reference to *Ips* spp., this means more than one species of *Ips*.

BIBLIOGRAPHY

Borror, D. J., and White, R. E. 1970. *A field guide to the insects of America north of Mexico*. Boston: Houghton Mifflin Company.

Borror, D. J., et al. 1976. *An introduction to the study of insects*. 4th ed. New York: Holt, Rinehart & Winston.

Daly, H. V., et al. 1978. *Introduction to insect biology and diversity*. New York: McGraw-Hill Book Company.

Ross, H. H. 1965. *A textbook of entomology*. 3d ed. New York: John Wiley & Sons.

Chapter 4

Reproductive Potential

The number of potentially injurious forest insects varies from season to season and from locality to locality. These fluctuations with time and place are of major concern to the forest entomologist. If insects can be prevented from increasing above the level of economically serious damage, we will attain our objective; if this is impossible, we may be able to anticipate outbreaks and be prepared to control them. In order to either predict or prevent, it is essential to know the capacity to increase that is characteristic of the more important insect species and to understand the forces that prevent the attainment of these rates of increase.

The actual abundance of a forest-insect species depends on its ability to multiply and to live in spite of the various destructive forces in the environment. Before we can ascertain what the actual abundance of any insect species will be, we must take into careful consideration (1) the ability of that insect to multiply in the absence of any destructive force and (2) the value of the sum total of all the environmental forces working toward its destruction. In other words, the rate of population growth is the result of a struggle between the forces of potential creation and the forces of potential destruction. Chapman

(1931) proposed apt terms for these two opposing forces: *reproductive potential* and *environmental resistance.*

The capacity of a species to increase at an exponential, logarithmic, or geometric rate is an old concept with many names. For example, Lotka (1925) used "intrinsic rate of natural increase," Price (1975) uses "the instantaneous rate of population increase," and Daly et al. (1978) discuss the "change in numbers."

The reproductive potential of an insect species, usually expressed in terms of a single individual, is its ability to multiply in a given time when relieved of all environmental resistance. The reproductive potential depends on: (1) the female's fecundity, (2) the sex factor, and (3) the length of the developmental period. At this point, we should make clear that reproductive potential is a theoretical concept developed to aid our understanding of the population fluctuations of a given insect species. The reproductive potential does not determine the population level a species will reach; rather, it indicates the rate of increase that will occur in the absence of environmental resistance (Fig. 4-1).

Fecundity The ability to produce young is one of the basic factors determining reproductive potential. Among insects, there is a wide range of variation in this ability. The number of eggs deposited by a single female may vary from a few to hundreds of thousands. In general, the fecundity of insects is relatively high. According to the definition of reproductive potential, fecundity can be observed only in the absence of environmental resistance. It is, not

Figure 4-1 Larvae of the redhumped oakworm, *Symerista canicosta,* clustered on an oak stem. Most trees in this forest contained similar clusters. *(Photo by T. C. Eiber.)*

the number of eggs a typical female of a species actually lays, but the number of eggs she is capable of laying. To determine this potential number, several methods may be used. The most accurate would be to rear a number of females from eggs to adults under ideal conditions, count the total egg production of each, and use the maximum number as the fecundity. Although this method has been used in several instances, it is difficult and time-consuming.

One method that is easier but not as accurate is to count the fully developed eggs in the ovaries of gravid females that have never laid eggs and to assume the maximum number found in any one female to be the fecundity for that species. The reason for taking the maximum rather than the average is that in nature every individual is subject to the action of environmental resistance, including factors that might affect the fecundity of the individual. For example, a food shortage for only a day or two during the larval period can affect the number of eggs produced by the adult developing from such a larva. Therefore, even the maximum number of eggs per female observed in nature is probably lower than the fecundity (Table 4-1).

Thus, counting the number of eggs laid under ideal conditions of confinement seems to be the only accurate method that can be used for determining the fecundity of an insect.

Sex Factor The next important factor determining the reproductive potential of a species is the proportion of males and females characteristic of the species. The sex factor is used to calculate the number of individuals in each generation of a species capable of producing new individuals. This proportion is determined in part by the habits of reproduction.

Sexual reproduction is the type most commonly found among insect species. The sexes are separate, and most females lay eggs.

Some insects possess an ability that materially increases their capacity to produce young. This is called *polyembryony.* A single egg deposited by a polyembryonic species produces from several to many individuals. Such a

Table 4-1 Fecundity of Some Major Forest-insect Pests, Determined by Various Means

Species	Eggs	Authority
Spruce budworm	359	Morris, 1963
Gypsy moth	1178	Brown and Cameron, 1979
Forest tent caterpillar	327	Hodson, 1941
Saratoga spittlebug	15	Ewan, 1961
Balsam woolly adelgid	248	Balch, 1952
White pine weevil	115	Graham, 1926
Pales weevil	107	Taylor and Franklin, 1970
Southern pine beetle	159	Clark et al., 1979
Engelmann spruce beetle	176	Knight, 1969

species, although it may produce comparatively few eggs, may still have a high rate of fecundity. Small hymenopterous parasites, notably the braconid wasps, commonly possess this ability.

Either partial or complete parthenogenesis is characteristic of some species. *Parthenogenesis* is the ability of females to produce young without fertilization. This habit of reproduction occurs very commonly among the Hymenoptera. For example, the larch sawfly, *Pristiphora erichsonii,* is partly parthenogenetic (Drooz, 1960). Complete parthenogenesis usually occurs within the Homoptera, especially the adelgids. The balsam woolly adelgid, *Adelges piceae,* gives birth parthenogenetically to many generations of agamic females. In this way, many more individuals are produced during a season than would be possible with sexual reproduction (Balch, 1952).

Usually, the females predominate in parthenogenetic species. This holds true in the case of the larch sawfly, *P. erichsonii.* Under conditions during outbreaks, as many as 96 females to 4 males have been observed. With such a high proportion of females, the number of progeny resulting from any one generation is much greater than would be the case with the half-and-half proportion of the sexes normally found in insects with sexual reproduction. We have yet to find a male balsam woolly adelgid, *A. piceae,* in North America.

The sex factor of a species is determined by dividing the number of females in a given group by the total number of individuals in that group. Thus, if the sexes occur in equal numbers, the sex factor will be 0.5. In a purely parthenogenetic species, in which no males occur, it will be 1.0. In the cases noted in the preceding paragraph, the sex factors are 0.96 for the larch sawfly, *P. erichsonii,* and 1.0 for the balsam woolly adelgid, *A. piceae.*

Length of Developmental Period The reproductive potential is dependent not only on fecundity and the sex factor but also on the length of time required for the completion of each generation. Many insects develop slowly. Certain species of wood-boring Cerambycidae, for example, the southern pine sawyer, *Monochamus titillator,* have two generations a year in the Piedmont Region of the South. In the North, the whitespotted sawyer, *M. scutellatus,* requires a full year to complete a generation under the most favorable conditions and as many as 4 years under adverse conditions. Some cerambycid beetles, such as the old house borer, *Hylotrupes bajulus,* may live as larvae and pupae for as long as 8 to 10 years. It must be admitted that such extremely long developmental periods are unusual, even among wood borers, and are the result of unusually adverse conditions. However, a few insects, for example, the periodical cicada, *Magicicada septendecim,* have a normal life cycle as long as 17 years.

In contrast, under favorable conditions, some of the fruit flies of the genus *Drosophila* may complete their development from egg to adult in less than 2 weeks and may, consequently, produce from 10 to 25 generations per year.

The length of the developmental period may also be influenced by variations in latitude and altitude. The Nantucket pine tip moth, *Rhyacionia frustrana,* has approximately four generations per year in Georgia, three per year in Virginia, two in Ohio, and one in Ontario, Canada (Yates, 1960).

CALCULATION OF REPRODUCTIVE POTENTIAL

Unless a specific numerical value for the reproductive potential of a species can be assigned, the term has little or no practical significance. In order to calculate the reproductive potential of a species, all the factors discussed in the preceding section must be taken into consideration. The maximum number of eggs that a single female is capable of producing, the number of generations that will be produced in a given time under optimum conditions, and the sex factor must all be known. The values for these factors are practically constant for any species of insect existing under ideal conditions, and they are sometimes called the *constants* for that species. If these constants are known, then the reproductive potential of any species can be calculated. In the calculation of reproductive potential, a mathematical expression is used to express the number of reproductive progeny that will result from a given population in any number of generations.

$$\frac{dN}{dt} = rN$$

In this expression, N represents the original population, which, in the calculation of reproductive potential, will be a single individual; r represents the product of the fecundity (number of eggs per female) and the sex factor (number of females in the population divided by the total population); and t represents the number of generations in a given time. Thus, the reproductive potential for n generations (a given number of generations) is Nr^n. The reproductive potential is based on the number of individuals within the population capable of reproducing but does not represent the total population. This expression is sound except when dealing with species that are polyembryonic, in which case it is necessary to include that factor also. For instance, if y represents the number of progeny arising from a single egg, the expression will then read $N(ry)^n$.

If it is assumed, for example, that the fecundity of the spruce budworm, *Choristoneura fumiferana,* is 359 eggs per female, the sex factor is 0.5, the number of individuals produced from an egg is 1, and the number of generations per year is 1, then the annual reproductive potential for the species expressed for a single insect is $Nr^n = 1 \times (359 \times 0.5)^1 = 179$. If, however, this were a polyembryonic species producing 4 individuals from 1 egg, and if it passed through 2 generations per season, the reproductive potential would be $N(ry)^n = 1 \times (359 \times 0.5 \times 4)^2 = 515{,}524$.

Table 4-2 Hypothetical Young Produced during Four Generations by Spruce Budworm and Balsam Woolly Adelgid

Species		Generation			
	I	II	III	IV
Spruce budworm					
Males	1	180	32,041	5,735,339	1,102,625,681
Females	1	179	32,041	5,735,339	1,102,625,681
Total	2	359	64,082	11,470,678	2,053,251,362
Balsam woolly adelgid					
Males	0	0	0	0	0
Females	2	496	123,008	30,505,984	7,565,484,032
Total	2	496	123,008	30,505,984	7,565,484,032

The important part that the sex factor plays in determining the reproductive potential of a species can be illustrated (Table 4-2) by a simple calculation comparing two insect species, the spruce budworm, *C. fumiferana,* which usually produces equal numbers of males and females, and the balsam woolly adelgid, *A. piceae,* which produces only females. Assume that the fecundity is 359 for the spruce budworm and 248 for the balsam woolly adelgid. If we start with two individuals of each species and calculate the reproductive potential for each through four generations of progeny, assuming that each female lays her full quota of eggs and that every egg produces an adult, we shall see a great difference in the final numbers for each species.

In the fourth generation, the balsam woolly adelgid, with the sex factor 1.0, has reached a total of more than 3.5 times the number of the spruce budworm, with a sex factor of 0.5. Thus, we see that the sex factor is of great importance in determining reproductive potential because it expresses the numbers of individuals in each generation of a species that are capable of producing new individuals.

These simple calculations demonstrate the importance of each of the various factors of reproductive potential. Fortunately, this stupendous potential of creation is held in check by the forces of destruction that are analyzed in Chapter 5.

BIBLIOGRAPHY

Balch, R. E. 1952. Studies of the balsam woolly aphid, *Adelges piceae* (Ratz.) and its effects on balsam fir, *Abies balsamea* (L.) Mill., *Can. Dept. Agric.* pub. 867.

Brown, M. W., and E. A. Cameron. 1979. Effects of disparlure and egg mass size on parasitism by gypsy moth parasite, *Ooencyrtus kuwani. Ann. Entomol. Soc. Am.,* in press.

Chapman, R. N. 1931. *Animal ecology with special reference to insects.* New York: McGraw-Hill Book Company.

Clark, A. L., et al. 1979. The fecundity of the southern pine beetle. *Ann. Entomol. Soc. Am.,* in press.

Daly, H. V., et al. 1978. *Introduction to insect biology and diversity.* New York: McGraw-Hill Book Company.

Drooz, A. T. 1960. The larch sawfly, it's biology and control, *USDA For. Serv. Tech. Bull.* no. 1212.

Ewan, H. E. 1961. The Saratoga spittlebug, *USDA For. Serv. Tech. Bull.* no. 1250.

Graham, S. A. 1926. Biology and control of the white-pine weevil. *Cornell Agr. Expt. Sta. Bull.* no. 449.

Hodson, A. C. 1941. An ecological study of the forest tent caterpillar, *Malacosoma disstria* Hbn., in northern Minnesota. *Univ. Minn. Agric. Exp. Sta. Tech. Bull.* no. 148.

Knight, F. B. 1969. Egg production by the Englemann spruce beetle, *Dendroctonus obesus,* in relation to status of infestation. *Ann. Entomol. Soc. Am.* 62:448.

Lotka, A. J. 1925. *Elements of physical biology.* Baltimore: Williams & Wilkins Company.

Morris, R. F. 1963. The dynamics of epidemic spruce budworm populations. *Mem. Entomol. Soc. Can.* no. 31.

Price, P. W. 1975. *Insect ecology.* New York: John Wiley & Sons.

Taylor, J. W., Jr., and Franklin, R. T. 1970. Biology of *Hylobius pales* (Coleoptera: Curculionidae) in the Georgia Piedmont. *Can. Entomol.* 102:729–735.

Yates, H. O., III. 1960. The Nantucket pine moth, *Rhyacionia frustrana* (Const.): A literature review. *USDA For. Serv. SE For. Expt. Sta. Pap.* no. 115.

Chapter 5

Environmental Resistance

Environmental resistance is the sum of all the factors in an environment that reduce the reproductive potential of an organism. The factors that are combined to produce this force can be divided into four principal groups: physical, nutritional, host resistance, and biological. Each group will be considered separately.

An understanding of insects is essential in evaluating the factors of environmental resistance. Although insect physiology, structure, and function are major courses in entomology curricula, a few generalizations here are helpful in interpreting the evolutionary success of the insects. The insects have been on our planet since the Carboniferous period, 300 million years ago. Their survival may be attributed to small size, an exoskeleton, flight, metamorphosis, adaptability, and reproductive habits (Borror et al., 1976).

Their small size relates well to a small food requirement and a microhabitat for protection. But smallness also means a large surface area relative to volume, with a disproportionate water loss and heat gain compared with larger animals. The insects have developed an almost waterproof exoskeleton enclosing an open circulatory system. This cylinder is the basis of the insect's strength and capability of flight, both of which are advantages when searching for food and a favorable microhabitat. Yet the rigid exoskeleton forces the insect to molt in order to increase its size, and the larger the cylinder, the weaker it becomes.

The survival of insects is also aided by metamorphosis, a change in structure and habits during normal growth. The metamorphosis of the spruce budworm, *Choristoneura fumiferana,* for example, begins with the egg stage, followed by a larval stage during which feeding occurs, a quiescent pupal stage, and then the winged adult stage, when mating, dispersion, and oviposition occur. In addition to metamorphosis, insect survival has been aided by high fecundity, the ability to produce young, and ability to store sperm after mating, thus delaying fertilization until conditions are optimum for oviposition.

PHYSICAL FACTORS

Of all the limiting factors in the insect's environment, the physical factors have received the most intensive study and are best understood. Temperature plays an important role in determining both the rate of insect development and the distribution of all insects, geographically and locally. There are comparatively few species whose temperature reactions are definitely known. Similarly, the effects of light, moisture, air movement, and all the other factors that are thought of in connection with climate and weather are far from fully understood. Nevertheless, much important progress has been made toward the understanding of these matters. In the following pages, some of the more important effects of the action of these physical factors on insect activities are discussed.

Temperature

One of the most important physical factors regulating insect activity is temperature. The insects are *poikilothermic;* that is, their internal temperature varies with the environment. The insects are often referred to as *cold-blooded* animals, and their activities are directly controlled by their internal temperature. Although the insects cannot physiologically regulate their body temperature, they attempt to maintain an optimum internal temperature by behavioral patterns. The internal temperature of an insect is usually higher than the ambient temperature. This is a result of the insect's small size and large surface area, muscular activity, intensity of solar radiation, and coloration. Conversely, insects readily lose heat when water evaporates from their spiracles and by convection from their body surface.

Each species of insect has a definite range of temperature within which it is able to live; temperatures above or below the limits of tolerance result in the insect's death. Near the upper and lower endurable limits are the dormant zones in which all movement ceases. When an insect is in the lower of these zones, it is said to be in *hibernation;* when in the upper, *estivation.* Between these dormant zones lies the zone of activity or, as it is sometimes called, the *zone of effective temperature* or the *temperature preferendum.*

The temperature requirements of insects vary with the species. Some species are active at a temperature only slightly above 0°C, the freezing point of water; whereas most species are active only at temperatures above 10° to 15°C. The optimum temperature for most insects is about 26°C, and estivation usually begins at 38° to 45°C. For many insects, 48°C is a fatal temperature.

Low Temperatures   The low fatal temperature varies with both the species and the season. It has been shown that the same species is able to endure much more cold in the fall and winter than in the spring and summer (Payne, 1926).

The lowering of an insect's body temperature causes a decrease in activity, and if the insect's temperature remains lowered long enough, movement stops. When the insect cannot move to obtain food, it will starve to death. For example, the larvae of the forest tent caterpillar, *Malacosoma disstria,* feed and grow slowly at 15°C. However, if temperatures do not increase, feeding stops in the fourth instar. The ability to feed at 15°C is important for survival because the average maximum temperature during this stage of development (in May in Minnesota) is only a little above 15°C (Hodson, 1941).

The freezing of insects is a complex matter that varies with species, cold hardiness, and the effect of snow cover. When freezing temperatures occur within the insect's body, the free water will form into ice crystals. The growth of the ice crystals damages the surrounding cells and tissues. The temperature at which ice crystals form is termed the *supercooling point,* a measure of the cold hardiness of a freezing-susceptible insect.

Why some insects can survive freezing is debatable. Perhaps the free water is eliminated from the body and the remaining water becomes bound with other molecular compounds; perhaps there is an increase in the concentration of electrolytes in the body fluids; or perhaps the insect possesses glycerol (Salt, 1961).

Many insects are protected from freezing by the insulating properties of snow. The alleviation of cold temperatures by snow is proportionate to the depth of the snow. For example, although the supercooling point of the European pine shoot moth, *Rhyacionia buoliana,* is −23°C and that of the balsam woolly adelgid, *Adelges piceae,* is probably −36°C, both Green (1962*b*) and Balch (1952) warn that these insects can survive when insulated by more than 8 in (20 cm) snow because the internal temperatures do not reach the supercooling point (Fig. 5-1).

High Temperatures   When the body temperature of an insect becomes abnormally high, there is usually hyperactivity followed by stupor and then death from denatured proteins, accumulated toxic substances, starvation, and/

or dessication. The lethal temperature varies with each species, its stage of development, the duration of exposure, and also the ambient humidity.

Some insects, those living in logs for instance, are unable to move quickly. If conditions become sufficiently unfavorable, the insects may be killed. In logs having comparatively thin, dark-colored bark, the side exposed to direct solar radiation may reach 60°C or more (Graham, 1924). Such temperatures are far above the fatal point for all insects, and so on the upper side of logs lying in the sun there is usually a sterile zone. On the sides where favorable temperatures prevail, the insects occur in the greatest numbers. On the bottom side, where uniformly cool conditions are found, there exist only those species that are able to live under cool conditions. It is possible to determine whether a fallen tree was infested before or after its falling. If infestation came first, insects of the same species will be found on all sides of the log; if infestation followed falling, there will be a zone on the upper side devoid of insects and a zone on the lower side containing insects with a preference for cool conditions.

Insect eggs are immobile and cannot escape lethal high temperatures. Remarkably high mortality of forest tent caterpillar, *M. disstria,* eggs occurred in Minnesota when temperatures reached 41.1°C during July 1937 (Hodson, 1941).

Figure 5-1 The insulation by snow prevented the freezing of overwintering larvae of the European pine shoot moth, *Rhyacionia buoliana,* within the buds of this red pine plantation (trees 1 to 2 m in height) at Elmira, Ontario, in February 1960. *(Canadian Forestry Services.)*

Insect Distribution   This brings us to a consideration of the influence of temperature on the distribution of insects. Several more or less successful attempts have been made to determine life zones on the basis of the sums of effective temperatures for different localities. Merriam (1898) used this basis for the construction of the life zones that are still widely used in biological work. Sanderson (1908) has shown, however, that insects do not always occur in every locality where the total of effective temperature is sufficient to make possible their complete development. Thus, even though there may be a sufficiency of warm weather in the summer, a species may be kept out of a locality as a result of extremes of cold in the winter. For instance, the European pine shoot moth, *R. buoliana,* succumbs to temperatures of –33°C or lower and therefore cannot spread to the northern limits of its host species except when snow cover insulates the insects from low-temperature extremes (Green, 1962*a*). Such insulation by snow has permitted this pest to extend its range throughout Michigan, reaching and at least temporarily persisting in localities where temperatures of –33°C are frequently encountered. Similarly, the gypsy moth, *Lymantria dispar,* which overwinters in the egg stage, is protected by snow cover. Sullivan and Wallace (1972) have established the supercooling point of the egg to be –30°C; they predict that the snow cover will protect the eggs and that this pest will eventually spread northward until limited by the abundance of food plants. When a winter occurs during which snow cover is absent and at the same time temperatures are low, the northern limits of these species may be expected to recede southward.

In transition zones of temperature where fatal low temperatures occur only in certain especially cold winters, insects such as those mentioned in the preceding paragraph may temporarily move northward, only to be extirpated during a cold year.

Some insects may be limited in their southward distribution by high temperature, but little positive evidence is available on this effect. We know the larch sawfly, *Pristiphora erichsonii,* may pupate prematurely when exposed to excessively high temperature (Graham, 1956*a*) and that animals other than insects are restricted in their southern distribution.

Temperature not only affects the geographic and local distribution of insects but also regulates the rate of insect development. In general, the rate of development increases in proportion to the increase in temperature until the point of optimum temperature is reached. The rate of this increase varies somewhat at different temperatures and with different species, but, in general, between the lowest effective temperature and the optimum point, an increase of 10°C doubles the rate of development.

The seasonal activities of many insects are regulated by temperature. For instance, the Nantucket pine tip moth, *R. frustrana,* has only one generation per year in the north and as many as four in the southern states. Both the number of generations and the time of emergence are correlated with temperature conditions. During a late spring, insects may be greatly retarded in emerging from hibernation. For instance, during the cold spring of 1956 in Michigan, the European pine shoot moth, *R. buoliana,* was at least 4 weeks later than usual in emerging from hibernation (Graham and Williams, 1958). An increase in the length of the larval stage usually affects an insect species adversely by exposing it for longer than usual to predators and disease-causing organisms.

Light

The reactions of insects to light are not very different from their reactions to temperature, and because these factors are usually closely associated with each other and generally vary synchronously, it is often difficult to determine whether the effects are produced by the one or by the other.

A brief explanation for the similarity of insect responses to light and heat is that *light* and *heat* are terms applied to adjacent yet overlapping portions of the electromagnetic (EM) spectrum (Fig. 5-2). The incoming waves of solar radiation can be measured in microns (μ). The wavelengths between approximately 0.4 and 0.7 μ we can see and have termed *light.* The longer wavelengths between 0.7 μ and about 500 μ have heating properties we can feel and have termed *infrared (IR).* The term *ultraviolet (UV)* refers to wavelengths less than 0.4 μ.

Foresters should remember that many insect species can sense in UV and that others can sense in IR. Also involved is the direction of the rays, their intensity (illuminance), time (photoperiod), and degree of polarization.

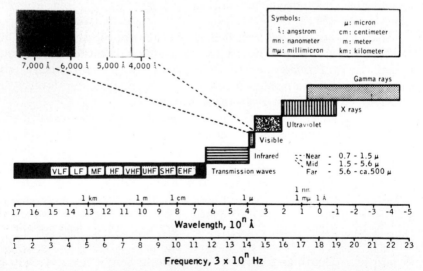

Figure 5-2 The electromagnetic spectrum. *(National Academy of Sciences.)*

Theoretically, it seems possible to divide light just as we do temperature: into optimum, effective, dormant, and lethal zones. Inasmuch as some insects respond positively and some respond negatively to light, it may be assumed the optimum must vary greatly with different species. The stimulating effect of light on certain species is well illustrated by the reactions of flying insects. Generally, species that fly in the day do not fly at low intensities of light or at night. Similarly, night insects do not normally fly during the day. The boundaries for flight can often be quite narrow, as with the males of European pine shoot moth, *R. buoliana,* which fly at dusk (Fig. 5-3).

The interactions of insect tropisms, taxes, and kineses with light are complex. *Tropism* is the orientation, either positive or negative, of an insect to a stimulus. *Taxis* is movement or orientation, either positive or negative, to a source of stimulation. *Kinesis* is movement without orientation but with velocity related to the intensity of stimulation (Chapman, 1971). For example, moths are not "attracted" to the light from a candle; rather, they evolved a positive phototaxis to a point of light—the sun, moon, or stars. The moths fly a straight line by keeping their eyes at the same angle to the parallel rays of the distant light source. However, when the moths fly at a constant angle to a nearby source of light emitting nonparallel rays, the result is ever decreasing circles and death.

Polarized Light The phenomenon of polarized light has a great influence on the taxes of some insects. Light waves vibrate in planes along their direction of movement. These planes are equally distributed throughout 360°. Occasionally, some planes are blocked by clouds, ice crystals, smoke, or fog and the remaining parallel planes are called *polarized light.* Polarized light,

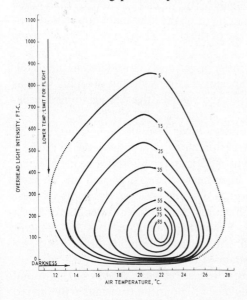

Figure 5-3 The combined effects of overhead light intensity and air temperature on mating flight of the European pine shoot moth, *R. buoliana.* The numbers in the isopleths are the percentages of males in flight. *(Canadian Forestry Service.)*

although invisible to humans, permits bees to navigate on cloudy days (Frisch, 1971) and also permits the orientation of defoliating hymenopterous and lepidopterous larvae (Wellington et al., 1951).

Insects do not react to light in the same manner at all stages of development or under all conditions. For instance, Wellington (1948), in a careful investigation of the photoresponses of the spruce budworm, *C. fumiferana,* points out that the first and second instars always react positively to light at room temperature. However, the light reaction is affected by feeding in the later stages, except for the third, which is indifferent to light even when starved. The later instars react negatively when hungry, and all stages at ordinary temperatures react positively when not hungry. At excessively high temperatures, all stages react negatively. The results of these experiments demonstrate that caution must be exercised in drawing generalizations from field observations.

From this discussion, we see that although the effects of light are difficult to separate from those of temperature, there is good evidence to indicate that light has an important regulating role in the life of insects.

Radiant Energy

The infrared portion of the electromagnetic spectrum lies between visible light for humans and radio waves (0.7μ to about 500μ). These longer wavelengths, although invisible to the human eye, can readily be detected by their heating qualities. The amount of radiant energy absorbed by an object depends on many factors. For example, a dark-colored object may absorb more energy than a light-colored object. Wellington (1950b) found the purple staminate flower buds of the balsam fir to be considerably warmer (5° to 8°C) than green vegetative buds. The increased temperature of the staminate flower buds enhances the development of young larvae of the spruce budworm, *C. fumiferana* (Shepherd, 1958).

The influence of radiant energy must be seriously considered in the development of all *cryptic* insects (those species that develop out of sight) such as the cone and seed insects, the tip moths and weevils, the web-spinning defoliators, and the bark beetles.

The taxes of insects to radiant energy have generated much interest and debate among entomologists. The attraction of wood-boring beetles to the heat and smoke of forest fires is well known, especially the buprestid genus *Melanophila* (Linsley, 1943). These insects have heat-sensing organs in pits on their middle legs. Evans (1971) has discovered that the spectral sensitivity of these organs ranges from 2.0 to 6.0μ, which coincides with the wavelengths of forest fire temperatures (425° to 1150°C). Evans has calculated the *Melanophila* beetles can readily detect a forest fire at distances ranging up to 3 mi (4.8 km). These beetles are often a nuisance to the fire fighters.

The heat-sensing organs of insects vary. Recently, Meyer (1977) found that the head capsule of many weevils has a region that transmits only far-red and near-infrared wavelengths. Callahan (1975), in his thought-provoking book, hypothesizes that the night-flying noctuid moths are attracted when the energy emitted by the female's thorax is chopped into a coded signal by her wing vibrations which is detected by sensilla on the antennae of the male moth.

Moisture

Another factor of the environment that plays an important part in insect activity is moisture. As with all other forms of life, so it is with insects: Both their distribution and their development are dependent on the presence of water in the environment.

The influence of moisture stress on terrestrial insects is closely associated with the insect's ability to prevent the loss of internal moisture. The immature, soft-bodied nymphs of the Saratoga spittlebug, *Aphrophora saratogensis*, feed at the base of plant stems and generate a mass of spittle around themselves (Ewan, 1961). Without an extremely high humidity, the nymphs die within hours. But as adults with hardened exoskeletons, they feed on the exposed twigs of young red pines. The effectiveness of heat in stimulating or retarding the rate of insect development is also influenced by the amount of moisture present. Under favorable moisture conditions, an insect is not as susceptible to extremes of temperature as it would be under unfavorable moisture conditions (Pierce, 1916).

With moisture, as well as with temperature, each species has definite requirements with optimum or effective zones. The extremes are less clearly marked than is the case with heat. Under ordinary conditions, an excess or deficiency of moisture does not result in an insect's immediate death, only in a disturbance of its activities. There are, however, certain outstanding instances of forest insects that are definitely limited in their activities by the moisture factor. The powderpost beetles, for instance, cannot live in moist wood (Snyder, 1926); whereas the ambrosia beetles cannot develop in dry wood. Bark beetles are definitely limited by moisture conditions in the phloem region. The southern pine beetle, *Dendroctonus frontalis*, is able to develop better under moist conditions than are *Ips* spp.; whereas *Ips* spp. can develop under conditions that are too dry for the southern pine beetle. Also, the smaller European elm bark beetle, *Scolytus multistriatus*, exhibits a marked preference for freshly cut logs and is unable to live in those that have dried to a point where the phloem is discolored (Martin, 1946). The rate of development of other wood-boring species may be greatly reduced by the desiccation of the wood. In fact, there are cases on record in which a wood borer having a normal life cycle of 1 or 2 years has required 20 years or more to reach maturity because of unusually dry conditions.

The activities of most insects are affected to varying degrees by moisture extremes. For example, the spruce budworm, *C. fumiferana*, larvae cease feeding when the air becomes saturated with water. Thus, during periods of moist weather, the larvae may be practically inactive, with the result that the length of the developmental period is extended (Wellington, 1950*a*). Excess moisture, for example, as when logs are floated in water or log decks are sprinkled, will stop insect development. Of course, there are interactions with low temperatures and oxygen deficiencies.

Weather

When the factors of temperature, light, and moisture are combined with still other factors, the resulting complex is termed *climate* or *weather.* Climate is the expression of the average physical conditions to be observed in a locality over many years. Climate may change, but such changes are slow. Weather, on the other hand, is the result of the combined action of all the physical factors of the environment at any given time and varies from hour to hour, day to day, and week to week. Weather influences the abundance of insects and the rate of development from year to year and from season to season in every locality. An understanding of the importance of climate and weather is fundamental in any consideration of forest-insect problems.

Weather Patterns   The assumption that all species of insects respond to the same or even similar weather conditions is unsafe. There is, however, a growing accumulation of data to indicate that outbreaks of numerous species occur in the same climatic area at about the same time. Carpenter (1940) summarized outbreak records for Europe and treated the data as if all species were part of a single population. He found a strong tendency for outbreaks of any group of species in the same vegetational formation to occur concurrently.

Although no similar study has been made in America, Graham and Knight (1965) suggested a comparable situation exists. In Minnesota, Wisconsin, and Michigan (the Lake States), outbreaks of the spruce budworm, *C. fumiferana,* and larch sawfly, *P. erichsonii,* occurred simultaneously between 1910 and 1920. At about the same time, the pine tussock moth, *Dasychira pinicola,* and the hemlock looper, *Lambdina fiscellaria fiscellaria,* were abundant. The study of tree rings indicates that both a budworm outbreak and a larch sawfly outbreak occurred about 1880; in the middle twenties, local outbreaks of larch sawfly and spruce budworm again occurred. Wellington et al. (1950) have studied the influence of weather on the outbreaks of the spruce budworm and determined that in any area outbreaks of the spruce budworm tend to follow a period when the number of atmospheric lows has been decreasing for 3 or 4 years.

The close association of insect outbreaks with weather fluctuations stimulates interest among forest entomologists in the so-called climatic cycles. Long-term records and evidence in tree rings provide considerable information on these trends (Keen, 1937). Alternating wet and dry periods occur, and along with them there are alternating periods during which certain insects are sometimes abundant and other times scarce. Although neither the time between wet and dry periods nor the amplitude of the variations appears to be uniform, there is a certain amount of consistency that is useful for general forecasts (Blais, 1965).

Thus, we are encouraged to hope that studies of the ways in which the physical factors of environmental resistance, especially the weather patterns, affect various species of forest insects may lead to better preparedness for outbreaks.

Abnormal weather   With insects of the forest, unusual weather conditions are exceedingly important in regulating abundance .For example, in northern Minnesota, abnormal spring weather has caused reductions in populations of the forest tent caterpillar, *M. disstria,* Witter and Kulman (1972) found high larval mortality when abnormally warm spring temperatures were followed by freezing temperatures ($-7°C$). Late frosts in the spring that kill the foliage on the host trees after the eggs of the forest tent caterpillar have hatched result in the starvation of almost all the young larvae (Blackman, 1918). The same situation may affect any leaf eater that feeds on trees in early spring.

The detrimental influence of abnormal weather is not limited to spectacular frosts; slightly reduced temperatures and increased precipitation can also cause declines in insect populations. Outbreaks of the lodgepole needleminer, *Coleotechnites milleri,* in the West, which have seriously influenced the recreational values of the national parks in western America and Canada, have declined significantly when pupation, emergence, and oviposition were delayed because late spring and early summer temperatures remained several degrees below normal (Struble, 1972).

The effect of abnormal weather should be considered relative to a given forest insect and the probability of occurrence during the outbreak of the insect. Unusually high summer temperatures can have a lethal effect on insects, for example, the summer egg mortality of the forest tent caterpillar, *M. disstria* (Hodson, 1941). Extremely heavy precipitation, usually during thunderstorms, is detrimental to many defoliating insects (especially the needleminers) because the needles or larvae are dislodged from the tree (Struble, 1972). To date, there is little (if any) information available to aid the forest entomologist in forecasting the occurrence of abnormal weather. However, when such weather fortuitously occurs, we should be cognizant of the event, document the evidence, and adjust or cease our control efforts accordingly.

NUTRITIONAL FACTORS

Whereas the physical factors of the environment do not usually lend themselves to regulation, the nutritional factors can often be controlled with a reasonable degree of exactness. If the food conditions can be so controlled that outbreaks of pests are improbable, even in weather favorable for insects, we shall be less concerned about the effects of weather on the rate of insect multiplication. For this reason, we should be particularly interested in the nutritional factors.

Quantity of Food

It is a biological expectation that if other conditions are favorable, an organism will eventually multiply to the limit of its food supply. As a rule, the more numerous the individuals of a tree species within a given area, the more abundant are its insect enemies. When there is an unlimited and convenient supply of a certain species of tree, the stage is set for an outbreak of the insect pests of that tree. Spruce budworm, *C. fumiferana,* outbreaks in the eastern United States and Canada are associated with an overabundance of mature balsam fir; forest tent caterpillar, *M. disstria,* outbreaks are associated with extensive stands of aspen or southern bottomland hardwoods; and some bark beetle outbreaks are associated with stands of high-density, overmature conifers.

Our forest and ornamental trees are often listed according to their susceptibility to insect attack. When these lists are examined, it is found that, with rare exceptions, the species listed as insect-resistant are those which do not occur commonly in large masses. Red pine was once classed as insect-resistant, but it cannot be accurately said that this and similar species are any more resistant than other species. The apparent resistance was due to their relative scarcity in the forest. The validity of this statement is supported by experience following the widespread planting of red pine in the Lake States after 1933 (Graham, 1956*b*). These plantations were attacked by the Saratoga spittlebug, *A. saratogensis,* several sawflies (especially the redheaded pine sawfly, *Neodiprion lecontei*), and in the more southerly plantations by the European pine shoot moth, *R. buoliana.* In the 1960s, when trees reached pole size, bark beetle outbreaks occurred.

In the southern forests during the 1960s, vast acreages were planted to slash pine. Many of these plantations were on shallow upland soils, although slash pine is usually found on wetland sites. The off-site plantations have been seriously damaged by tip moths (Bethune, 1963) and pitch canker (Dwinell and Phelps, 1977). Both slash and loblolly, when planted on shallow, sandy soils, are extremely susceptible to root rots (Morris and Frasier, 1966). Instead of scattered stands of slash pine on lands too wet to support other species, we now have vast areas over which this species is predominant. Thus, the quantity of food available becomes an important factor in regulating insect abundance.

Kind and Quality of Food

The abundance of forest insects and the length of the developmental period are both limited by the kind and quality of food that the insects are able to use. For instance, an insect species that feeds only on the succulent tissues of the phloem must complete its feeding period quickly, while the perishable material that it eats is still in usable condition. Although perishable, this food is comparatively high in nutritional value. We should expect, therefore, that the phloem region would be filled with many kinds of insects, each with a short developmental period. This is exactly the condition that exists.

By the nature of their food, the development of almost all leaf-eating species is limited to a single season or even to a short part of one season. Few leaf eaters can pass the winter as partly grown larvae; consequently, they must complete their development before the leaves drop in the autumn. In some cases, the length of the developmental period is limited to a few weeks during which the foliage is soft and succulent.

The success of an insect species is determined not only by the kind of food available when it is needed but also by the quality of food as affected by soil nutrients or moisture. The lack of some element in the soil may affect the plants adversely but at the same time improve nutritional conditions for some insects, such as aphids (Haseman, 1950). The availability of soil moisture has long been associated with numerous genera of forest insects, especially the terminal weevils, tip moths, defoliators, and bark beetles. The interactions among soil moisture, trees, and insects will be discussed as host-resistance factors in this chapter.

Wood borers, on the other hand, feed on a medium that with time will decay, yet may change very slowly. It is, therefore, not at all uncommon for their life cycle to extend over several years. During the process of wood disintegration, there is, in many instances, a close relationship between fungi and insects. In some cases, particularly in the early stages, a true symbiotic relationship can be demonstrated (Baumberger, 1919). Just as the fruit flies can develop normally only where bacteria or fungi are present to aid in the elaboration of raw food, so, in the case of many wood borers, fungi must be present to alter the character of the wood, thereby making the food materials available for the insects.

Many of the wood-boring species, such as the Cerambycidae and Buprestidae, feed for a time in the phloem before entering the solid wood. In this way, the young larvae are provided with more nourishing and more easily digestible food than the larger larvae are. This suggests that the powers of digestion of these species are better in the later than in the earlier stages. In some cases, such as that of the ambrosia beetles, the fungus furnishes the entire food for the insect, and the galleries that are cut into the wood by the insects serve only for shelter and a place for the fungus to grow. These interactions are discussed in great detail in Chapter 19, "Wood Destroyers (Wood in Use)."

Forest insects have widely varying food requirements. Some require leaves; some, the phloem region; and still others, the solid wood. As the wood is consumed by insects and fungi, its chemical and physical characters change, and with this change come new species of insects to replace those that attacked the living tree or recently cut log. These, in turn, are replaced by others, until in the last stage of decomposition, the population is identical with that of the duff stratum of forest soils. Thus, there is a continuous succession of insect species inhabiting the tree from the fresh green condition, the newly felled or killed tree, to the completely decomposed condition.

Host Selection

Most species of forest insects are limited in their feeding either to one species of tree (*monophagous*) or to a more or less prescribed group of related species (*oligophagous*). The locust borer, *Megacyllene robiniae,* for example, feeds on black locust; the sugar maple borer, *Glycobius speciosus,* attacks maples; the larch sawfly, *P. erichsonii,* defoliates larches; the Nantucket pine tip moth, *R. frustrana,* infests only pines, and the spruce budworm, *C. fumiferana,* is confined to a comparatively limited group of conifers, particularly the firs and spruces.

When an insect attacks only one species of tree, its control presents a comparatively simple problem. When it is a general feeder, like the gypsy moth, *L. dispar,* the problem becomes more complex. Fortunately, with many species that feed on a variety of hosts, the problem is simplified by the fact that there is a strong tendency for an insect to oviposit on the host upon which it was reared. This is called *Hopkins' host-selection principle* (Craighead, 1921).

The application of this principle has, in the past, given rise to the naming of new species based solely on their host associates. A definitive study of host selection by *Ips* beetles led Wood (1963) to suggest that the principle should be considered a hypothesis. Wood's suggestion has been supported by Smith and Sugden (1969) following cytological studies of speciation in the terminal weevils (*Pissodes* spp.).

HOST-RESISTANCE FACTORS

Trees possess certain physiological and genetic characteristics that result in producing, in vary degrees, a certain real ability to resist insect attack. These characteristics, by making environmental conditions less favorable for the insect than would otherwise be the case, increase the force of environmental resistance working against the insect. The environmental, physiological, and genetic factors influencing the resistance of a tree to insects and the simultaneous effects of the tree on the population dynamics of the insect can be discussed from the standpoint of either or both. Inasmuch as foresters are primarily concerned with the tree, this discussion of the host tree factors of environmental resistance will emphasize the tree.

Resistance

The resistance of a plant to insect attack was defined by Painter (1951, p.15) as "the relative amount of heritable qualities possessed by the plant which influence the ultimate degree of damage done by the insect." Painter classified the degree of the resistance of a plant to insect attack at levels ranging from immunity, through high, medium, and low resistance, to highly susceptible. The immune plant is never infested under any condition, the various levels of resistance reflect a lesser amount of damage than expected for a given set of conditions, and a highly susceptible plant receives more damage than would have been expected.

The resistance of trees to insects involves the tree, the insect, and their environments. The interactions are numerous and complex, acting either directly or indirectly on either the tree, the insect, or both. A good classification of the factors involved was developed by Painter (1951) (Fig. 5-4).

The phenomenon of resistance was divided by Painter into three interrelated mechanisms: *nonpreference, antibiosis,* and *tolerance.*

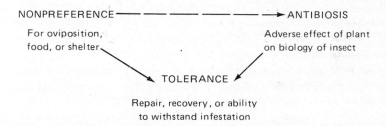

These three mechanisms can be demonstrated by questioning why the gypsy moth, *L. dispar,* attacks oak but not walnut (nonpreference), why gypsy moth larvae die when force-fed on walnut leaves (antibiosis), and why young, vigorous oaks survive gypsy moth outbreaks yet suppressed and overmature oaks are killed (tolerance).

Preference The mechanism of preference, why one given tree species and not another provides food and shelter for a given insect, involves the directed responses (taxes) of insects to chemical stimuli produced by the tree. There is evidence that many forest insects exhibit positive chemotaxes to volatile compounds found in trees. The monoterpene α-pinene is attractive to the Douglas-fir beetle *D. pseudotsugae* (Heikkenen and Hrutfiord, 1965), and the pales weevil, *Hylobius pales* (Thomas and Hertel, 1969); ethanol produced when a dead tree ferments attracts the ambrosia beetles (Cade et al., 1970, and Moeck, 1970). Numerous wood borers are attracted to turpentine (Wickman, 1969), and a coneworm, *Dioryctria abietella,* responds to the volatiles isolated from slash pine cones (Asher, 1970).

A positive flight taxis is perhaps the initial stage in the preference mechanism.

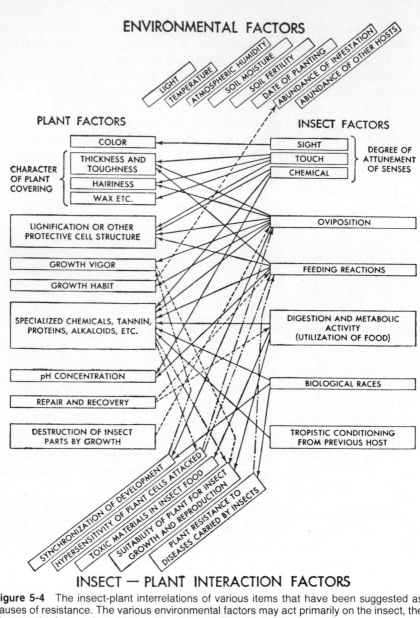

Figure 5-4 The insect-plant interrelations of various items that have been suggested as causes of resistance. The various environmental factors may act primarily on the insect, the plant, or the insect-plant relationship. *(Kansas State University.)*

Nonpreference The mechanism of preference has the corollary of nonpreference; that is, a tree emits no chemicals that attract an insect or masks the attractant with other chemicals.

Although the occurrence of a nonpreference is commonplace, there have been few studies of this resistance mechanism in forest entomology. We will seldom (if ever) see a coniferous bark beetle infesting hardwoods, cone and seed insects damaging seedlings, or termites attacking sound, dry wood. The phytophagous insects have co-evolved with plants and have apparently developed taxes to locate suitable sites for oviposition, shelter, and food.

The bark beetle complex associated with Virginia and loblolly pines exhibits a nonpreference for normal, healthy trees; a positive flight taxis does not exist. The insects do not attack until after the pines have been severely stressed (Hines and Heikkenen, 1977; Heikkenen, 1977).

The feeding stage is of utmost importance to insects that randomly locate their host. For example, the *Scolytus* beetles, which disperse through the crowns of host and nonhost trees, are inhibited in feeding on nonhost trees. Gilbert and Norris (1968) established that the smaller European elm bark beetle, *S. multistriatus,* will not feed on nonhost hickory because of the chemical juglone, yet juglone had no deterrent effect on the hickory bark beetle, *S. quadrispinosus.*

Nonpreference for shelter may also occur. For example, the Douglas-fir beetle, *D. pseudotsugae,* is sensitive to the moisture content of the sapwood. When the moisture content drops below 90 percent (ovendry weight), the beetles will leave the gallery, seek fresher material, and construct a new gallery (Johnson, 1963). Similar emergence has been observed for the southern pine beetles, *D. frontalis,* and termites when the moisture content of their shelter and food drops below their preference zone.

The nonpreference mechanism is often observed when resistant trees are planted adjacent to susceptible species. Such a case exists with the European pine shoot moth, *R. buoliana,* in the Lake States. When numerous species of resistant pines (Virginia, pitch, jack, and eastern white pine) are interplanted with highly susceptible red pines, these pines are infested but at significantly lower levels (Miller and Heikkenen, 1959). If the resistant trees are planted separately, the insect attack will probably not occur (Fig. 5-5).

Antibiosis The resistance mechanism, antibiosis, is the detrimental effect a tree has on an insect. Certain trees apparently possess immunity from insects because they are actually distasteful or poisonous to the insects. Exactly what these characteristics are is beginning to be known for a few species. The existence of antibiosis cannot be doubted.

Antibiosis may not be possessed to the same degree by all the individuals of a species. Some trees may be immune to attack, whereas a neighboring tree may be highly susceptible. A striking example of variation in this respect is

Figure 5-5 In this plantation infested with the European pine shoot moth, *R. buoliana,* the highly susceptible red pine *(left)* growing adjacent to the medium-resistant Scotch pine *(right)* masked the nonpreference mechanism of highly resistant jack and eastern white pine. *(U.S. Forest Service photo.)*

to be observed in Norway spruce when attacked by the eastern spruce gall adelgid, *A. abietis.* Most of the trees are heavily infested, but others are absolutely untouched. Even when the branches of an immune tree interlock with those of a heavily infested one, the immune tree will remain uninfested. The antibiosis mechanism may be the presence and abundance of a phenolic compound in the needles (Tjia and Houston, 1975).

The operation of the antibiosis mechanism of an immune or resistant tree on an insect usually occurs when the insect begins to feed. Either the insect will die, or its development will be severely curtailed, resulting in a prolonged development, reduction in size, and reduced fecundity.

The resins of nonhost pines are toxic to the western, *D. brevicomis,* mountain, *D. ponderosae,* and Jeffrey pine beetles, *D. jeffreyi* (Smith, 1972). Laboratory tests have demonstrated that selected monoterpenes (β-pinene) are repellant to the Douglas-fir beetle, *D. pseudotsugae.* Beta-pinene does not occur in the parts of the tree inhabited by the beetle; however, it is concentrated in the resin exuded from wounds (Heikkenen and Hrutfiord, 1965).

Terpene differences having insect- and fungus-repellant properties also occur in the western true firs (Russell and Berryman, 1976).

Chemical variation within species of the same tree genus will also affect antibiosis. The spruce budworm, *C. fumiferana,* is deterred by the chemical pungenin (Heron, 1965). The concentration of pungenin is low in the tissues of white spruce preferred by the larvae, yet the larvae avoid the high concentrations that exist in the mature needles of white spruce and in the tissues of the resistant black spruce. Even within a given tree species, seasonal changes of chemicals will affect insect preference. Sawflies, *Neodiprion* spp., which normally feed on old needles of jack pine, will not feed on the new needles in the early summer because of the high concentrations of resin acids. However, the concentrations of resin acids diminish later in the growing season, and the next generation of sawflies will feed on the new needles (Ikeda et al., 1977).

The English oak, *Quercus robur,* is attacked by numerous defoliators in the early summer; yet after mid-June, the insects are relatively scarce. Feeny (1970) attributes the decline of the defoliators (while an apparent food supply exists) to the seasonal increase in the toughness of the leaves and a decline in the availability of nitrogen as increasing concentrations of tannins combine with the proteins. The larvae of the gypsy moth, *L. dispar,* do not feed on mountain laurel, dogwood, holly, red spruce, tulip poplar, or black walnut (Mosher, 1915; Doskotch et al., 1977). If the gypsy moth spreads southward through the Appalachian forests, the antibiotic effect of these tree species should be considered when projecting the insect's economic importance (Fig. 5-6).

Tolerance   The resistant mechanism of tolerance is the ability of a tree to grow, reproduce, or repair injury in spite of an insect population that would damage a susceptible tree. The degree of tolerance is strongly influenced by the age and vigor of the tree and is subject to variations in the environment.

Vigorous, rapidly growing trees exhibit more tolerance to insect attack than do trees of the same species that are older, growing more slowly, or under stress. For instance, some of the buprestid wood borers may deposit their eggs on trees that are fairly healthy. These eggs will hatch, and the larvae will penetrate the bark. In order to develop, these larvae must feed for a period of time in the inner phloem and cambium, but they can feed only on the phloem of decrepit trees or freshly cut logs. Therefore, when they reach the inner phloem of a healthy tree, they either die immediately or are forced to turn back into the bark, where they eventually die. These small buprestid larvae have frequently been found starving in the bark of healthy jack pine, hemlock, or ponderosa pine.

A characteristic that is often associated with rapid growth is the copious flow of resin and sap. This flow will drive out or overwhelm and kill almost any of the phloem borers before they can become established in the tree. Only

Figure 5-6 This tulip poplar *(right)* was not defoliated during an outbreak of the gypsy moth, *Lymantria dispar*. Tulip poplar and black walnut have an antibiotic mechanism that stops larval feeding. Note the defoliated eastern white pine *(left)*. *(Pennsylvania State University.)*

when an especially vigorous tree is attacked simultaneously by large numbers of these insects is it unable to resist attack. Illustrations of resistance because of higher vigor may be seen in any forest infested with *Dendroctonus* beetles. When resistant trees are attacked, they usually become covered at each point of attack with pitch tubes formed by hardened resin. Examination of these trees will show that some beetles are entombed in their resin-filled galleries and that other tunnels are deserted (Belluschi et al., 1965) (Fig. 5-7).

The galleries of phloem insects are often found overgrown by many years of woody growth, indicating that an old attack was aborted. Galleries of pine engravers (Kulman, 1964) and the fir engravers (Berryman, 1969) are frequently thus overgrown.

The tolerance of a tree is often associated with site. When red pine is planted on sites that have adequate soil moisture, the tree exhibits a tolerance to attack by the European pine shoot moth, *R. buoliana.* The larval population collapses during the needle-mining stage (Heikkenen, 1963). Similarly, eastern white pine in New York is tolerant to attack by the white pine weevil, *Pissodes strobi,* when the depth of rooting exceeds 90 cm (Connola and Wixson, 1963). Conversely, loblolly pine has no tolerance to the southern pine beetle, *D. frontalis,* when growing in shallow, rocky soils when droughts occur (Brender, 1955) or in the flatwoods when the root systems are flooded (Heterick, 1949).

Pseudotolerance When potentially susceptible trees have an apparent tolerance to insect attack, the possibility of pseudoresistance by evasion, induced resistance, or escape should be considered.

Vigorous, rapidly growing trees are damaged less from insect injury than are trees of the same species that are growing more slowly. Rapid growth has a double effect: (1) it produces in the tree more tolerance both to attack and to injury, and (2) it shortens each developmental stage of the tree, thereby reducing the period of susceptibility to the insects attacking each stage. Throughout the United States and Canada, pine plantations are considered susceptible to the tip moths, *Rhyacionia* spp., until the stands close. When vigorous seedlings are planted on good sites with no brush and weed competition, the trees may evade the tip moths by quickly passing through the susceptible stage.

Sometimes a tree species may evade insect attack because of phenological variation in the foliage. The lack of synchronization between the spruce budworm, *C. fumiferana,* and black spruce illustrates this kind of resistance. When the larvae emerge in the spring, they require tender, fresh foliage for food. Their time of emergence is about 2 weeks prior to the opening of the black spruce buds; consequently, in a stand of black spruce, the larvae are faced with a food shortage in the early part of the season. Some of them find the green tissues they require by boring into a bud or by mining a needle, but many of them fail to do this and perish. Thus, the presence of black spruce in a forest reduces budworm abundance.

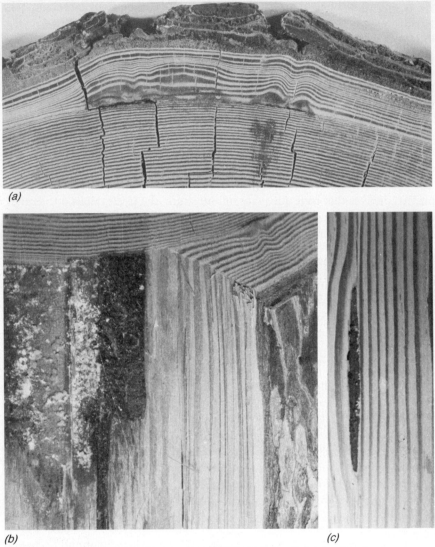

(a)

(b)   (c)

Figure 5-7   Douglas-fir can abort the attack of the Douglas-fir beetle, *Dendroctonus pseu-dotsugae.* The cross section *(a)* reveals 13-year-old attacks aborted during the 1952 out-break. The pitch pockets as seen in the flat-grain *(b)* and vertical- grain board *(c)* are lumber defects. *(Virginia Polytechnic Institute and State University.)*

The white spruce also reduces the number of budworm larvae. Unlike the black spruce, the buds of the white spruce open at a suitable time, but the needles harden and become unpalatable before all the larvae can reach full growth. Many of them perish when they are forced to leave the trees to seek more suitable food (Graham and Orr, 1940).

A tree species may escape an insect pest when it is planted outside the range of a native insect or an introduced pest whose distribution has yet to include the tree. For example, red pine planted south of the 40th parallel is not damaged by the European pine shoot moth, *R. buoliana*; the tree is beyond the insect's present range. In Connecticut, red pine planted south of its range is now being destroyed by the introduced red pine scale, *Matsucoccus resinosae.* [1] Red pine is highly susceptible to this scale, and noninfested trees beyond the insect's range will perhaps escape if the pest does not continue to spread. A similar situation exists for white fir in California if the balsam woolly adelgid, *A. piceae,* spreads southward from Oregon. And the Appalachian hardwood forests may have only escaped the gypsy moth, *L. dispar,* for the present.

An incomplete insect outbreak will, of course, miss certain trees. If an insect such as the white pine or Sitka spruce weevil, *P. strobi,* randomly damages 20 percent of 1000 trees per acre each year for 20 years, you could expect 14 trees ($0.8^{20} \times 1000$) to have escaped any damage in a 20-year-old plantation.

BIOLOGICAL FACTORS

Biological factors include all interactions among organisms that result in the limitation of insect multiplication. The biological factors that play important roles in environmental resistance are competition, parasites, predators, and pathogens.

Competition

Competition between insects may occur either within the same species (intraspecific) or between individuals of different species (interspecific). In either case, the result is a reduction in the rate of increase of the competing organisms. Competition occurs *only* when a requisite resource—that is, something desired or needed, such as food, space, or shelter—is in short supply.

Under normal conditions, when forest insects are present in moderate numbers, the things they need are in excess of their demands. As a result, there can be no competition. Food is an example. In order for a continuous supply of food to be available, only a small part can be used at any one time. Food must be produced more rapidly than it is consumed; otherwise, the supply will eventually be cut off. This means the forest trees can afford to supply other organisms with only that amount of food which they are able to produce in excess of their own requirements for maintenance and growth. Thus, when other environmental forces are maintaining the insects at normal numbers, the insects are using only this surplus or less. Therefore, there is no scarcity of food and no competition.

[1] Personal communication, R. L. Talerico.

At a time when some insect species temporarily increase in number to a point where an outbreak exists, the situation is changed. Then the insects are no longer feeding on the surplus but are actually using up the capital. Under such conditions, competition occurs not only among the insects of the species in the outbreak state but also with other species that require the same food. There can be only one result: Sooner or later, if the outbreak is not checked by the action of some outside controlling influence, the supply of food will become exhausted. The insects dependent on it will either starve or be forced to migrate. Thus, competition for food reduces the number of the excessively abundant species and possibly some other species dependent on the same food.

Similarly, two different species of bark beetles may compete directly with each other for closely similar food. When windfallen spruce trees are attacked simultaneously by the spruce beetle, *D. rufipennis,* and *Ips* beetles, direct competition results. *Ips* beetles develop more rapidly, and they may soon destroy the habitat for the *Dendroctonous* brood (Fig. 5-8). Only in trees or

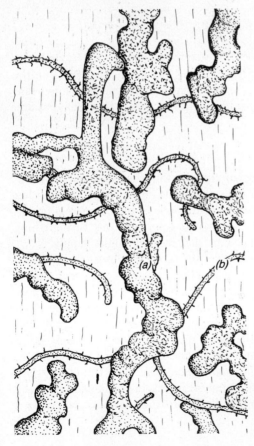

Figure 5-8 Cerembycid *(a)* and *Dendroctonus (b)* tunnels in the p hloem of pine, illustrating competition between these insects. Increase of *Dendroctonus beetles* is often prevented by interspecific competition with cereambycid larvae. *(From Graham and Knight, Principles of Forest Entomology, 4th ed., Fig. 13, p. 61.)*

logs too moist for *Ips* or uninfested by them can the *Dendroctonus* succeed. Evidence is plentiful in support of the contention that species with closely similar needs and habits cannot live together indefinitely.

Competition for space is an important limiting factor among some groups of forest insects. Competition of this type is sometimes very keen among the phloem insects and also among the wood borers. In nature, this competition for space is usually difficult to evaluate because other kinds of competition operate simultaneously. Only by carefully controlled experiments, in which space is the single limiting factor, can the true effects be demonstrated. Whenever competing individuals are limited by restricted space, a definite population equilibrium develops. The population level is characteristic of the species and is attained regardless of the initial number of individuals (Crombie, 1946). If individuals are too numerous, the population will be reduced; if individuals are too few, the organisms will increase until the number characteristic of the species and the space is attained. In the case of forest insects, the number of young produced is usually in excess of the equilibrium number. Competition for space causes the elimination of part of the juvenile population (Fig. 5-9).

Sometimes the abundance of forest insects is limited by competition for suitable shelter. The abundance of termites or of carpenter ants is definitely limited in certain instances by a scarcity of places in which to build their nests.

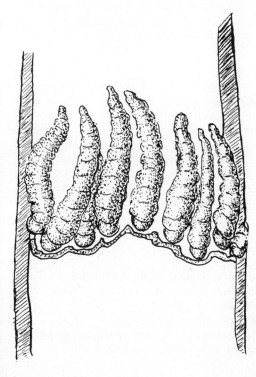

Figure 5-9 The larvae of the white pine weevil, *Pissodes strobi*, occasionally arrange themselves in a ring as they feed downward in the terminal shoot of eastern white pine. This results in intraspecific competition for food and space. *(From Graham and Knight, Principles of Forest Entomology, 4th ed., Fig. 14, p. 62.)*

With most forest insects, however, if food is sufficient, shelter is also available, for the insect's shelter is usually made from the same material that is used for its food, from its body secretions, or from a combination of both.

Predators, Parasites, and Pathogens

The effects of predators, parasites, and pathogens in producing environmental resistance are so nearly identical that it seems advisable to discuss them together in order to avoid repetition. These three closely allied factors constitute an important part of environmental resistance. Without the control exercised by these beneficial organisms, almost all insect pests would be more serious than they now are. The effect of predators, parasites, and pathogens is the subject of Chapter 9, "Biological Control."

BIBLIOGRAPHY

Major Texts

Andrewartha, H. G., and Birch, L. C. 1958. *The distribution and abundance of animals.* Chicago: University of Chicago Press.

Borror, D. J., et al. 1976. *An Introduction to the study of insects.* 4 ed. New York: Holt, Rinehart & Winston.

Chapman, R. F. 1971. *The insects: Structure and function.* New York: American Elsevier Publishing Co.

Painter, R. H. 1951. *Insect resistance in crop plants.* Lawrence: University Press of Kansas.

Literature Cited

Asher, W. C. 1970. Olfactory response of *Dioryctria abietella* (Lepidoptera: Phycitidae) to slash pine cones. *Ann. Entomol. Soc. Am.* 63:474–476.

Balch, R. E. 1952. Studies of the balsam woolly aphid, *Adelges piceae* (Ratz), and its effects on balsam fir, *Abies balsamea* (L.) Mill., *Can. Dept. Agr. Publ.* no. 867.

Baumberger, J. P. 1919. A nutritional study of insects with special reference to microorganisms and their substrata. *J. Exptl. Zool.* 28:1–81.

Belluschi, P. G., et al. 1965. Douglas-fir defects caused by the Douglas-fir beetle. *J. For.* 63:252–256.

Berryman, A. A. 1969. Responses of *Abies grandis* to attack by *Scolytus ventralis* (Coleoptera: Scolytidae). *Can. Entomol.* 101:1033–1041.

Bethune, J. E. 1963. Pine tip moth damage to planted pines in south Florida. *U.S. For. Serv. Res.* note SE-7.

Blackman, M. W. 1918. The American tent caterpillar. *J. Econ. Entomol.* 9:432.

Blais, J. R. 1965. Spruce budworm outbreaks in the past three centuries in the Laurentide Park, Quebec. *For. Sci.* 11:130–138.

Brender, E. V. 1955. Drought damage to pines. *For. Farmer* (July):7, 15.

Cade, S. C., et al. 1970. Identification of a primary attractant for *Gnathotrichus sulcatus* isolated from western hemlock logs. *J. Econ. Entomol.* 63:1014–1015.

Callahan, P. S. 1975. *Tuning in to nature.* Old Greenwich, Conn.: Devin-Adair Co.

Carpenter, J. R. 1940. Insect outbreaks in Europe. *J. Animal Ecol.* 4:108–147.

Connola, D. P., and Wixson, E. C. 1963. White pine weevil attack in relation to soils and other environmental factors in New York. *New York St. Museum and Sci. Serv. Bull.* no. 389.

Craighead, F. C. 1921. Hopkins' host-selection principle as related to certain cerambycid beetles. *J. Agr. Res.* 22:189–220.

Crombie, A. C. 1946. Interspecific competition. *J. Animal Ecol.* 16:44–73.

Doskotch, R. W., et al. 1977. Feeding responses of gypsy moth larvae, *Lymantria dispar,* to extracts of plant leaves. *Environ. Entomol.* 6:563–566.

Dwinell, L. D., and Phelps, W. R. 1977. Pitch canker of slash pine in Florida. *J. For.* 75:488–489.

Evans, W. G. 1971. The attraction of insects to forest fires. *Proceedings, Tall Timbers Conference on Ecological Animal Control by Habitat Management* 3:115–127.

Ewan, H. E. 1961. The Saratoga spittlebug. *USDA For. Serv. Tech. Bull.* no. 1250.

Feeny, P. 1970. Seasonal changes in oak leaf tannins and nutrients as a cause of spring feeding by winter moth caterpillars. *Ecology* 51:565–581.

Frisch, K. von. 1971. *Bees, their vision, chemical senses, and language.* Ithaca: Cornell University Press.

Gilbert, B. L., and Norris, D. M. 1968. A chemical basis of bark beetle (*Scolytus*) distinction between host and non-host trees. *J. Insect Physiol.* 14:1063–1068.

Graham, S. A. 1924. Temperature as a limiting factor in the life of subcortical insects. *J. Econ. Entomol.* 17:377–383.

———. 1956a. The larch sawfly in the Lake States. *For. Sci.* 2:132–160.

———. 1956b. Forest insects and the law of natural compensations. *Can. Entomol.* 88:45–55.

———, and Knight, F. B. 1965. *Principles of forest entomology.* 4 ed. New York: Mc-Graw Hill Book Company.

———, and Orr, L. W. 1940. The spruce budworm in Minnesota. *Univ. Minn. Agric. Exp. Sta. Tech. Bull.* no. 142.

———, and Williams, C. B. 1958. Forecasting spring activation of the European pine shoot moth larvae from phenological observations. *Univ. Mich. For. Note* 21.

Green, G. W. 1962a. Flight and dispersal of the European pine shoot moth, *Rhyacionia buoliana. Can. Entomol.* 94:282–314.

———. 1962b. Low winter temperatures and the European pine shoot moth, *Rhyacionia buoliana* (Schiff.) in Ontario. *Can. Entomol.* 94:314–336.

Haseman, L. 1950. Controlling insect pests through their nutritional requirements. *J. Econ. Entomol.* 43:399–401.

Heikkenen, H. J. 1963. Influence of site and other factors on damage by the European pine shoot moth. University of Michigan School of Natural Resources. Ph.D. Dissertation (University Microfilms, Ann Arbor, No. 64–6689).

———. 1977. Southern pine beetle: A hypothesis regarding its primary attractant. *J. For.* 75:412–413.

———, and Hrutfiord, B. F. 1965. *Dendroctonus pseudotsugae:* A hypothesis regarding its primary attractant. *Sci.* 150:1457–1459.

Heron, R. J. 1965. The role of chemostatic stimuli in the feeding of spruce budworm larvae on white spruce. *Can. J. Zool.* 43:247–269.

Heterick, L. A. 1949. Some overlooked relationships of southern pine beetle. *J. Econ. Entomol.* 42:466–469.

Hines, J. W., and Heikkenen, H. J. 1977. Beetles attracted to severed Virginia pine. *Environ. Entomol.* 6:123–127.

Hodson, A. C. 1941. An ecological study of the forest tent caterpillar, *Malacosoma disstria* Hbn., in northern Minnesota. *Univ. Minn. Agric. Exp. Sta. Tech. Bull.* no. 148.

Howard, L. O. 1926. The parasite element of natural control of insects and its control by man. *J. Econ. Entomol.* 19:271–282.

Ikeda, T., et al. 1977. Chemical basis for feeding adaption of pine sawflies *Neodiprion rugifrons* and *Neodiprion swainei. Sci.* 197:497–499.

Johnson, N. E. 1963. Factors influencing the "second-attack" of the Douglas-fir beetle. *Weyerhaeuser Co. For. Res. Note* 53.

Keen, F. P. 1937. Climatic cycles in eastern Oregon as indicated by tree rings. *USDA Weather Bur., Weather Rev.* 65:175–188.

Kulman, H. M. 1964. Pitch defects in red pine associated with unsuccessful attacks by *Ips* spp. *J. For.* 62:322–325.

Linsley, E. G. 1943. Attraction of *Melanophila* beetles by fire and smoke. *J. Econ. Entomol.* 36:341–342.

Martin, C. H. 1946. Effect of condition of phloem moisture on the entry of *Scolytus multistriatus. J. Econ. Entomol.* 39:481–486.

Merriam, C. H. 1898. Life zones and crop zones in the United States. *USDA Biol. Survey Bull.* no. 10.

Meyer, J. R. 1977. Head capsule transmission of long-wave length light in the Curculionidae. *Sci.* 196:524–525.

Miller, W. E., and Heikkenen, H. J. 1959. The relative susceptibility of eight pine species to European pine shoot moth attack in Michigan. *J. For.* 57:912–914.

Moeck, H. A. 1970. Ethanol as the primary attractant for the ambrosia beetle, *Trypodendron lineatum* (Coleoptera: Scolytidae). *Can. Entomol.* 102:985–995.

Morris, C. L., and Frasier, D. H. 1966. Development of a hazard rating for *Fomes annosus* in Virginia. *Plant Dis. Rep.* 50:510–511.

Mosher, F. H. 1915. Food plants of the gypsy moth in America. *USDA Bull.* no. 250.

Payne, N. M. 1926. Freezing and survival at low temperatures. *Quart. Rev. Biol.* 1:270–282.

Pierce, W. D. 1916. Relationship of temperature and humidity to insect development. *J. Agr. Res.* 5:1183–1191.

Russell, C. E., and Berryman, A. A. 1976. Host resistance to the fir engraver beetle. I. Monoterpene composition of *Abies grandis* pitch blisters and fungus-infested wounds. *Can. J. Bot.* 54:14–18

Salt, R. W. 1961. Principles of insect cold-hardiness. *Ann. Rev. Entomol.* 6:55–74.

Sanderson, E. D. 1908. Relation of temperature to hibernation. *J. Econ. Entomol.* 1:56–65.

Shepherd, R. F. 1958. Factors controlling the internal temperatures of spruce budworm larvae, *Choristoneura fumiferana* (Clem.). *Canad. J. Zool.* 36:779–786.

Smith, R. H. 1972. Xylem resin in the resistance of the pinacene to bark beetles. *USDA For. Serv. Gen. Tech. Rep.* PSW-1.

Smith, S. G., and Sugden, B. A. 1969. Host trees and breeding sites of native North

American *Pissodes* bark weevils with a note on synonymy. *Ann. Entomol. Soc. Am.* 62:146–148.

Snyder, T. E. 1926. Lyctus powder-post beetles. *USDA Farmers' Bull.* no. 1477, pp. 1–13.

Struble, G. R. 1972. Biology, ecology and control of the lodgepole needleminer. *USDA For. Serv. Tech. Bull.* no. 1458.

Sullivan, C. R., and Wallace, D. R. 1972. The potential spread of the gypsy moth, *Porthetria dispar* (Lepidoptera: Lymantriidae). *Can. Entomol.* 104:1349–1355.

Thomas, H. A., and Hertel, G. D. 1969. Responses of the pales weevil to natural and synthetic host attractants. *J. Econ. Entomol.* 62, no. 2: 383–386.

Tjia, B., and Houston, D. B. 1975. Phenolic constituents of Norway spruce resistant or susceptible to the Eastern spruce gall aphid. *For. Sci.* 21:180–184.

Wellington, W. G. 1948. The light reactions of the spruce budworm. *Can. Entomol.* 70:56–82.

———. 1949. The effects of temperature and moisture upon the behavior of the spruce budworm, *Choristoneura fumiferana* (Clem.) (Lepidoptera: Tortricidae). I. The relative importance of graded temperatures and rates of evaporation in producing aggregations of larvae. *Sci. Agric.* 29:201–215. II. The responses of larvae to gradients of evaporation. *Sci. Agric.* 29:216–229.

———. 1950*a*. Variations in the silk-spinning and locomotor activities of larvae of the spruce budworm, *Choristoneura fumiferana* (Clem.), at different rates of evaporation. *Trans. Royal Soc. Canada* 44:89–101.

———. 1950*b*. Effects of radiation on the temperatures of insectant habitats. *Sci. Agric.* 30:209–234.

———, et al. 1950. Climate and spruce budworm outbreaks. *Can. J. Res. D.* 28:308–331.

———, et al. 1951. Polarized light and body temperature level as orientation factors in the light reactions of some hymenopterous and lepidopterous larvae. *Canad. J. Zool.* 29:339–351.

Wickman, B. E. 1969. Wood borers attracted to turpentine. *USDA For. Serv. Res. Note* PSW-195.

Witter, J. A., and Kulman, H. M. 1972. Mortality factors affecting eggs of the forest tent caterpillar, *Malacosoma disstria*. *Can. Entomol.* 104:705–710.

Wood, D. L. 1963. Studies on host selection by *Ips confuses* (LeConte). *Univ. Calif. (Berkeley) Publ. Entomol.* 27:241–282.

Chapter 6

Population Levels

The actual level of insect populations at any time or place is determined by the interaction of reproductive potential and environmental resistance. This chapter considers some of these interactions and their results. The importance of insect numbers lies in the fact that the capacity of an insect species to injure trees is in direct proportion to the number of that insect in the forest. When its population is low, a potential pest is innocuous; whereas when its population is high, the same insect may cause tremendous devastation.

The student is aware that insects are the dominant group of animals on the earth (Borror et al., 1976). Their populations may number in the millions per acre, and their evolution has resulted in great diversity and adaptability. The early insects may have been mostly scavengers, but evolution has resulted in many other adaptations, so that today, though many insects are still scavengers, there are others that feed on living plants and animals.

BIOTIC BALANCE

When the astonishing ability of insects to reproduce is considered, it is amazing that our forests are not damaged far more than is actually the case. But the fact remains that our natural forests usually develop to maturity and are not destroyed by insects. Insects kill trees, but there is a maintenance of a relative equilibrium, so that over long time periods the forest persists, and, on the average, insects are prevented from increasing in number to a destructive level. This condition of equilibrum is called *biotic balance*.

Insect numbers fluctuate from year to year and from season to season, but their varying numbers and the elements of the environment are so equalized that survival of both trees and insects continues. Thus, in the struggle for supremacy among different forms of life, there is a deadlock. Biotic balance is, therefore, an oscillating, not a static, equilibrium. It is a condition wherein, on the average, reproductive potential and environmental resistance are equal.

Weather conditions, the number of parasites and predators, and available food all change continuously. As a result, the number of individuals in each forest environment varies from time to time and from place to place, but usually the amplitude of the variations is not great because the numbers oscillate about a mean point. The importance of this oscillating equilibrium cannot be overemphasized. It is only under such a balanced state that long-lived organisms exist at all. The life of a tree extends over 100 years or more; sexual maturity is attained after many years of growth. Such lengthy processes, obviously, presuppose generally favorable environmental conditions extending through centuries of time.

It should not be inferred from this discussion, however, that the environmental conditions remain constant or even relatively constant throughout the life of any individual forest tree. Obviously, the ecological picture gradually changes as trees pass from infancy to old age. It has already been pointed out that trees of different ages are attacked by different insects. Similarly, different plants and animals are associated with each age-class of trees. Thus, in considering any tree or small group of trees, we are dealing with only a segment of the complete environmental complex. Only in a large forest area, with a normal distribution of age-classes and their associated flora and fauna, is the environmental complex complete. Then each element of the complex is operative but in different spots at different times. As one factor ceases to operate in one spot, it is replaced by another.

ENVIRONMENTAL FACTORS INFLUENCING POPULATIONS

The balance is maintained by a wide variety of environmental forces that act on insects with different reproductive potentials. For instance, if an insect produces 100 eggs per female and the sex factor is 0.5, then the reproductive potential of 50 must be matched by an environmental resistance of 49 if the population is to remain constant. Similarly, if a species has a reproductive potential of 200 individuals, the environmental resistance must be sufficient to destroy close to 200 individuals, including parents, for each individual of the parent generation. If the forces are not nearly equal, the population will either increase or decrease. The difference between the number of individuals in one generation and the next, after the force of environmental resistance has taken its toll, determines the rate at which a species will increase or decrease. Thus, the number of each succeeding generation will equal the number of individuals at the beginning of the generation times the reproductive potential minus the environmental resistance.

Compensation among Factors

If a species is to remain in relatively constant numbers in successive genera-tions in spite of changes in the force of different factors of environmental resistance, it is apparent that the changes in resistance factors must be compen-sating (Graham, 1956). This is proved by the fact that more than 95 percent of the forest-insect species in a locality continue to be present year after year in relatively low numbers in spite of variations in the individual factors of environmental resistance.

An excellent specific illustration of this compensation of resistance factors was observed by S. A. Graham (Graham and Knight, 1965) in studying the jack pine sawfly, *Neodiprion pratti banksianae.* For 6 years, under favorable food conditions, the population remained approximately static. During two seasons, there was a combination of unfavorably wet weather and a disease; during another year, there was little disease, but a late frost killed a large proportion of the young larvae; and during 3 different years, heavy storms washed many of the larvae from the trees. Thus, there was no increase during the period in spite of the favorable food situation.

In the example just cited, the compensation among the environmental resistance factors occurred in the same locality but in different seasons. Similar compensatory effects are apparent when geographic areas are compared. For example, in the moist tropics, physical resistance is practically inoperative. There the increase of insects is prevented by the biological factors (competi-tion, predators, and parasites). In arctic and subarctic regions, on the other hand, physical resistance is high, and biological resistance is correspondingly low. In other words, increased biological resistance compensates for lack of physical resistance in the tropics, and physical resistance compensates for lack of biological resistance in the colder localities.

Life Systems

Clark et al. (1967) presented their definition of an ecological unit termed the *life system.* It is defined as the part of the ecosystem that determines the existence, abundance, and evolution of a population. This idea provides a new terminology for the general population level in relation to the environment in which it exists. The entire ecosystem then is made up of a large number of interlocking life systems (Price, 1975).

The components of the life system are simply stated but have far-reaching implications for the entire system. The genotype must be first because this is the inherited ability of the individuals to survive and multiply as affected by environmental factors. This gives us the individuals in the population (phenotypes), which together form the existing group having statistics such as numbers and rates of birth, death, and dispersal. All these qualities of the subject species and those of the environment are the determinants of population numbers. Those genotypes that result in vigorous phenotypes will be favored in the environment, assuring that the population will continue a slow change molded by genetic and environmental elements.

Some authors prefer to use the name *balance of nature,* rather than biotic balance; this is a matter of preference. A comment is needed on the terms *feedback* and *homeostasis,* [1] which are commonly used in a discussion of the basic concepts of biotic balance. Feedback is merely those loops or control mechanisms that prevent a population from effectively destroying itself. Such feedback processes, either positive or negative in effect, lead to a general homeostasis in the environment, a biotic balance. Homeostasis, then, is a balance brought about by feedback loops and control mechanisms that prevent total loss of stability. We are all aware that stability is a basic characteristic of all natural populations.

The idea of zones of abundance has considerable validity. This concept divides the range of a species into zones based on the most favorable environment for the species. In the most favorable portion of the range lies the *zone of normal abundance,* within which physical conditions are invariably favorable for the species and within which outbreaks are controlled or prevented by the action of nutritional and biotic factors of environmental resistance.

Outside this zone lies the *zone of occasional abundance,* within which the physical conditions are always sufficiently favorable to permit the species to survive but within which favorable variations will only occasionally permit them to become abundant.

Next comes the *zone of possible abundance,* within which conditions are usually so unfavorable that the insect cannot maintain itself except when favorable weather variations permit invasion from a more favorable zone. When weather conditions return to normal, the insects that have invaded the zone are destroyed. Another zone might well be added, called the *zone of possible occurrence,* within which physical conditions are never sufficiently favorable to permit the insect to become numerous but within which it might occasionally be found in favorable habitats.

[1] Homeostasis, derived from the Greek *homoios,* meaning like or similar, and *stasis,* meaning to stand.

Physical factors are especially important in determining the distribution of a species by permitting it to occupy or by preventing it from occupying a given place. The success or failure of a species outside the zone of normal abundance depends on the distribution and size of the favorable habitats. It is apparent that an insect is not likely to become economically important throughout its entire range and that outbreaks will seldom (if ever) occur outside the zones of normal and occasional abundance.

The zone concept is only one illustration of the workings of a life system. It does demonstrate the entire process of population performance, including feedback and homeostasis. The population numbers may fluctuate considerably, but the life system as a whole is protected from extinction by the many interacting factors of the environment in which it exists.

Population Regulation

Frequently, the value of a certain factor of environmental resistance may vary with the population density of the insect against which it is working. When the population of the insect is at one level, a factor of this kind may be relatively ineffective in reducing the insect's numbers; whereas when the population is at another level, it may be very effective. Smith (1935) has given the name *density-dependent* to such a factor, in contrast with the physical factors previously discussed, which he terms *density-independent.* Density-dependent factors generally operate most effectively against relatively high populations and correspondingly less effectively against low populations. For instance, insect species that are parasitic or predaceous on other insects and entomogenous disease-causing organisms usually increase their efficiency at high host density and decrease their efficiency at low host density. The rate at which this efficiency increases with host density determines the population level at which parasite and host will reach equilibrium. Thus, a parasite that is able to produce a high rate of mortality in its host at low density will, if other things are equal, be economically effective; whereas another parasite that increases more slowly in efficiency with increased host density may be relatively ineffective.

Birds are density-dependent, as are other vertebrate predators. Unlike parasites and disease-causing organisms, birds are most efficient in retarding insect increase when the prey population is relatively low but still numerous enough to offer an attractive reward for hunting. This is especially true during the breeding period, when the birds are confined to their breeding grounds, where their rate of increase is limited by their seasonal and spatial requirements. Territorial requirements limit to a certain number the nesting pairs that can live and breed in an area; during the breeding period, birds cannot increase their numbers at a rapid rate to take advantage of an unusually abundant food supply. Therefore, birds are effective as resistance agencies up to the point at which their nutritional requirements are satisfied by the insect population on which they are feeding. However, when the insect population exceeds that amount which the birds can eat, these vertebrates cease to exercise any important restrictive influence during the breeding season.

There have been numerous theories on the subject of population regulation, and these have led to a considerable argument between those who believe that factors external to the population are responsible and those who propose that factors within the population produce regulation. Clark et al. (1967) and Price (1975) have included brief summaries of the various points of view as expressed by Andrewartha (1961), Milne (1957), Nicholson (1958), Solomon (1957), and others. The arguments in general revolve around the interpretation of specific resistant mechanisms and genetic adaptability. The theories are different ways of regarding and evaluating the same things. Some may wish to relate regulation directly to density-related processes within life systems; others may prefer to view the life system based on differences in favorableness of environmental conditions. Some insect populations may clearly express the results of density dependence; others may seem to be regulated by independent factors. The forest manager is probably not overly concerned about the theoretical aspects of the problems but may be more interested in knowing those key factors that can be adjusted to reduce losses in the forests being managed.

The term *key factor* is used frequently in the literature and is of value to the study of forest entomology (Morris, 1959). A key factor is that factor (or factors) in a population which causes mortality, closely related to total changes in the population from generation to generation. Close correlations between population changes and a specific cause of change may lead to the conclusion that that factor was indeed the cause and therefore the key factor. But such conclusions must be made with caution because correlation does not always mean there is a direct and independent relationship.

Life tables have become a very useful tool in population ecology because the various factors influencing populations may be presented in a logical display that can be readily interpreted. Morris and Miller (1954) summarize the procedure used in developing the tables and explain their use. Many papers have been published since 1954 that also utilize life tables to express population changes in an understandable and systematic form. Such life tables reveal the accumulative effects of all mortality factors throughout the insect's life cycle. In insect studies, the survival of a large group (born at about the same time) is measured at fairly close intervals throughout its existence. Ideally, the same individuals should be observed at each sampling; but in field studies of forest insects, this is generally impossible. Thus, survival is estimated, not by observing the same individuals, but by sampling the population. It should be pointed out that life tables are not an end point. They are merely a format in which survival data can be systematically presented.

The format for a life table is illustrated in Table 6-1, in which hypothetical

Table 6-1   Format for a Life Table, with Possible Sampling Stages

x	$1x$	dxF	dx	$100qx$
Sampling interval	No. alive at beginning of x	Factor responsible for dx	No. dying during x	Percent mortality $dx/1x$
I. Eggs (fall)	—	Parasites	—	—
		Other	—	—
		Total	—	—
II. Larvae (fall)	—	Parasites	—	—
		Temperatures	—	—
		Other	—	—
		Total	—	—
III. Larvae (spring)	—	Parasites	—	—
		Predators	—	—
		Other	—	—
		Total	—	—
IV. Pupae (summer)	—	Parasites	—	—
		Predators	—	—
		Other	—	—
		Total	—	—
V. Adults (summer)	—	Sex ratio	—	—
		Predation	—	—
		Weather	—	—
		Other	—	—
		Total	—	—
Generation			—	—

where x = Age interval at which the sample was taken
         $1x$ = Number living at the beginning of the stage noted in the x column
        dx = Number dying within the age interval in the x column
       dxF = Mortality factor responsible for dx
     $100qx$ = Percent mortality = $dx/1x$

sampling intervals and mortality factors are presented. The table headings have been utilized by numerous authors.

This discussion of life tables has related to forest insects, but the reader should realize the broader usefulness of such presentations. Life tables were originally used by people interested in human populations, but the usefulness of the procedure was recognized by many other population ecologists. Harcourt (1969) provided the first review of the life table work on insects.

FACTORS INTRINSIC TO POPULATIONS

Most insect species are provided with protective devices that assist them in overcoming the effects of excessive environmental resistance; thus, they are able to maintain their level of population under temporarily adverse circumstances. The influence of these protective forces varies directly with environmental resistance. Under optimum conditions, when environmental resistance is inoperative, their value in maintaining numbers is zero; when the factor of environmental resistance against which the device is operative is at its peak, the protective value is also highest. The various protective factors are discussed individually in this section.

Insurance of Mating

It has previously been pointed out in our discussion of reproductive potential that the sexual type of reproduction is most commonly found among insects and that mating of the two sexes is therefore necessary to the survival of most species. It has also been shown that insects which reproduce sexually have a lower reproductive potential than equally fecund species that are parthenogenetic. Therefore, if the former type is to maintain any equality with the latter, mating must be assured. The very necessity for mating, demanding as it does close proximity of the sexes, means that the physical factor of space may result in raising the value of environmental resistance for a bisexual species more than for a parthenogenetic one. Consequently, insects reproducing sexually need something to offset their relatively low values for reproductive potential and high values for environmental resistance.

This they accomplish by means of a number of modifications and adaptations for bringing the sexes together. Many are attracted to the opposite sex by odors given off by that sex. These materials, called *sex pheromones,* have been found in numerous insects from many orders. The practical application of these materials in integrated pest management will be discussed in sections on survey techniques and control. Stridulation[2] is another method of attracting the opposite sex. This method is common among the Orthoptera and other orders of insects, but there are many other special adaptations, such as the flashing of fireflies, that aid the male in locating the female.

The insects may be attracted to host materials and thus will be drawn into proximity of the opposite sex; in such cases, there may be no specific means of direct attraction from a long distance.

Care of the Young

After mating, insects continue their struggle against the forces of the environment. They have developed a number of devices for the protection of their eggs and young by which they offset to a certain extent the destructive forces of environmental resistance. Care of the eggs or young by the adults is one of the means by which this is accomplished. Care of the young is usually associated with the higher animals, but nevertheless it manifests itself in primitive ways among some insects. The highest development of this protective function in insects is usually found among the social forms, particularly among the Hy-

---

[2]The production of sounds by rubbing one body surface against another.

menoptera and the Isoptera; even among the solitary species, such as the carpenter bee, *Xylocopa virginica,* it is strikingly developed.

When an adult insect selects and deposits eggs on the host tree on which the larvae naturally feed, she is showing an involuntary type of care for the young. Such care is exhibited by *Chrysobothris* beetles and other buprestids. In the cracks where they are placed, the eggs are inconspicuous and therefore are not likely to be disturbed before they hatch. To this extent, potential care is characteristic of most species, although there are some, such as walking-sticks, that drop their eggs without regard to location. Species such as these must have a very high reproductive potential if they are to survive because many of their young will almost certainly starve before finding suitable food. Such insects are very seldom serious pests.

Some moths, such as the gypsy moth, *Lymantria dispar,* cover their eggs with a material that may protect them from excessive evaporation and may make them less attractive to predators. Others may provide a hygroscopic covering such as that on the tent caterpillar egg mass. This protective covering collects water on the surfaces to keep the eggs from drying.

One effective device for protecting the welfare of the young is illustrated by some of the Diptera that are larviparous. These insects hold their eggs within their bodies until after they hatch. Then they deposit the larvae onto the host. This is true of some of the Tachinidae, many of which are parasitic on the larvae of forest insects. The young of larviparous species have a much better chance of entering an active host than the oviparous flies do because the eggs of oviparous flies may be rubbed off or molted off the host's body before they have had time to hatch.

With the exception of the social species, parental protection of the young beyond the selection of a proper feeding place is rather rare among forest insects. In most instances, the larvae, having been placed near a supply of food, must shift for themselves. The mother aphid and the mother tingid are usually found near their young and are sometimes said to guard them, but it is doubtful that this proximity is an example of parental care. More likely, it is a question of convenience. In social species, however, there are many striking examples of most meticulous care for all immature stages. It is, not the mother, but definitely designated workers that usually care for the young in these social groups. The mother, meanwhile, attends strictly to the process of oviposition. The young of both carpenter ants and termites are fed and cared for in this manner. In the ant nest, the young are carried from place to place in order to keep them under the most favorable conditions possible.

Defense and Camouflage

Some species are provided with disagreeable odors or flavors that make them distasteful to predaceous animals and serve as means of defense (Fig. 6-1). The pentatomids and other plant bugs, certain beetles, and some butterflies are notable examples of insects possessing this means of defense.

Many insects are protected by their inconspicuous coloring or form. Few observers of nature have failed to notice how most insects blend into their backgrounds. Even some of the most gaudily colored species are comparatively inconspicuous in their natural surroundings. Protective coloring and protective form are so characteristic of insects as to be obvious to the most casual observer.

There are three types of protective appearance: (1) The color pattern of the insect's markings blends into the surroundings. Leafhoppers, aphids, underwing moths, and many other insects exemplify this protective coloration. (2) The insect resembles some part of the plant on which it lives. There are many well-known examples of this sort of protective resemblance: treehoppers, walkingsticks, and spanworms. (3) The resemblance is to some other insect that is either distasteful or feared by insect predators. This last method of protective resemblance is known as *mimicry* (Rettenmeyer, 1970), one of the most common examples being that of the viceroy butterfly, *Limenitis archippus,* which is similar to the monarch butterfly, *Danaus plexippus,* a species that is said to be distasteful to birds.

Figure 6-1 Larva of the Douglas-fir tussock moth, *Orgyia pseudotsugata.* Persons working in close proximity where contact is made with hairs from this insect may develop serious skin irritations. *(U.S. Forest Service photo.)*

In addition to the various means of defense already mentioned, many insects are provided with mechanical contrivances for defense. The heavy exoskeletons of many beetles make them less vulnerable than they would be if their bodies were soft. The sharp and sometimes venomous spines possessed by many caterpillars serve as a defense against insectivorous birds. Other mechanical means of defense may be mentioned, for instance, the strong mandibles of many forest insects. These are ready weapons of defense and are frequently used effectively. A defense mechanism of sawfly larvae feeding in colonies involves a simultaneous arching of abdominal segments as a defense reaction that seems to frighten the predator. Another powerful weapon with which some of the Hymenoptera are provided is the venomous sting. By means of the various methods of defense, insects are often able to prevent the normal reduction of their numbers as a result of the action of environmental resistance.

Use of Shelter

Some forest insects are able to evade at least partially the force of environmental resistance by retiring into shelters of various sorts. Shelter protects insects to some degree against changes of weather and also makes them less easily found by their enemies. These shelters may be constructed of silk or wax secreted by the insect or of other materials such as sticks, stones, wood, earth, or combinations of these materials.

One type of insect shelter with which everyone is more or less familiar is the lepidopterous cocoon (Fig. 6-2). By means of this structure, the pupa is protected during its periods of quiescence. In the larval stage, some insects build cases of silk and other materials for their protection throughout the developmental period. The bagworm, *Thyridopteryx ephemeraeformis,* for in-

Figure 6-2 The cocoons of the Douglas-fir tussock moth, *Orgyia pseudotsugata,* are spun by the larva to shelter the defenseless pupal stage from many potential enemies *(U.S. Forest Service photo.)*

stance, carries a case about with it from the time it hatches until it is fully grown. As the larva grows, the case is enlarged to accommodate it. The female bagworm never leaves the case but lays her eggs and therein dies.

The spruce budworm, *Choristoneura fumiferana,* and some others of the same group construct another kind of shelter. During the larval period, they work under a light web of silk that they spin. The tent caterpillars and the fall webworm, *Hyphantria cunea,* build nests of silk into which the larvae retire. Other species, such as the armored scales and the woolly adelgids, are sheltered by their own waxy secretions.

Some species of insects make use of their food material for shelter. Leaf-miners are sheltered between the upper and lower surfaces of the leaves in which they feed. Various species of flies, wasps, aphids, and mites stimulate the development of plant galls that supply both food and shelter. The bark-mining and wood-boring insects are well sheltered in their tunnels and galleries. From these examples, it becomes evident that insects make use of many different types of shelter to aid them in overcoming environmental resistance (Fig. 6-3).

Locomotion

When an insect is threatened by a parasitic or predaceous insect or by unfavorable conditions of moisture or temperature, it has two choices: (1) It may remain quiescent, trusting that its enemy will not notice it or that the tempera-

Figure 6-3 The larva of the northern pitch twig moth, *Petrova albicapitana,* lives within a hollow blister of pitch. The larva feeds on the phloem and is protected from its enemies and from the elements by the structure. *(From Graham and Knight, Principles of Forest Entomology, 4th ed., Fig. 75, p. 319.)*

ture or moisture conditions will remain within its range of tolerance, or (2) it may try to escape.

Adult insects are usually well provided with the necessary means of locomotion to permit escape. Most of them possess both legs and wings, which they use effectively. By means of these appendages, they are able to escape either from their enemies or from locally unfavorable physical conditions of the environment. Furthermore, the ability to move about helps to bring together the sexes, thus ensuring successful mating. The power of locomotion also helps the insect in its search for food, often makes possible effective distribution, and helps the insect to place its eggs in a location favorable to the larvae.

Larvae, on the other hand, are generally much less able to move about. None of the immature stages of insects have functional wings except the Mayflies, which have a subimago winged stage. Therefore, larvae must depend primarily on their legs for moving from place to place except as they are carried by birds, other animals, vehicles, or wind. Some of them have well-developed legs and can move about rapidly, but others, such as the caterpillars and grubs, can move only slowly. Even this poorly developed ability to move makes it possible for many of them to secure more favorable conditions of moisture or temperature or to improve their food conditions. This is true of practically all insects that feed on the surfaces of plants.

In addition to the use of their legs, many lepidopterous larvae have another means of locomotion useful for protection. The larvae are able to spin very rapidly a thread of silk on which they can drop out of harm's way. Later, when it is safe to do so, a larva hanging by its thread can return to its original position. This silk is also a means of dispersal for many insects.

The miners and borers, on the other hand, are greatly restricted in their movements. Only a few species, such as the larvae of the carpenterworm, *Prionoxystus robiniae,* can move back and forth in their galleries. The others must take what comes. Such insects, having a limited sphere of activity, are confined to those locations in which the extremes of environmental conditions always lie within their zone of tolerance. The motile species, because they are better able to adjust themselves to changing conditions, have a much wider range of activity; and when conditions become unfavorable in one place, they move to another.

Migration and Dissemination

At some time during the year, almost every insect species passes through a period of local redistribution. Then the individuals move out in various directions in search of favorable new locations. This movement usually occurs during the adult stage, but in some instances, it may also occur during the larval stage. As a result of this movement from centers of concentration, almost every suitable breeding place will receive its share of insects. Sometimes this dissemination is very limited in distance. For instance, Graham (1937) determined that the walkingstick, *Diapheromera femorata,* because it is wingless and not especially active, moves out from centers of infestation at the rate of only ⅛ mile (about 1/5 km) per year; whereas the western pine beetle, *Dendroctonus brevicomis* (Miller and Keen, 1960), may fly miles (kilometers) from the trees in which it developed. In a few instances, insects may migrate from north to south in the autumn, as migratory birds do. Also, some insects move northward in the spring as the season advances and feed far north of the areas in which they can winter (Johnson, 1969).

The efficiency of insect distribution can be well illustrated in any large coniferous forest area that has been killed by fire in the spring of the year. Even though the area of killed trees may be hundreds of acres in extent, practically every suitable tree will be infested with its complement of borers: buprestids, cerambycids, scolytids, certain siricids, and other subcortical and xylophagous insects. There could not possibly have been present before the fire enough adults to reproduce the multiple of borers found in the area after the fire. It is evident, therefore, that they flew in from surrounding territory (Evans, 1972). The omnipresence of certain leaf eaters in every suitable location is still further proof of the general distribution of at least the more common species. The search for food or for a suitable place for oviposition is the chief cause of dissemination, but apparently, even when food and places for oviposition are abundant, insects may still move about.

The desire to move to new locations seems to be inherent. Dissemination is a means by which the intensity of local environmental resistance may be reduced. The value to the species of the ability to move out from centers of concentration is inestimable because the more widely a species is distributed, the less likely it is to be wiped out by some local adversity. Some insects move from one host to another within the same locality. Frequently, this involves the alternation of hosts, the insects transferring their activities from one host to another and later returning to the original host species. This sort of migration is common among aphids. Some insects, for example, some lady beetles, move from low to high altitudes in late summer and autumn, returning to lower elevations in the spring.

Dissemination of insects is often aided and directed by air movements. Winged insects may be carried upward by convection currents and drift for many miles before returning to the earth. The influence of the prevailing wind in North America accounts for the more rapid spread of insects from west to east than in other directions. The occasional flights of insects observed far from possible breeding centers are likely to be the result of air movements combined with peculiar atmospheric conditions that stimulate the insects to take wing.

Johnson (1969) points out that strong wind inhibits takeoff but light wind can stimulate it. One cannot generalize on this subject because of the great variation among species. It does seem that large insects may be more susceptible to the effects of wind than smaller ones partly because of cooling effects and partly because smaller insects are more protected by being closer to the surface of the substrate.

Many wingless insects are carried aloft in air currents and may be transported long distances. Many have special adaptations that aid in air transportation. For example, some young caterpillars are provided with long hollow hairs that increase the surface exposed to air currents without materially increasing their weight. Others spin a silken thread on which they are capable of ballooning for long distances.

Tropic Responses

Some insects possess other characteristics that are advantageous in the struggle for existence. These abilities relate to reactions called *tropic responses,* which result in the orientation of the insect's body in relation to some physical or chemical stimulus. For instance, when one of the hover flies habitually heads into the wind, it is exhibiting a tropic response: anemotropism. By responding anemotropically, species living near the sea may prevent themselves from being blown away from land and thus becoming reduced in numbers.

Examples of other types of tropic responses are numerous. Lady beetles, when seeking a place to hibernate, are positively thigmotropic; that is, they respond positively to the sense of touch and, by creeping into cracks where their bodies are in close contact with surrounding surfaces, they ensure their safety. Many insects respond either positively or negatively to chemical stimuli and are then said to be either positively or negatively chemotropic. Insects that turn toward the light are positively phototropic; whereas those that turn away are negatively phototropic. Those that seek out sunny places are positively heliotropic; whereas those that shun the sun are negatively heliotropic.

These terms are descriptive of how insects respond to external stimuli. They do not in any way explain *why* insects react in certain ways. In order to avoid the common error of trying to explain the cause of certain reactions merely by giving those reactions a name, it must be kept clearly in mind that the terms are descriptive rather than explanatory.

Through many generations, natural selection has brought about a survival of those insects that have come to respond in ways that are favorable to the species. Thus it is that tropic responses have had, during the course of evolution, an important influence in determining how successful an insect will be in overcoming environmental resistance. The insect that orients itself to environmental forces so that it is most likely to be favored is much more likely to survive than one that is erratic in its reactions or one that fails to react at all. Some authorities think that practically all the activities of insects are the results of tropic responses; even if this is not entirely true, it cannot be doubted that tropisms play an important part in the life of insects and that they determine to a large degree whether or not an insect will thrive.

Effect of These Protective Devices

It is evident that an insect's abundance depends not only on the reproductive potential of the species and the amount of environmental resistance but also on the inherent ability of each insect species to meet and reduce in various ways the force of environmental resistance. Like other inherent qualities, this ability varies with different species. Its value also differs with varying conditions. Under optimum conditions, the ability of insects to overcome environmental resistance is not needed and is, therefore, inoperative. But when pressure is so increased that this ability is needed, its operation becomes more and more important until finally, under conditions of extreme environmental pressure, it may be so important as to determine whether or not a species may survive. Thus, by acting with increasing power as environmental resistance increases and with decreasing power as resistance decreases, the ability to overcome environmental resistance tends to stabilize insect numbers and to reduce extreme fluctuations in abundance that otherwise might occur as a result of fluctuations in the environment.

HIGH POPULATION LEVELS

Damage by insects is almost always the result of high population levels. Usually, these are temporary in character, occurring for a season or a few seasons and then subsiding. When a decided upswing in the population level of an insect occurs and injury to trees or products results, we say that outbreak conditions prevail. Very often an insect outbreak is referred to as an *epidemic.* Although that term should be applied only to diseases of human populations, its use has become so common in forest entomological literature that it seems (to some) acceptable to apply the term to insect outbreaks.

Continuous Abundance

Damage by insects is usually associated with outbreaks, but there are a few instances when injury is not associated with any marked upswing in abundance of the insect concerned. Such insects (termites, for example) are always present in sufficient abundance to cause injury to materials that are exposed to them. In instances of this sort, the biotic balance has been established above the point at which economically important damage will result.

Examples of continuous abundance may not be common among insects that attack living trees, although insects such as the white pine weevil, *Pissodes strobi,* may be judged by some as being of that type over reasonably short spans of time. They may be more common among insects that function in the deterioration of dead material, including wood products. The process of deterioration is a natural one that must take place at a reasonably rapid pace in the environment. The process is ecologically necessary and desirable in the natural environment, but it may cause severe problems to the products of the forest and to human habitations. We would not expect many persons to consider termites, ambrosia beetles, powderpost beetles, or seed and cone insects among those that develop outbreaks in the typical sense.

Low Population Numbers

In some forests, conditions are favorable for certain insect pests, so that their activities are always evident, although their numbers are not increasing and are not sufficiently high to cause enough damage to justify the application of control measures. This situation is referred to by some forest entomologists as being *endemic.* Endemic infestations are best illustrated in overmature pine stands in which a moderate number of trees are succumbing each year to bark beetle attack. An endemic infestation has been defined as one in which the annual losses are less than the annual growth of the trees, as expressed in merchantable volume (Salman and Bongberg, 1942).

This definition is suitable when applied to bark beetle infestations but will not fit some other conditions. For example, in pine plantations in the Lake States, when a few trees are infested with sawflies, an endemic condition is said to prevail. Or when any defoliating insect is present in usual low numbers, the infestation is endemic. This situation should not be confused with incipient conditions that exist when the insect population is on the increase but is below the numbers required to cause economically serious injury.

It is unfortunate that terminology varies in usage from one profession to another. The use of *endemic* here is quite different from the definition used in zoological circles. Endemic, by definition, means the native inhabitants of a specific region; Webster states that it means "native; not introduced." The concept is introduced here because the word will be used in both senses and the forester should understand both. It is advisable to avoid the term altogether and use *low population numbers* as a possible substitute.

Outbreak Conditions

An *insect outbreak* is a temporary condition characterized by excessive insect numbers and injury to valuable materials. Outbreaks may be either of two kinds: *sporadic* or *periodic.* A sporadic outbreak is one that appears suddenly in a small or restricted area, lasts a short time, and then subsides. Sporadic outbreaks are usually associated with changed conditions, which may be the result of human activities.

These changes are usually temporary in character and result in the brief flare-up of an insect population, followed by a return to normal. A good illustration of this type of outbreak can frequently be observed when coniferous trees are cut to clear a right-of-way or when logs are decked temporarily in the woods. Bark beetles will be attracted to the freshly cut material. If more are attracted than can find space in the cut wood, they may attack and kill nearby green trees. Also, beetles emerging from the cut materials may kill a few trees. Usually, little or no brood is produced in the attacked trees, so the beetle population disappears.

Periodic tree-killing outbreaks are far more serious. They are characteristic of many defoliating insects that become injuriously abundant at more or less regular intervals. Examples of these are the spruce budworm, *C. fumiferana,* the hemlock looper, *Lambdina fiscellaria fiscellaria,* and the tussock moths. These outbreaks may occur at susceptible stages of forest development as a result of certain weather conditions or as a result of changed biotic resistance. The length of time between periodic tree-killing outbreaks is often determined by the time required for the forest to recover from a previous outbreak and develop once more to a susceptible stage. Periodic outbreaks are usually catastrophic in character and often result in the destruction of a considerable part of the attacked forest. They constitute some of the most important and difficult forest-insect problems. Their control depends on the ability of people to maintain conditions in the forest that are generally unfavorable to large populations of harmful insects.

There are other periodic outbreaks on more resistant trees, especially hardwoods, that are not tree-killing outbreaks because the trees in the majority may survive the serious infestations. The forest tent caterpillar, *Malacosoma disstria,* in Minnesota may not kill aspen even after 3 successive years of defoliation. The caterpillar populations subside, and most of the trees survive to become subject to another outbreak some years later. This particular scenario of 1 to 3 successive years of moderate to heavy defoliation, followed by a population crash, fits many of the hardwood defoliators.

Causes of Outbreaks

Outbreaks of insects are the result of some environmental conditions that temporarily disturb the biotic balance. If, instead of the usual compensatory fluctuations in the factors of environmental resistance, a change should occur that would lower the combined effect of the resistance factors for any insect, an outbreak of that insect would be inevitable. Relieved of the resistance pressure that had held its numbers down, the insect would reproduce rapidly and soon come to dominate its environment.

The numbers attained by insects when they reach outbreak proportions are almost beyond belief. For instance, the larvae of the forest tent caterpillar, *M. disstria,* were so abundant during an outbreak in Minnesota that they almost covered the ground when they dropped from the trees. Similarly, in Michigan, the same species wandering across highways were crushed in such numbers by passing automobiles that the road became slippery and highly hazardous. Often, thread-spinning caterpillars in outbreak numbers will cover ground, stumps, and trees with a mantle of silk built from the threads that the larvae spin as they move about. During the flight period of the spruce beetle,

D. rufipennis, in Colorado, the numbers were so large that windrows of beetles were found in small lakes between timbered areas and lights in the towns within the infested areas were almost completely blacked out. Such examples are common to those with experience working with forest insects. They serve to illustrate the tremendous numbers of individuals involved. Once started, an outbreak will continue until the forces of environmental resistance destroy the pest and the balance is restored at a low population level.

In the following chapters, the causes leading to the development of specific outbreaks will be discussed. We shall see that outbreaks are sometimes the result of overmaturity or decadence of the trees in the forest. Sometimes, they are the result of our attempts to grow forests without sufficient knowledge of actual needs of the species being managed. Sometimes they are the result of poor tree health caused by excessive competition, poor site conditions, and various other undesirable situations too numerous to mention at this point. Almost invariably, outbreaks follow the development of unsteady environmental conditions. After the outbreak, when the excessive insect numbers have subsided, the environmental complex will have become more steady than it was immediately prior to the outbreak. An outbreak is a part of the natural process by which a forest may be restored to a biotic balance, but although the action of the insects may be logical from an ecological viewpoint, it may be extremely wasteful from an economic standpoint. A special comment is needed regarding the introduction of exotic insects. In such cases, outbreaks may develop without significant changes in environmental conditions because the insect pest is in itself the changed factor in the environment.

Insects are useful in the natural order of the forest as a means of eliminating overmature or decadent trees, as a means of eliminating species from sites to which they are not suited, as a means of thinning stands that are too dense, and as a means of improving stand composition.

Although the insect outbreak may be effective in correcting an undesirable condition, it will be more sensible for foresters to manage the forest so that the undesirable conditions that lead to outbreaks will not develop. In the following chapters we shall show how this can often be accomplished and how losses occasioned by insect outbreaks may be avoided.

BIBLIOGRAPHY

Andrewartha, H. G. 1961. *Introduction to the study of animal populations.* Chicago: University of Chicago Press.

Borror, D. J., et al. 1976. *An introduction to the study of insects.* 4 ed. New York: Holt, Rinehart & Winston.

Clark, L. R., et al. 1967. *The ecology of insect populations in theory and practice.* London: Methuen & Company.

Evans, W. G. 1972. The attraction of insects to forest fires. *Proceedings, Tall Timbers*

Conference on Ecological Animal Control by Habitat Management, Tallahassee, Fla.: Tall Timbers Res. Sta., pp. 115–127.

Graham, S. A. 1937. The walkingstick as a forest defoliator. *Mich. Univ. Sch. of For. and Conserv. Circ.* no. 3.

──────. 1956. Forest insects and the law of natural compensations. *Can. Entom.* 48:45–55.

Graham, S. A., and Knight, F. B. 1965. *Principles of forest entomology.* 4th ed. New York: McGraw-Hill Book Company.

Harcourt, D. G. 1969. The development and use of life tables in the study of natural insect populations. *Ann. Rev. Entomol.* 14:175–196.

Johnson, C. G. 1969. *Migration and dispersal of insects by flight.* London: Methuen & Company.

Miller, J. M., and Keen, F. P. 1960. Biology and control of the western pine beetle. *USDA Misc. Publ.* no. 800.

Milne, A. 1957. Theories of natural control of insect populations. *Cold Spring Harbor Symp. Quant. Biol.* 22:253–71

Morris, R. F. 1959. Single factor analysis in population dynamics. *Ecology* 40:580–588.

──────, and Miller, C. A. 1954. The development of life tables for the spruce budworm. *Can. J. Zool.* 32:283–301.

Nicholson, A. J., 1958. Dynamics of insect populations. *Ann. Rev. Entomol.* 3:107–136.

Price, P. W. 1975. *Insect ecology.* New York: John Wiley & Sons.

Rettenmeyer, C. W. 1970. Insect mimicry. *Ann. Rev. Entomol.* 15:43–74.

Salman, K. A., and Bongberg, J. W. 1942. Logging high risk trees to control insects in pine stands of northeastern California. *J. Forestry* 40:533–539.

Smith, H. S. 1935. The role of biotic factors in population densities. *Econ. Entomol.* 28:873–898.

Solomon, M. E. 1957. Dynamics of insect populations. *Ann. Rev. Entomol.* 2:121–142.

Chapter 7

Detection and Evaluation

Success in preventing and controlling outbreaks will depend to a great extent on preparedness. If we are forewarned of an outbreak, it may be checked in the incipient stage; whereas if an outbreak develops without our knowledge, it may reach proportions beyond any practicable possibility of control. There are numerous cases on record in which easily obtainable information could have led to control measures and the prevention of major insect outbreaks. In control of insect outbreaks, as in fire control, success depends on early detection. The reason for this is that an incipient condition can grow to serious proportions in only a season or two.

The task of protecting 460 million acres of commercial forests, urban trees, homes, and the health of forest recreationists is a monumental one. Obviously, it would be impossible to employ enough forest entomologists to carry out such a huge job. All people working in and enjoying the forest who are well trained and on the alert for the early detection of insect problems must aid in the effort.

The realization that prompt detection is an essential preliminary to effective control has led to the organization of various types of surveys to determine the trends of potentially damaging forest insects. Some of these surveys have been organized and directed by state or provincial governments. Others have been conducted by federal agencies such as the U.S. Forest Service and the Canadian Forestry Service. Private agencies have seldom organized forest-insect detection under any formal administrative setup, but they have cooperated enthusiastically and become effective participants in surveys directed by government agencies.

No matter what the administrative organization may be, the objectives of all forest-insect surveys are the same: the detection of potential outbreaks and the evaluation of both insect abundance and timber damage or potential damage.

INVOLVED AGENCIES

In this section, brief descriptions of a number of the kinds of surveys are presented. They illustrate how the various agencies have approached the problem of forest-insect detection.

U.S. Department of Agriculture's Animal and Plant Health Inspection Service

A general insect-pest survey has been conducted since 1920 by the U.S. Department of Agriculture. This survey is now performed by the U.S. Department of Agriculture's Animal and Plant Health Inspection Service (APHIS) and includes reports on all noxious species, among them forest pests. Weekly, monthly, and annual *Cooperative Plant Pest Reports*[1] are published regularly; and from time to time, special articles on insects of current interest have been prepared and distributed. Information for these reports comes from cooperators who send in reports of conditions in their localities. Most of the personnel are trained entomologists in the employ of the various states or the federal government.

The various survey reports published by APHIS provide a general picture concerning the annual abundance of important agricultural and forest pests. Usually, the reports on forest insects are made only when outbreaks prevail.

U.S. Forest Service

The detection of incipent outbreaks requires much more complete information than is provided by the APHIS insect-pest survey. Therefore, in the nine national forest regions of the U.S. Forest Service, more detailed information is collected on various important forest insects. The U.S. Forest Service is also responsible for surveying all federal lands and works in close cooperation with the Bureau of Land Management, the Bureau of Indian Affairs, the National

[1]Supersedes *Cooperative Economic Insect Report,* which ended in 1975 with vol. 25.

Park Service, and the Department of Defense. Although much of this information never finds its way into print, it is used to guide local federal insect control activities.

Each region publishes an annual summary of forest-pest conditions. The national summary is published annually under the title *Forest Insect and Disease Conditions in the United States.*

The U.S. Forest Service also cooperates with the various state and private agencies. In this way, unnecessary duplication is avoided, and coordinated plans are made for detection and later for control. The normal detection surveys are a standard contract, based on the amount of forest land in the state. However, when surveys are conducted for suppression of outbreaks, the cooperation must be initiated by the state, and the cost-sharing percentages are then negotiated.

Canadian Forestry Service

Far more detailed and comprehensive than any survey thus far developed in the United States is the Forest Insect and Disease Survey of Canada made by the Forest Protection Branch, Canadian Forestry Service.

The Canadian survey attempts to obtain information on most of the species of forest insects. Forest biology rangers attached to each of the six regional laboratories make intensive examinations of locations where infestations of important forest insects occur. These well-trained rangers are the backbone of the Canadian survey. Their collections are representative of every forest condition. Each ranger is carefully instructed in how to make and report observations and is provided with supplies to facilitate collecting and shipping materials to the nearest regional laboratory. When the collections are received, the insects are counted, reared, and identified. Records are then published showing population trends and the proportion of the various stages attacked by parasites or diseases. Thus, the survey provides information not only on pests that are known to be important but also on many other species.

States

In the United States, forest-insect surveys are conducted by the majority of states. The most common survey relies on the states' district foresters. For example, in Virginia's Division of Forestry, the chief of insect and disease investigations is responsible for the training and updating of division personnel. All federal and private foresters are invited when training sessions are held periodically on a regional basis (coastal plain, piedmont, mountains). The programs emphasize forest insects of regional importance. Attendance at, and interest in, these programs are good. Bimonthly and annual reports are published and are available to all interested parties.

An expansion of the Virginia type of survey has developed in the Wisconsin Department of Natural Resources. Each of Wisconsin's six forest districts has a trained forest entomologist-pathologist who reports to both the district forester and the Office of the State Entomologist. This type of survey emphasizes close professional cooperation at both the local and the state levels and coordination of policy and reports. Also of great importance is the fact that in each district, an annual survey is conducted on specific forest insects and diseases. Wisconsin now has more than 20 years of data on about 40 forest-insect and disease problems. These data provide a base from which to predict and interpret specific insect outbreaks.

Private

Private pulp and paper and timber companies are seriously concerned with forest-insect problems. Companies send their foresters to training sessions and cooperate by promptly reporting unusual insect situations. The most outstanding example of the well-trained, interested forester being the mainstay of a detection program was the discovery of the outbreak of the balsam woolly adelgid, *Adelges piceae,* in the Pacific Northwest during 1954 by M. M. Grobin of the Harbor Plywood Corporation, Hoquiam, Washington.

Private companies believe in early detection. In Arkansas, the Weyerhaeuser Company surveys each plantation, at monthly intervals after planting, for pales weevil, *Hylobius pales,* damage. During outbreaks of the southern pine beetle, *Dendroctonus frontalis,* many companies fly over their holdings to locate fading trees. The Weyerhaeuser Company uses helicopters at biweekly intervals, and its foresters attempt to have the trees salvaged within three weeks.[2] The Kirby Lumber Company in Texas has developed an extremely thorough aerial detection plan to cope with bark beetle outbreaks (Stanley, 1972).

In the Southern states, the private industries have organized the Southern Forest Disease and Insect Research Council. This council supports graduate research assistantships, awarded annually to Southern universities on a competitive basis.

In the Western states, various pest action councils have been formed. These councils are composed of private and public forest managers, entomologists, pathologists, and other people interested in protecting forests from damage by animals, insects, and diseases. The councils are participating members of the Western Forest Pest Committee of the Western Forestry and Conservation Association. The councils are extremely effective in coordinating and lobbying for Western problems, and they assist in assembling the annual reports on forest-insect conditions.

Throughout the United States and Canada, the forest entomologists have formed work conferences that meet annually to discuss professional matters within their geographic region. The Southern Forest Insect Work Conference,

[2]Personal communication from S. C. Cade and R. L. Hedden at Weyerhaeuser Company.

organized in 1956, publishes research results in its annual minutes and, perhaps most important, is the only organization publishing a summary of all
damage caused by forest insects in the South.

DESIRABLE PRACTICES

The ideal detection procedure has still to be devised. There are, however,
certain desirable practices that might well be incorporated into any survey.

Cooperation

The trend in the direction of cooperative programs is certainly desirable. By
cooperation, the number of people watching forest-insect activities is multi-
plied many times. However, the effective use of cooperators depends on their
ability to observe accurately and to report in a manner that will permit correct
interpretation. Reports based only on impressions of untrained people are so
inadequate that reexamination is necessary before conclusions may safely be
drawn.

Training

One of the preliminary requirements for successful survey work, therefore, is
an adequate training program. This training cannot be expected to enable
everyone to recognize all potentially dangerous pests, but it can teach a person
how to recognize those species most likely to be encountered and to report in
a meaningful manner. A printed description of the important insects and their
work should be furnished. Information should also be provided on such matters as how, when, and where to look for evidence of each important species.
To encourage the making of observations at the proper time, foresters should
be sent various memoranda during the course of the season advising appropriate action.

Foresters who have had the benefit of such a training program are
equipped to make reports that are valuable and trustworthy. Their data will
be accurate and inclusive. A forester who finds an unknown insect in large
numbers should send the agency's entomologist not only specimens of the
insect and its work but also a report describing the abundance of the insect
and the characteristics of the locale.

Ideally, reports should contain sufficient information about the conditions
of host and insect to permit correct interpretation by forest entomologists.
Foresters with the U.S. Forest Service must report insect problems to the Pest
Control Division of the Regional Office. State and private foresters should
report to their state's forest entomologist. Each agency will have its own forms
and procedures. The forester must assume the responsibility for comprehend-
ing the employer's forms and the changes.

Insect Identification

Two important factors must be known from the start in the detection and evaluation of a forest-insect problem: (1) the name of the tree species and (2) the name of the insect species. The correct identification of a species is the key to the literature pertaining to the species. The pertinent literature should be reviewed by the forester concerned with an insect problem so that advice may be better evaluated. It is assumed that the forester knows the trees, but the insects are another matter. Although the more common insect species can be readily identified, the forester should submit specimens for identification by entomologists specializing in a species' taxonomic family.

When a suspicious insect is sent to a specialist for identification, there is a minimum amount of evidence that should be included:

1 The name of the host (scientific name if known) and a sample of damage to the host
2 The insect
3 The location, specifying at least the county or parish within the state in which the insect was collected
4 The date of collection (day, month, year)
5 The name of the collector

Insects should not be shipped alive. Hard-bodied insects can be killed by placing them in a full vial of alcohol; moths and butterflies, by pinching their thorax; larvae, by boiling them in water for a few moments before placing them in alcohol; and scale insects, by drying them in an oven.

The alcohol-filled vials should be capped, placed in boxes or mailing tubes, and surrounded with crushed newspapers. The moths and butterflies should be placed between pieces of writing paper within boxes. Do not use cotton; the fibers become entangled with the insect's body parts, making identification difficult. Also avoid plastic bags because they retain moisture and the specimens may become covered with molds (Knutson, 1976).

Foresters who are interested in collecting insects are advised to study the publications of Oman and Cushman (1948) and Ross (1966).

METHODS OF DETECTION

In practice, detection and evaluation are closely related. From the time an outbreak is detected, the evaluation begins. Detection surveys are designed to locate threatening populations of forest insects before severe outbreaks can develop. The type of survey used depends on many factors, both biological and economic. The detection survey is usually extensive, leaving until later the job of careful evaluation and appraisal. Extensive surveys have been conducted in various ways, but at present, they are most frequently made from the air. The objective is to determine where insects are located as evidenced by their damage.

Aerial Surveys

The aircraft is the major means of conducting planned detection surveys (U.S. Forest Service, 1970). Outbreaks of various kinds can be easily detected from the air because evidence of excessive insect activity such as defoliation or change of foliage color is plainly discernible. An example of this is the yellow or reddish hue that characterizes coniferous trees infested with most bark beetles. Survey by airplane for detection purposes is both rapid and effective over wide areas of forest.

The personnel selected for aerial surveys should be volunteers who like to fly, do not readily succumb to motion sickness, and can aerially orient their position relative to the terrain. The usual procedure is for one or sometimes two observers to ride with the pilot so that during the flight, the infested area can be spotted accurately on maps or aerial photographs where the nature and extent of the infestation is recorded. The plane should be high-winged for good visability of ground conditions (Fig. 7-1). The height at which the plane flies is determined by the character of the infestation. The usual minimum altitude is 1000 feet above the ground. Flying below this altitude is dangerous.

Flights must be carefully timed to correspond to the season of the year in which injury is most conspicuous. Because of the seasonal nature of conspicuous injury, observations during any flight are usually aimed at a single species

Figure 7-1 The airplanes used in aerial surveys must be of high wing type because of the need for forward and lateral vision, as in this Cessna 206. *(U.S. Forest Service photo.)*

of insect. This simplifies the problems for both pilot and observers because it limits the examination to forest types susceptible to one specific insect.

Improvements are being made continually in aerial survey procedures through the better timing of examinations, greater accuracy of observations, the use of photographic recording equipment, and improved flying techniques. However, it is unlikely that sufficient detail can be secured from the air to serve as a basis for the detailed planning of most control projects. Therefore, follow-up examinations from the ground will be necessary in most cases.

Ground Surveys

Extensive observations are regularly made by trained personnel whenever they travel from place to place. In the course of a season, these incidental observations may cover much of a locality. The combined reports of all trained personnel in a region will provide information on the more conspicuous species present and over a series of years will indicate trends.

The general reconnaissance becomes more specific and usually more valuable when observations are made on a series of definite plots or areas. Conspicuous evidence of insect injury in these areas can be observed from the road; thus, rapid coverage of extensive areas is obtained (Fig. 7-2). Roadside observations are open to criticism because they are not always representative of conditions throughout a locality. Nevertheless, they are useful indicators of

Figure 7-2 Surveys may sometimes be accomplished by viewing from higher elevations. Groups of trees killed by the mountain pine beetle, *Dendroctonus ponderosae,* viewed from a hillside. *(U.S. Forest Service photo.)*

extensive conditions. Through an examination of the same plots year after year, trends become evident.

In some mountainous regions, it is possible to view large areas of forest from elevated points. When such lookouts are readily accessible, they make possible extensive surveys of insect injury at very low cost. Field glasses can be used, and infested areas can be spotted on a map.

Problem Areas

In all detection surveys, it is desirable to keep costs at a minimum without reducing effectiveness. One way to accomplish this is to concentrate efforts where insects have been known to cause problems in the past.

Foresters must know the timber types, sites, and past histories of their geographic region. Also important is a knowledge of the region's serious insect pests. Some forest conditions are more subject to attack by specific insect species than others are. The forester must know what these are and where they are located and must pay particular attention to these areas in developing management plans.

The conditions where forest-insect problems have often occurred are: (1) trees planted outside of their range, (2) pure stands, (3) stands on shallow soils, (4) areas where there has been recent cutting, (5) stands with high basal areas and/or decreasing growth rates (Fig. 7-3), and (6) stands disturbed by wind,

(a)

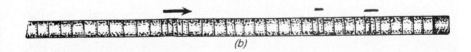

(b)

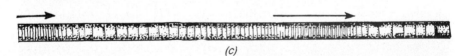

(c)

Figure 7-3 Increment cores illustrating three types of reduced radial growth are useful in evaluating conditions that have affected a tree. A wider than normal ring *(a)* is often laid down on the lower part of the bole during the first year of very severe defoliation, when transloca-tion of food is disturbed. Effects of unfavorable weather *(b)* are limited to the years during which the unfavorable conditions prevail. Suppression *(c)* results from root and crown com-petition; reduction of growth from year to year is gradual, but recovery requires only a few years. *(From Graham and Knight, Principles of Forest Entomology, 4th ed., Fig. 19, p. 92.)*

lightning, fire, drought, excessive rain, land drainage, dam impoundments, home construction, or road building. Specific examples of these conditions will be discussed in Chapter 10, "Preventive Control by Silvicultural Practices."

Where such conditions occur, detection surveys should be made regularly to determine where the hazard is greatest and where conditions are relatively safe. On the basis of such a survey, maps can be prepared to show hazard conditions in various places. With this information, foresters can direct logging or protection efforts into those areas that are of greatest value. The degree of hazard in any area may change over a period of years. Therefore, repeated surveys are necessary and should be prepared in advance. Once an outbreak is under way, time becomes a commodity in short supply.

Problem areas also exist regarding human health and recreational enjoyment. Foresters are well aware of the association of flies, waste material, and disease. But of equal importance are the problem areas associated with stinging and sucking insects.

Tree Symptoms

The first symptom a forester will usually encounter is an abnormal condition of the tree's crown. The ability to recognize an abnormal condition presupposes that that forester knows the normal appearance of a tree. In the West, Keen (1943) developed an excellent system for rating the risk that a ponderosa pine would be attacked by the mountain pine beetle, *Dendroctonus ponderosae.*

Keen established that the following symptoms are especially useful indicators of poor tree health: (1) fading or off-color foliage, (2) top dying, (3) thin, open crown throughout, (4) dying lateral twigs, and (5) short needles above, longer below. Good tree health, associated with low risk, is indicated by crowns with full, dense foliage, long needles, dark green foliage, and almost no dead twigs.

Arbitrary values have been given to the various symptoms, the combination of all indicating the risk rating for an individual tree. The trees are divided into four risk classes: (1) low risk, (2) moderate risk, (3) high risk, and (4) very high risk. Therefore, the proportion of high-risk trees in the stand can be determined, and the relative degree of hazard can be evaluated.

Risk rating of individual trees is now applied generally in mature ponderosa and Jeffrey pine types in California and ponderosa pine types in Idaho and Montana (Johnson, 1972). The rating is used not only in connection with surveys but also in certain silvicultural practices that are discussed in Chapter 10, "Preventive Control by Silvicultural Practices."

However, the forester cannot always rely on abnormal crown coloration. Many tree species that die in the late summer remain green throughout the winter and drop their needles in the spring without the needles having faded. This is especially true with Douglas-fir and the firs in the Pacific Northwest and occasionally with the southern pines.

Many insects that feed on the terminal and twigs will initially cause drops of resin to form, and these will glisten in the sun. This symptom is followed by a drooping of the terminal; then comes needle discoloration. Many trees attacked by bark beetles first produce streaks of resin, and the tubes of pitch may form on the tree long before the needles begin to fade (Fig. 7-4).

The early detection of defoliating insects is next to impossible. The forester should rely on sampling for egg masses, beating the foliage to dislodge larvae, or traps for the collection of flying adults or larval frass.

Trapping Insects

Adult insects are often attracted by lights or odors and so can be captured in traps and counted. As a means of estimating the abundance of species that are attracted to them, traps are a valuable aid. With traps, it has been possible,

Figure 7-4 Pitch tubes produced by a pine infested by the black turpentine beetle, *D. terebrans,* in the Osceola National Forest. *(U.S. Forest Service photo.)*

for instance, to detect the spread of some exotic insects. The presence of the Japanese beetle, *Popillia japonica,* has often been detected by trapping in newly invaded localities before it has become sufficiently numerous to be detected by other means. Similarly, traps have been used to detect the spread of the gypsy moth, *Lymantria dispar.* Traps have also been used later to verify the success of control programs and when bark beetles are attracted to stressed pines (Fig. 7-5).

The chief objection to the trapping method for general survey work is the time required to operate the traps. The insects captured must be sorted, preserved, and counted, and the numbers must be tabulated. The operation of traps will usually be limited to the detection of special insects. Occasionally, they may be used for more general collecting where the traps can be easily attended.

BIOLOGICAL EVALUATIONS

After an insect-caused problem has been detected, a biological evaluation of the situation may be immediately imperative. Two evaluations of the problem

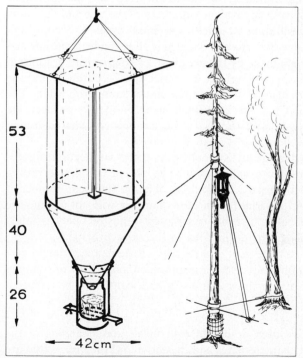

Figure 7-5   A flight trap for insects, placed at the midbole of stressed pine trees. The flying insects hit the four-way glass and drop into the jar at the bottom of the funnel. *(From Hines and Heikkenen, 1977.)*

must be made: (1) the extent and severity of damage that has occurred and (2) the future trend of the insect population.

Damage

The extent and severity of damage is related to the value of what is being damaged. There are differences in the value of a cubic foot of wood as stump-age, pulpwood, firewood, lumber, joists, or an objet d'art. The differences in value will be reflected in the measurements and sample size used to determine the damage.

The evaluation of damage will also be influenced by the extent of the damage, the insect causing the damage, and the accuracy desired by the manager. For example, you would have to use different methods to evaluate overmature trees infested with bark beetles, a region-wide outbreak of a defoliator, 1-year-old plantations infested with pales weevil, *Hylobius pales,* campgrounds swarming with mosquitoes, or termites in your sills and joists.

The methods used to estimate insect damage in the forest are usually similar to those used in timber cruising. This is natural because the objective is to determine the volume of timber, the number of trees, and the acreage or the value of the materials involved. The measurement of the tree and the forest will follow standard mensuration practices. On small areas or where high values are at stake, a 100 percent cruise of damage is often made. If the infested areas are large, estimates are made by the use of either strip cruises or random sample plots. Damage on these sample areas is measured accurately, and the total damage to the entire area is assumed to be in the same proportion. The percentage of an area that must be examined in order to give an accurate estimate of damage will vary with the degree of homogeneity characterizing the population and with the size of the area involved.

In the evaluation of damage for computing control costs, accurate figures are essential. These figures are usually obtained by taking samples according to a systematic pattern, either fixed or random. When a sufficient number of samples have been taken, the addition of more samples will not change the average. Thus, with an adequate number of samples, it may be safely assumed that the results will apply to the entire area in the same proportion.

The final statement of the extent and severity of the damage should give the quantity for the sampling universe, for instance, so many trees or so many board feet of timber per acre or how many sills and joists must be replaced.

Insect Populations

The evaluations that have been described rest on an understanding of the biological and ecological characteristics and the activity of the insects that are being surveyed. The characteristics of insect populations must be measured in terms of this biological knowledge and usually for successive generations. Only by reliable measurements can we determine with certainty whether we are dealing with increasing or decreasing numbers. The decision to use or not to use some direct control procedure is, or at least should be, based on the facts thus disclosed. This section examines some specifications for making biological evaluations in such a manner that the results will command confidence.

Insect evaluations are designed to measure and predict the trends in forest-insect populations. When such evaluations are being made, many inter-related factors must be known. The insect population must be carefully measured, and the susceptibility of the host product or area must be rated, with the recognition that two major biological organisms, the host and the insect, are involved. Each is dependent on the other. If the host is resistant, the insect will do no serious damage; if the insect is not present in sufficient numbers, a susceptible host will remain practically undamaged.

Effective evaluation procedures for predicting insect population trends have not been developed for most of the important forest insects because of the vast amount of information needed. This lack of technique does not mean that these evaluations are of minor importance. On the contrary, effective insect evaluations can help greatly in reducing the number of mistakes due either to not acting when outbreaks are imminent or to conducting needless control when outbreaks are subsiding.

Counting Insects

The procedure to use in sampling is determined by the insect species and the intensity and size of the infestation. There is no standard procedure applicable to all species or situations; however, similar techniques may be used for closely related species.

The forester should remember that insects pass through various stages of development in varying microhabitats at different times of the year. Thus, the direct sampling of insect populations can be quite technical, especially the timing and duration of the sampling period. Because host damage is often proportionate to the insect population, it is possible to express population trends in terms of injury. There are numerous examples of how variations in host damage can be used to indicate the abundance of some forest insects. For instance, the rate of bark beetle increase or decrease may be determined by comparing the number of entrance tunnels with the number of emergence holes. The relative abundance of a defoliator may be determined by the percentage of defoliation. The numbers of sucking insects, such as the spittlebugs and cicadas, may be indicated by the number of *flags,* that is, twigs on which the needles or leaves have turned red.

The direct counting of insects is possible. Usually, a specially trained person will be required to do this, and the method must be adapted to the specific insect. In many instances, the actual number of insects may be determined by counting those present on a certain unit of shoot, twig, or stem. The counting of flying adults is difficult and expensive. They must first be captured in light or odor traps that have been maintained for consecutive years before population trends can be evaluated. The direct count of insect eggs, larvae, and cocoons is less difficult because of reduced insect mobility. Direct counts are frequently used in evaluating population trends.

Indirect estimates of insect numbers can also be made by measuring the amount of boring dust or *frass,* the term used for the waste materials excreted by insects. Also, a reflection of insect numbers is the number of complaints received from people who have been stung or otherwise annoyed by insects.

The terminal-infesting weevils and moths cause annual reductions in the branches per whorl and also cause stem deformities. These symptoms permit estimates of annual variation in past insect-caused damage.

Types of Populations

Morris (1955) describes five ways in which population data may be expressed. This classification is helpful in understanding some of the problems involved. The first four expressions are based on direct population measurement; that is, insects are located and counted. The fifth is a result of indirect observation based on measuring some index related to insect numbers. Morris's classification is as follows:

1   Population Intensity   Insect populations may be expressed in relation to the food supply available to them. A defoliator might be described as a certain number of individuals per shoot or leaf group; a bark beetle, in terms of number of beetles per square foot of bark surface. Such an expression is perhaps the most useful for biological evaluations because the number of insects is described in relation to the host and can readily be subdivided into various degrees of damage.

2   Absolute Population   In this expression, the insect counts are related to some absolute unit such as insects per acre. Such measurements can be very useful in expressing changes in total insect population from year to year or generation to generation. Because of changes in the quantity of food available, the population-intensity expression may reveal only apparent trends that are not accompanied by absolute changes in insect numbers.

3   Basic Population   The insect population may be measured relative to some specific unit in the forest stand, an individual tree for example. Such a unit then can be converted to absolute population if supporting data on the structure of the tree and the forest stand are known. Such measurements may be used as an indicator of absolute population if the normal characteristics of the forest stand are not disturbed. The important difference between basic population and population intensity is that an expression of basic population can be directly related to absolute population; whereas population-intensity measurement may not represent valid expressions of change in absolute population.

4 Relative Population Some surveys utilize gross measurements unrelated to any specific unit of the forest or tree as a means of expressing trends. Examples are the *beating method* used in forest-insect surveys and many of the *sweeping procedures* used in fields. These methods are useful in keeping track of the presence or absence of insects but are not precise enough for use in accurate evaluations.

5 Population Indexes In some situations, an index can be used more effectively than a count of the insect population. In a survey of defoliators, the index may be an ocular estimate of defoliation intensity; in bark beetle work, the index may be the numbers of dead trees. Indexes are extremely useful in population evaluations and in many cases are the only feasible method because of the expense entailed in other, more precise procedures.

Sampling Statistics

The specifications for sampling must be based on a knowledge of statistics. There are numerous texts that cover the subject. If all biological populations were distributed according to the normal curve, the development of insect procedures that are statistically sound would be relatively simple. Unfortunately, among forest-insect populations, other distributions are generally more common than the normal. Thus, considerable statistical knowledge is required for measuring forest-insect populations or conditions. The forester and forest entomologist must have a basic knowledge of statistics so that they can express their ideas clearly to statisticians and, in turn, understand the advice received.

Insect evaluations must be based on soundly conceived plans. Steele and Torrie (1960) list five basic steps in the planning and execution of a sample survey. The steps are as follows:

1 The objectives must be clearly stated.
2 The sampling unit and sampling universe must be defined.
3 The sample must be chosen.
4 The survey must be conducted.
5 The data must be analyzed.

These steps are self-explanatory, but two terms should be defined. The *sampling universe* is chosen by the surveyor for a specific purpose. The universe may be in terms of trees or portions of trees, or it may be in acres or other portions of the forest floor. The *sampling unit* is selected as the basic unit within the universe on which the counts are made. The sampling unit should be such that all the units in the universe have an equal chance of selection. It should be stable in number and size but should be small enough so that an ample number of samples can be taken to estimate variance adequately.

The method by which the sampling units are selected is important. Generally, a random selection is advisable, although there are specific situations in which other methods are more practicable. Stratification of samples into groups having some characteristic in common is often necessary to obtain the maximum information for the time available. Timing is a major problem in insect evaluations; the sampling period may be relatively short in duration.

Distributions

As we have noted, the distribution of data from biological populations does not generally approximate the normal curve. This matter deserves further consideration Two questions arise:

1 How can we analyze such data?
2 Are there mathematical distributions that do fit?

No set of biological data is likely to fit any mathematical form exactly. But the data may satisfactorily approximate one of several mathematical functions that may be represented by either a straight or a curved line. Discussions of the various types of distribution can be found in most statistics texts, and good discussions in relation to forest insects can be read in the papers by Waters (1955) and Waters and Henson (1959). Insect data will often approximate one of the following three distributions.

1 **Binomial** The binomial distribution may be applicable when the data concern merely success or failure. In forest-insect work, we might record only insects present or insects absent or damage present or absent. An assumption is made that every sample unit in a lot has an equal chance of containing insects or damage.

2 **Poisson** The Poisson distribution is a random distribution that is often approximated when insect populations are at low levels. When the populations are higher, a binomial distribution, either positive or negative, is more likely. The Poisson is of doubtful reality in nature because the following basic assumptions are seldom met:

a Each individual has the same chance of occurring on any unit.
b Each sampling unit has the same chance of having an individual land on it.
c The presence of an individual or individuals on a sampling unit neither attracts other individuals to nor repels them from the unit.

3 **Negative Binomial** The negative binomial is one of several "contagious" distributions and is applicable to biological data of many types. This distribution is nonrandom. There is a definite tendency among biological organisms toward *clumping;* that is, if one insect of a species is present in a unit, there is an increased chance that others of the same species will be present in the same unit. This may happen in nature when larvae hatch from eggs laid

in masses, when insects are attracted to especially favorable spots, or when they exhibit gregarious tendencies.

The three distributions as well as the normal and others that biological data may fit are discussed in books and journals on statistics. Our objective here is merely to impress on the forester the facts that there are many possible types of distributions and that insect populations do not always occur randomly.

Transformations may be effectively used on biological data to convert data to approximate a normal distribution. Once the transformation has been made, the transformed data may be analyzed using statistical procedures based on the normal curve. Transformations that are often effective are the log, the square root, the reciprocal, and the arc sine.

Nonparametric Statistics

The use of nonparametric statistics should be mentioned briefly because of the usefulness of the techniques. In insect surveys and in research as well, many data are collected for which the underlying distribution is not known. To handle such data, procedures are needed that are not dependent on a specific parent distribution. The nonparametric procedures supply ways in which such data can be analyzed. These procedures are not often used because of lack of knowledge of their usefulness. The individual should investigate the possibility of a nonparametric test before accepting parametric procedures based on weakly supported assumptions. Generally, the procedures in such tests as the sign test, Wilcoxon's signed rank test, or the rank correlation tests are quick and easy to learn and apply. Those who desire to read a short discussion on nonparametric methods should consult a statistics text such as Steele and Torrie (1960), *Principles and Procedures of Statistics.*

Sequential Sampling

Sequential sampling procedures have been extremely useful in biological evaluations. The method involves a flexible sample size, in contrast with the conventional methods, which utilize fixed numbers of sample units to be examined. In sequential sampling, the basic units are examined, and cumulative counts are recorded until the sample total fits one of the classes defined in advance. The least sampling is required where the insect is very sparse or very abundant. The savings in sampling time can be considerable when using this method. Waters (1955) presents detailed procedures for computation of sequential plans for four mathematical distributions: binomial, negative binomial, Poisson, and normal.

The method is not useful in population work, where counts consisting of numbers of insects are desired, because it yields only classes by previously defined abundance groups. The development of a sequential plan requires a considerable amount of knowledge about the insect. Its biology must be known. Also the mathematical distribution of counts for the basic sampling unit is needed, and the effects of the insects on the host trees must be known. If the sequential sampling procedure is to be used to distinguish between light versus moderate defoliation, we need the class limits in terms of insect numbers. What is the maximum number of insects per sampling unit that will cause no more than light defoliation, and what is the minimum number that will cause no less than moderate defoliation? This information must be obtained through research before a sequential plan can be developed. Generally, no more than three classes are needed, and sometimes two will suffice. A sequential system could be devised with two classes, for example, treatment needed or treatment not needed. Confidence limits are a part of the equations for the sequential method, and in most forest-insect plans, a 10 percent probability has been set.

There are numerous plans that could be used as examples. Here we will use one from Cole (1960) for the western spruce budworm, *C. occidentalis,* in Douglas-fir. This plan was developed for use in postcontrol surveys to class the control as either satisfactory or unsatisfactory. Such a procedure is much more useful than conventional methods, which state merely that 95 percent or more mortality is satisfactory. Percentage mortality values are often misleading. In a very heavy infestation, 95 percent mortality could result in more survivors than are normally present in a light, increasing infestation. The sequential plan avoids this problem by basing success on the number of living insects remaining, rather than on the percentage surviving.

Postcontrol population data were found to approximate the Poisson distribution closely. Cole set the following class limits for the two classes (Table 7-1):

M_0 (satisfactory): the average number of larvae per 15-in twig is 0.35 or fewer

M_1 (unsatisfactory): the average number of larvae per 15-in twig is 0.50 or more

Cole's limit was set at 100 twigs. If after 100 twigs had been sampled, the cumulative total was between 36 and 48, the statement was made that the control was "equal to 95 percent reduction." Thus, the maximum sampling for borderline cases is 100 twigs; but for clearly satisfactory or clearly unsatisfactory control, far fewer samples would be examined. The procedures followed in selecting sampling points within spray blocks are described in Cole's paper.

The example cited is not an insect evaluation, but it illustrates the method. Plans have been developed for insect evaluations. Knight (1960) developed such a plan for mountain pine beetle, *D. ponderosae,* surveys. In this plan, predictions are made that the number of trees killed will increase, remain static, or decrease in the next generation by making counts of beetles develop-

Table 7-1   Sequential Table for Field Use in Postcontrol Sampling of Spruce
Budworm Larval Populations

No. of twigs examined	Cumulative no. of budworm larvae satisfactory vs. unsatisfactory	
15	—	12
20	2	14
25	4	17
30	6	19
35	8	21
40	11	23
45	13	25
50	15	27
55	17	29
60	19	31
65	21	34
70	24	36
75	26	38
80	28	40
85	30	42
90	32	44
95	34	46
100	36	48

Source: Cole, 1960.

ing to the adult stage in presently infested trees. The sequential method has
its limitations, but it is soundly based mathematically and is proving to be
extremely useful in the evaluation of forest-insect abundance.

This chapter on detection and evaluation has served only to introduce this
broad subject. We have not attempted to do more than present some of the
ideas that are currently being used in this field. The forester will find these
methods useful in entomology and in other phases of forestry. New methods
are being developed and will be adopted in place of many that are in use at
present. The forester should adopt new methods as they are developed and are
proved effective.

BIBLIOGRAPHY

Bliss, C. I., and Fisher, R. A. 1953. Fitting the negative binomial distribution to
    biological data and note on the efficient fitting of the negative binomial. *Biometrics*
    9:175–200.
Canadian Forestry Service. 1976. *Annual report of the forest insect and disease survey,
    1975.* Ottawa: Canadian Forestry Service.
Cole, W. E. 1960. Sequential sampling in spruce budworm control projects. *Forest Sci.*
    6:51–59.
Hines, J. W., and Heikkenen, H. J. 1977. Beetles attracted to severed Virginia pine
    (*Pinus virginiana* Mill.). *Environ. Entomol.* 6:123–127.

Johnson, P. C. 1972. Bark beetle risk in mature ponderosa pine forests in western Montana. *USDA For. Serv. Res. Pap.* INT-119.

Keen, F. P. 1943. Ponderosa pine tree classes redefined. *J. Forestry* 41:249–253.

Knight, F. B. 1960. Sequential sampling of Black Hills beetle populations. *USDA Forest Serv. Rocky Mt. Forest Range Expt. Sta. Res.* Note 48.

Knutson, L. 1976. Preparation of specimens submitted for identification to the systematic entomology laboratory. *USDA Entomol. Soc. Am. Bull.* 22:130.

Morris, R. F. 1955. The development of sampling techniques for forest insect defoliators with particular reference to the spruce budworm. *Can. J. Zool.* 33:225–294.

———, and Miller, C. A. 1954. The development of life tables for the spruce budworm. *Can. J. Zool.* 32:283–301.

Oman, P. W., and Cushman, A. D. 1948. Collection and preservation of insects. *USDA Misc. Publ.* no. 601.

Ross, H. H. 1966. How to collect and preserve insects. *Il. St. Nat. Hist. Survey* circular no. 39.

Stanley, G. W. 1972. Kirby's southern pine beetle control program. Houston, Tex.: Kirby Lumber Corp.

Steele, R. G. D., and Torrie, J. H. 1960. *Principles and procedures of statistics.* New York: McGraw-Hill Book Company.

Stevens, R. E., and Stark, R. W. 1962. Sequential sampling for the lodgepole needle miner, *Evagora milleri. J. Econ. Entomol.* 55:491–494.

U.S. Forest Service. 1970. Detection of forest pests in the Southeast. *USDA For. Serv. Southeastern Area St. and Priv. For.* no. 7.

———. 1977. Forest insect and disease conditions in the United States, 1977. Washington, D.C.: U.S. Forest Service Div. For. Insect Disease Management.

Waters, W. E. 1955. Sequential sampling in forest insect surveys. *Forest Sci.* 1:68–79.

———, and Henson, W. R. 1959. Some sampling attributes of the negative binomial distribution with special reference to forest insects. *Forest Sci.* 5:397–412.

Chapter 8

Control

Insect control may be defined as the "regulation of insect activities in the interest of human beings." It includes practices of all sorts that are directed toward that end. Some of these activities will be corrective and some preventive, but all will aim at the same objective. All control efforts must provide a benefit to people; thus, thorough cost-benefit studies are essential. In this chapter, some basic ideas on pest management, direct control, and prevention of spread will be discussed. Biological, silvicultural, and chemical controls are discussed in more detail in Chapters 9, 10, and 11.

CERTAIN ASPECTS OF CONTROL

Purpose of Control

The purpose of control is to avoid economically important injury, and the success of any control operation must be measured in terms of the value of the products protected. The objective is not the eradication of the insect, but the reduction of its numbers below some level of injuriousness. Eradication can best be attempted when the pest is newly introduced and dangerous. One conspicuous reason for this attitude toward eradication is the cost. To eradicate a native insect would be prohibitively expensive. This extreme cost coupled with the very low probability of success should be enough to prevent eradication attempts against forest insects.

Eradication of a native insect should not be attempted for another, less obvious reason. There is a fundamental biotic principle that any niche left vacant will soon be occupied by some other organism and that that organism is likely to be a species similar to the original occupant. Therefore, if we eliminate one insect pest, another, similar one, previously present in small numbers and now relieved of competition, may become a pest in place of its former associate. Thus, we might solve one problem by eradication only to create another.

If we can hold the pests within reasonable bounds but still permit them to be present in such numbers that they can continue to compete successfully with their subordinate associates, the biotic balance will be maintained at an innocuous level. By so doing, we should accomplish control in an ideal manner.

Outbreaks of forest insects result from fluctuations of the biotic balance. Conversely, as long as the balance can be maintained at low levels without major fluctuations, outbreaks are impossible. In its simplest terms, therefore, the purpose of control is to maintain the balance between the pests and the trees at a level below the threshold of serious injury.

Although natural disturbances may upset the balance, causing a species to attain injurious levels, the activities of humans are the most far-reaching and important. The cutting of trees, the building of trails and roads, the construction of buildings, the drainage of land, the damming up of water—all these, in fact, practically all human activities, tend to modify or change environmental conditions, thereby increasing or decreasing the probability of outbreaks.

The value of the forest to people depends on the use made of it. Obviously, people cannot use it without causing some disturbance. Nevertheless, every possible effort should be made to carry on activities so that the balance will be disturbed as little as possible. Through understanding and care, it is possible to make all the necessary changes in conditions and yet to a very large extent maintain the biotic balance. We have seen that that balance is maintained only when the two opposing forces of reproductive potential and environmental resistance are equal. Therefore, whenever one or more factors of resistance are either removed or reduced in intensity, other factors of equal force must replace them if the equilibrium is to be maintained. This is what is done when we carry out control measures. We merely substitute one set of resistance factors for another set that has been removed. The change in environmental resistance may be either the result of human activities or the result of natural forces.

This brief discussion on biotic balances should have caused some doubts to arise. One becomes aware of situations that have apparently become pest problems only after interest in management has increased. There is clear evidence that outbreaks of species such as the spruce budworm, *Choristoneura fumiferana,* or the western pine beetle, *Dendroctonus brevicomis,* were causing severe losses (from our present viewpoint) long before we did anything to the forest. Biotic balance does not necessarily mean that organisms must always exist together in harmony with only minor fluctuations in numbers. There may be very great changes while a general balance is maintained over long time periods. Thus, the economic injury level is dictated by the value placed on the forest and the need for the products and services derived from it, and not always by some theoretical biotic balance that people might have influenced by their misguided management applications.

There are many reasons why the land manager may decide to control the damage done by pest populations, and although we attempt to relate these to economic values, these may not always apply. Management objectives may be extremely varied. The National Academy of Sciences (1975) attempted to place forests in categories that illustrate this diversity.

1 Noncommercial forests
2 Extensively managed forests
3 Intensively managed forests
4 Christmas tree plantations
5 Seed orchards
6 Forest tree nurseries
7 Suburban and urban

These categories are not exhaustive, nor are they even comparable; several could be further subdivided. For example, the noncommercial forests (U.S. Forest Service, 1972) of the United States total 253 million acres, including much nonproductive forest and over 19 million acres of productive forest lands withdrawn from commercial use for various reasons.

Cost of Control

Like every other forest operation, the application of forest-insect control must be looked at from the business point of view. The material saved by the application of control measures must justify the expense involved. The cost of control must be less than the loss that would have occurred had no protective measures been applied. The lower the value of the trees or wood products to be protected or the smaller the margin of profit, the smaller will be the amount that can justifiably be expended for protection.

The forest entomologist must keep these economic requirements in mind when called upon to decide whether or not control is justified in a specific instance. As a rule, the nearer the ultimate product, the more valuable the materials become. In the early years of a forest rotation, the value of the trees is comparatively small. At that time, even small injudicious expenditures may easily wipe out the entire expected profit. The value of mature timber justifies greater expenditure for its protection than the value of younger trees does, the value of partially manufactured products warrants a greater expenditure than standing timber does, and the protection of manufactured wood products justifies a still greater expenditure.

However, strictly economic values are not the only considerations. Trees may have very real values in checking soil erosion or for aesthetic purposes. The value of the forest as a habitat for wildlife should be recognized by any person interested in resource management. People are interested in these animals for both their game and their aesthetic values. As plans for forest protection are developed, these values and other resource considerations require analysis. Their value may be greater than that of the timber produced, and the combined values in multiple-use management could make a real change in a cost-benefit analysis. For example, where trees are used to prevent the erosion of soil on steep slopes and watersheds, their destruction means not only the loss of a certain amount of wood but also damage through erosion of hillsides and silting of reservoirs. Such damage may be so great that, by comparison, the loss in wood may appear infinitesimal. Under such conditions, larger expenditures for protection than in the timber forests are justifiable. Similarly, parks, recreation areas, roadside forests, and ornamental trees have an aesthetic value far in excess of the value of the wood produced. The real value of ornamental trees is difficult to determine and seems to rise continually. The cost of tree removal and replacement is only a part of the value.

The forest entomologist will frequently be making evaluations for control purposes. It is not difficult to calculate the approximate cost of a certain control operation, but when an attempt is made to decide whether or not the expenditure is justifiable in terms of the values involved, the problem is not always simple. On watersheds and where the aesthetic considerations are paramount, the values are so great that almost any practicable control expenditure may be justified, but placing a value on aesthetics has not yet been satisfactorily accomplished. In commercial forests, on the other hand, the manager must be able to determine the actual monetary worth of the products saved before deciding with confidence to apply a certain control procedure. The difference between the total operational expenditure and the monetary value of the loss expected without control represents the amount saved. Naturally, control should not be recommended if this difference is a negative value. In such a case, some less expensive control method must be used, or the insect loss must be accepted as unavoidable.

Estimating the expected loss is not always easy and must be based largely on experience. Not infrequently, the needed experience is lacking. Then one can only judge from what is known about closely related species. But a well-trained forest entomologist should, in most cases, be able to judge approximately the amount of injury that a certain infestation is likely to cause. In many instances, the experienced professional should be able to estimate this with a high degree of accuracy. The amount of injury will depend on the kind of insect, the forest type, and the site conditions. This subject is discussed further in the various chapters dealing with special insect groups.

Another important matter bearing on control is the length of time that treatment will protect trees from the particular pest in question. How many times during a rotation must treatment be repeated? Usually, this question cannot be answered with certainty, but there are some principles that will serve as guides in securing a reasonably correct answer. For instance, if the control operation changes the forest composition so that it is no longer favorable for the pest, the control measure will not need to be repeated. Conversely, if the treatment leaves the forest in the same susceptible condition as before, another outbreak can be expected within a few years, the number depending mostly on the insect's reproductive potential and weather conditions. For example, spraying a plantation to control sawflies does not change the character of the stand, and another treatment may be necessary within a few years. Repeated treatments in the same tree rotation are likely to be prohibitively expensive.

From this discussion, we see that the forest entomologist must know not only how a forest insect can be controlled but also when it is profitable to apply a certain measure. Decisions on the latter require knowledge of the expected severity of injury, the values involved, whether or not control will be permanent, and the cost of control operations (Fig. 8-1).

Figure 8-1 A truck-mounted mist blower can be used in treating plantations. Expensive control measures may be necessary in growing trees for special purposes such as Christmas trees. *(Photo courtesy Dr. W. E. Wallner.)*

control of the spruce beetle, *Dendroctonus rufipennis* (Nagel et al., 1957). In contrast with the European experience, the beetles are not particularly attracted to girdled trees, but populations in felled trees become extremely high. Trees are felled in small groups at intervals of ¼ to ½ mi (0.4 to 0.8 km) so that they lie in partial shade. After the logs are infested, the beetles are killed. If enough trees are felled to absorb a large proportion of the beetles in the area, control can be achieved with this method alone. However, the method may also be used advantageously as a supplement to chemical control.

In the control of certain of our ornamental-tree pests that tend to congregate in secluded places during some portion of the day, loose bands of burlap or other material placed around the tree trunks are sometimes used. The insects will gather under the protection of the band, where they can easily be destroyed in large numbers. This method is, of course, applicable only to ornamental and shade trees. The spring cankerworm, *Paleacrita vernata,* and the fall cankerworm, *Alsophila pometaria,* may be trapped in sticky bands on shade trees when the wingless females attempt to crawl up the tree trunks to lay their eggs.

Destroying Infested Materials

In forest entomological literature, one of the most frequent recommendations for the control of bark beetles and wood borers is to cut and burn the infested portion of the tree. Doubtless, this is an effective means of control if practically all the infested trees in a community are treated in this way. Consequently, cutting and destroying the infested portions of trees should become a community control method. One insect that might be controlled by this method is the smaller European elm bark beetle, *S. multistriatus,* the chief vector of the Dutch elm disease. If, by community cooperation, all dying trees were cut before the beetles emerged, an outbreak of this elm disease could be checked.

The destruction of infested material is an extremely important part of the planned control of this disease. The sanitation procedure of removing and destroying all dead and dying elm trees or branches of trees is a necessity if control is to be successful.

The method is not confined to protecting ornamental trees and forest products. In the West, the spruce beetle, *D. rufipennis,* infests logging slash and as a result often becomes a serious problem in the adjacent standing timber. It is common practice to wait until the slash has been attacked by beetles and then pile and burn all the infested material, destroying the developing broods in the process.

Application of Heat

The use of heat to kill wood-boring insects is occasionally applied in the forest, but it is more frequently employed to kill forest product insects in furniture factories and other wood manufacturing plants. In the woods, solar energy provides the source of heat.

Temperatures beneath the bark on the top side of logs lying in full sunlight frequently exceed the fatal temperature for log-inhabiting insects. These temperatures exceed the temperature of the surrounding air in direct proportion to light intensity. If logs lying in bright sunlight are turned every week or two during warm, bright weather, all insect life in the surface layers may be destroyed. For this solar treatment to be successful, the logs must be exposed on all sides to the full force of the sun's rays (Fig. 8-2).

In manufacturing plants, kiln treatment is the simplest and most convenient method for destroying insects in wood. In standard practices, the kiln temperatures practically always exceed the fatal point for wood-boring insects.

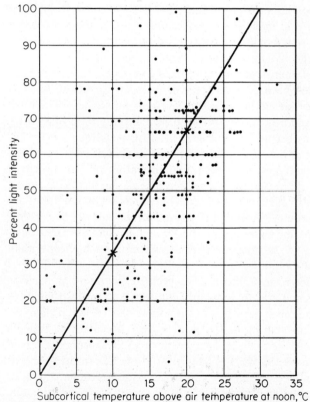

Figure 8-2 The influence of shade on the subcortical temperature of similar logs. The extreme temperatures when high light intensities are achieved are fatal to the insects in the surface layers. *(From Graham and Knight, Principles of Forest Entomology, 4th ed., Fig. 25, p. 120.)*

USE OF BIOTIC METHODS AS DIRECT APPLICATIONS

Biotic methods will be discussed in Chapter 9 in more detail in relation to their effectiveness in regulation of natural populations and their possible use in prevention of outbreaks. Numerous attempts to apply biotic methods directly have been made, but most of them have never passed the experimental stage. In a few instances, however, these efforts have been successful. Some of the successful cases have been so striking in their results that we shall probably see a much wider application of this kind of control in the future. Stinner (1977) points out that mass releases of insects have given mixed results but that there have been notable successes.

Sex attractants (pheromones) have been discussed elsewhere as a survey technique, but they may also have potential in control. Roelofs et al. (1976) considers the use of pheromones in a disruption procedure promising. Pheromones are a means of communication (Shorey, 1973) between two organisms that might be useful to pest managers if well-enough understood. Theoretically, any manipulation that would prevent the communication from reaching its objective might be effective in control of pest populations.

The practical use of pheromones in control of forest insects has not been shown yet, but testing has been done on some problems. Coppel and Mertins (1977) describe the use of the method and the developments to date. The work of Lanier et al. (1976) and others on utilization of pheromones in mass trapping of forest insects has been developed into a useful technique.

A striking example of the successful direct application of biotic methods to the control of an insect pest is to be found in California. In the citrus orchards of that state, the mealybug, *Pseudococcus gahani,* has been a most serious pest. From 1918 to 1929, it was effectively controlled by the direct use of a predatory beetle, a coccinellid named *Cryptolaemus montrouzieri.* According to Smith (1925, 1926), the use of this method passed beyond the experimental stage and proved itself to be not only effective but also much cheaper than chemical or mechanical methods.

The procedure followed in California has been to rear these predaceous beetles in large numbers and liberate them in the infested orchards in sufficient quantities to destroy the mealybugs before they have caused much damage. Special insectaries for the rearing of the beetles were operated in strategic locations from the proceeds of the sale of the beetles reared. During the season of 1929, over 14 million beetles were produced by these laboratories; and as a result, the mealybugs were effectively controlled over an orchard area of about 75,000 acres (33,600 ha). The total cost of this work was approximately $125,000. The beetles were not liberated in every orchard in the infested region but were concentrated in those that were in danger of serious injury. If it is assumed, as Smith did, that one-third of the orchards were actually treated, the cost per treated acre was about $5. If, however, the cost was spread over the entire area affected, the cost per acre was only $1.66. These costs have, of course, increased with rising labor and other expenses, but no later figures are available. When compared with spraying, which cost, in 1929, from $25 to $30 per acre for one application, the expense of the biotic method was very low indeed. The student is advised that these cost figures are merely illustrative and that not all of them have increased over time. The present cost of aerial spraying is considerably below the 1929 prices of $25 to $30 per acre. The price today is dependent on many factors, but on very large acreages, it may be as low as $3 per acre.

At present, *Cryptolaemus* is not being used as extensively as formerly because the citrophilus mealybug has been brought under excellent indirect control by the introduction of two very effective internal parasites, *Coccophagus gurneyi* and *Tetracnemus pretiosus,* introduced from Australia in 1928. The use of *Cryptolaemus* is now mostly limited to the control of another, somewhat less injurious mealybug. Nevertheless, it still remains one of the most striking illustrations of how a predatory species may be used directly for the control of a noxious species.

DeBach (1974) discussed the possible use of virus and bacterial materials as sprays for the control of insects. Insect virus materials can be used effectively in the control of forest insects. Two such materials have been registered, one for use on the Douglas-fir tussock moth, *Orgyia pseudotsugata,* and the second for use on the gypsy moth, *Lymantria dispar.* There is evidence that others will receive similar treatment. The bacterium now available for use in forest entomology is *Bacillus thuringiensis,* which is available from several commercial firms and has been registered for use on a number of agricultural and forest pests.

The ecological advantages of pathogens are evident. Microorganisms form a vital part of natural forests by being a part of the process of recycling. Some of these organisms are *pathogenic,* meaning that they may cause disease in living plants and animals. We make use of these characteristics in our control efforts. The high specificity of organisms such as the nuclear-polyhedrosis viruses makes them especially valuable. But by 1978, only two viruses had been registered for use on forest insects. The cost of production, availability, and research required have all been drawbacks to rapid development. *B. thuringiensis* is not as specific, but it is expensive and has given erratic results in many tests.

The success of fungous diseases in reducing insect abundance depends so much on favorable weather conditions that the use of these organisms offers comparatively little promise. Similarly, airborne bacterial infections are influenced by weather conditions. Other organisms not so dependent on the weather may ultimately be used. Soil-inhabiting bacteria and viruses seem to offer the greatest promise.

The successful use of insect parasites and insect predators in inundative releases depends on the possibility of rearing them on a large scale at a reasonable cost. If it were not for difficulties in the techniques of rearing these organisms, we should now be able to use them extensively in direct control. Another obstacle to be overcome is the difficulty of providing a sufficient quantity of suitable food to make wholesale rearing possible. This is no easy task. Artificial or partly artificial food has been used for rearing various insects. Therefore, the development of artificial food formulas for parasitic or predatory insects seems possible.

Although no obstacle presents so great a difficulty as the provision of suitable food, other difficulties stand in the way of large-quantity production of parasites or predators. For instance, it may be difficult to provide conditions suitable for mating, or transportation of the insects to the localities where they are needed. The problem of synchronizing laboratory production of parasites with the development of the pest in the forest so that a maximum supply of the beneficial species will be on hand at the proper time must be solved.

The use of biotic agencies for the direct control of forest insects, therefore, has real possibilities, even though at present they have not been fully explored. The future is almost certain to witness an increased use of biotic methods for direct control.

PREVENTION OF SPREAD

A list of the forest pests of the United States and Canada includes numerous enemies that have been introduced from foreign lands. Outstanding among them are the gypsy moth, *L. dispar;* the poplar-and-willow borer, *Cryptorhynchus lapathi;* the European spruce sawfly, *Gilpinia hercyniae;* the elm leaf beetle, *Pyrrhalta luteola;* the larch casebearer, *Coleophora laricella;* the smaller European elm bark beetle, *S. multistriatus;* the European pine shoot moth, *Rhyacionia buoliana;* the balsam woolly adelgid, *Adelges piceae;* the Formosan subterranean termite, *Coptotermes formosanus;* and the old house borer, *Hylotrupes bajulus.* The great majority of these unwelcome immigrants reached America prior to 1912.

In addition to these immigrants, some of our native species have expanded their range to include localities from which they were originally absent. In an effort to check this invasion and spread of pests, regulations have been adopted by both federal and state governments providing for quarantines and embargoes to prevent invasion of new pests and inspection and certification to ensure that products in commerce are not infested with dangerous insects.

Legislative Approach

The first quarantine act was passed by the Congress of the United States in 1912. Since that time, other federal and state acts, complemented by similar legislation in Canada and Mexico, have implemented national and state inspection and quarantine systems until a high degree of efficiency has been attained.

Prior to 1912, plant products of all kinds could be moved freely both into and within the United States. As a result of low costs in Europe, nursery stock of all sorts was propagated overseas later to be shipped to American nurseries as lining-out stock. Some conifers, such as eastern white pine and Colorado blue spruce, were grown in Europe to the height of a foot or more before being shipped with the roots balled in earth. Without inspection or any restrictions on movement of stock from infested to uninfested areas, the number of introduced pests grew at an alarming rate, and action became imperative. The shipment of logs with bark intact is believed to have been the means by which many bark-, phloem-, and wood-destroying insects have been dispersed throughout the world.

Laws regulating the movement of plants may be grouped into two general categories: (1) quarantines and embargoes and (2) inspection and certification. No attempt will be made here to discuss the specific legislative acts. This has been effectively covered by Popham and Hall (1958). In the following two sections, we shall present only the objectives and the results that may be expected.

Quarantines and Embargoes

Quarantines are designed to regulate the movement of products from one country or locality to another in such a manner that the introduction of pests into uninfested localities will be either prevented or retarded. Quarantines prohibit movements except under specific restrictions that are deemed adequate to prevent spread of undesirable pests. For example, a shipment may be placed in quarantine at port of entry until it has received treatment, such as fumigation or dipping, in order to destroy any pests contained therein. Or a shipment may be planted temporarily in a quarantined area until close inspection demonstrates the presence or absence of pests. If they are free from infestation, the plants may then be distributed into uninfested territory. If they are infested, they are destroyed or treated to eliminate the pests.

Embargoes prohibit entirely the movement of products from one locality to another. They are applied to products that cannot be effectively inspected and that grow in localities known to be infested. For example, there is a strict embargo against importing any plants with roots balled in earth. Embargoes on localities are usually temporary and are lifted when it is considered safe to do so.

Inspection and Certification

Plant products entering the United States and Canada are all subject to government inspection at port of entry. If they are found free of dangerous pests, they are then permitted to enter. If they are infested, they are destroyed or, if practicable, treated to eliminate the infestation. Similarly, importations into states are subject to state inspection. The task of inspecting all forest products moving in commerce would require a force of inspectors entirely out of reason. In order to reduce the task, certification may be made at the point from which a shipment originates. After nurseries have taken all prescribed precautions and made all required treatments, they are given an examination and, if the products are free from dangerous pests, a certificate for one season is issued covering their stock. A copy of this certificate must be attached to each individual shipment. In order to implement this law, common carriers are forbidden to accept any living plant materials that do not bear a certificate of inspection unless such shipment is routed through an inspection office.

Pros and Cons of Regulation

The combined enforcement of all these regulations has resulted in tremendous benefits. One of the most important results has been the elimination of infestations at the source, both within the United States and abroad. The probability that an infested shipment will be detected, confiscated, and destroyed has greatly discouraged carelessness on the part of nursery managers and importers. Undoubtedly, the regulations have prevented the introduction of many pests and have retarded the spread of those already here.

The enforcement of regulations has been efficient and for the most part effective in both national and local services. Nevertheless, we cannot expect that any pest can be permanently excluded from any locality where suitable conditions prevail. No matter how careful an inspector may be, there is the ever present danger that some pest may escape (Fig. 8-3).

When human frailty is taken into consideration, the quality of the inspection services deserves the highest commendation; nevertheless, we must recognize the weaknesses inherent in the system. We must expect every potential pest ultimately to reach every suitable point on the earth, regardless of our efforts. However, if, through the support and enforcement of intelligent regulations, the spread of pests into new areas can be retarded, the enforcement services will have justified themselves.

There are certain precautions that foresters should observe: (1) The movement of planting stock outside the natural range of a species should be discouraged. (2) Within their natural range, living trees should never be moved from an area infested with a dangerous pest into an uninfested area. (3) Whenever a species is to be introduced into a new locality, it should be grown there from treated, sound seed. (4) Local nurseries and seed orchards should be established to serve each general locality. (5) Insofar as is practicable, nursery windbreaks should be of species not grown in the nursery. These precautions, supplementing the enforcement of statutory regulations discussed above, will help prevent the introduction and spread of pests.

Figure 8-3   A small fumigation chamber attached to a green-house. This is an example of an inexpensive means of pre-venting the spread of pest organisms in a locality. *(Photo courtesy Institute of Paper Chemistry, Wisconsin.)*

MODIFICATION OF FOOD SUPPLY

For control purposes, the food supply of an insect may be modified in three different ways: (1) It may be made inaccessible by erecting either chemical or mechanical barriers. (2) It may be made less available by reducing its actual quantity. (3) It may be made unavailable by changing its composition.

Inaccessibility Through Barriers

The use of chemical barriers in the control of wood and tree insects has not been developed to a high degree of efficiency, but it has been demonstrated that many common spray materials may be used in this way. A number of them have a decided repellant effect on insects, for example, Bordeaux mixture, lime sulfur, iron sulfate, and to a certain extent, even whitewash. All these repel insects for a time after they are applied. Caution should be exercised in select-ing proprietary repellants because some of them are valueless and some are actually injurious. In general, the use of paints should be avoided on living trees.

In the protection of trees from insect attack, mechanical barriers are often used. Sticky bands encircling the trunk are used to prevent defoliating insects from climbing from the ground to the foliage. This method is widely used on ornamental trees to control defoliators such as cankerworms, tussock moths, the gypsy moth, *L. dispar,* and other insects that spend a part of their lives on the ground and move to the treetops by creeping up the trunks. The commercial materials called "tree tanglefoot," "Stickem," and "Tack Trap" are convenient to use for this purpose because they can be purchased ready for use in any desired quantity. To be effective, all bands should be watched closely to ensure renewal whenever they lose their sticky quality.

In the protection of forest products, particularly seasoned materials, the use of mechanical barriers is important. For instance, one of the best barriers against insect infestation is an unbroken coat of paint or varnish covering the surface of susceptible wood. This is particularly effective against dry wood insects, for example, the powderpost beetles. Because these beetles deposit their eggs in the open pores of seasoned hardwood lumber, a coat of paint or varnish that closes these pores prevents infestation. The pressure impregnation of wood with chemicals such as creosote and pentachlorophenol serves as a barrier to both insects and diseases.

Reducing Availability of Food

Reduction in the food available for insect pests offers an effective means of insect control, especially in the protection of forest products. This may be accomplished by prompt utilization, by barking freshly cut logs, and by the disposal of waste materials.

Prompt utilization of logs and other forest products is one of the most obvious methods of preventing insect injury to this class of materials, but it is surprising how often the importance of this simple means of protection is overlooked. A tree is most susceptible to insect attack shortly after it has been cut or killed. As the phloem region and sapwood dry, the wood becomes less and less attractive to a large proportion of wood-inhabiting insects. One of the most effective ways of preventing both insect injury and the succeeding fungous injury is to utilize the logs as soon as possible after they are cut or to salvage without delay trees killed by fire, insects, wind, or other causes. This calls for prompt action. When trees have been killed by a spring fire, the problem of prompt utilization is almost hopeless because the trees will be infested with borers within a few weeks after the fire. Unless it is possible to begin salvage immediately, heavy losses cannot be avoided. On the other hand, when the trees are killed in the summer or fall, the wood will not be seriously injured until the following spring. There is, therefore, more time to plan and execute salvage operations.

Prompt salvage also prevents the increase of harmful species that might attack standing timber. Disastrous outbreaks of bark beetles have resulted from their excessive multiplication in windfalls. The tremendous outbreak of the spruce beetle, *D. rufipennis,* that reached a peak about 1950 after destroying vast quantities of spruce in Colorado illustrates well the danger of neglecting wind-thrown material. The Douglas-fir beetle, *D. pseudotsugae,* and *Ips* spp. frequently build up to outbreak proportions in fire-killed or wind-thrown materials.

Logging is only the first step in the process of salvage. If losses are to be avoided, the logs should be either sawed immediately or treated with water or insecticides. Prompt utilization will also protect green logs cut in regular logging operations. The shorter the time between felling and manufacture, the smaller the amount of food available for the insects and the less chance of loss from insect attack.

Food for some wood-boring insects can be eliminated by removing the bark from freshly cut wood. For example, many wood borers require phloem during their early larval stages. The removal of the bark destroys this succulent tissue on which the young larvae are dependent. It has one objectionable feature that is sometimes difficult to overcome. As the barked log or stick of pulpwood dries, the outer portions shrink more rapidly than the inner parts, with the result that checks develop, which may reach almost to the center of the logs.

In the forest, the proper handling of logging waste or slash is sometimes important from the standpoint of insect control. The ordinary methods of slash disposal are directed primarily toward fire prevention; as a result, only the smaller and more inflammable portions of the slash are taken care of. This type of disposal has little or no deterrent influence on dangerous forest pests because the potentially injurious species are found only in the larger parts such as the big branches, broken logs, and stumps. Disposal of the larger parts by utilization, barking, burning, or some other suitable method may be necessary. As long as logging is going on and fresh slash is continually being supplied by successive operations, the insects breeding in this material will find adequate feeding places for each generation. When logging operations end in a locality, the slash insects may attack standing trees because of a scarcity of food, injuring advance growth and even killing some trees. Such outbreaks are usually sporadic and seldom occasion great losses.

The consensus is that hardwood slash is almost never a breeding place for insects that attack living trees. Even coniferous slash is not as serious a menace as has been sometimes stated.

The *Ips* beetles are attracted to dying and dead pines and spruce and commonly infest slash and thinnings in the manner described. The problem may be alleviated by thinning stands in the fall and winter and by removing wood from the harvest areas as quickly as feasible. The forester must remove the wood within a few days if thinnings are done in the summer and must be prepared for possible losses. The multiple generations of these insects, especially in the South, result in a long summer period during which hazard is continually high.

Another group of insects associated with pine slash is the reproduction weevils. These insects breed in green pine stumps, and the emerging adults kill small seedlings by chewing the bark. The pales weevil, *Hylobius pales,* is the most important of these pests. The standard recommendation is to delay replanting of harvested areas, but this delay could be undesirable (Warren, 1965). Delay may result in the development of undesirable competing species and will mean the addition to a rotation of the number of years involved.

Changing Food Composition

Because each species or group of species has its own special food requirements, another means of insect control is made available. If we can change the character of the insect's food so that it is no longer suitable for consumption, we can reduce the growth and development of the insect correspondingly. Fortunately, this modification is often possible; and consequently, some of the most effective control methods are based on this plan. These methods are especially applicable to the control of wood borers working in freshly cut wood. At best, the phloem region is suitable for the development of these insects for only a comparatively short time. If this period can be shortened still further so that there is not time for development, or if a change in composition can be brought about before the wood is exposed to insect attack, injury by these insects can be reduced or eliminated.

One of the ways by which wood can be protected from the attack of borers is to give an opportunity for changes to take place in the inner bark before there is time for the insects to begin work. This can be accomplished by cutting in autumn and early winter because no insects are flying in those seasons. Not until the following spring, 8 or 9 months later, will these insects be on the wing. By that time, changes in the inner bark will have taken place that reduce its attractiveness. It is not known exactly what changes occur during the fall and winter, although they are evidently both physical and chemical in nature. When the phloem darkens, it no longer furnishes suitable food for the borers; even when no apparent change has occurred, timber cut in the autumn is not as susceptible to attack as the same tree species cut in late winter, early spring, or summer.

When it is impractical to cut wood during a certain period of year, it may still be possible to change food conditions in other ways. The period of susceptibility can be shortened by rapid drying. This can be secured in thick-barked pieces by removal of the bark, which accomplishes a double effect: It reduces the total quantity of food available for the borers, and it makes possible the rapid drying of the sapwood, with the consequent food changes. Thus, the wood soon becomes unsusceptible to the attack of those insects that feed in green sapwood. Rapid seasoning of thin-barked pieces can be accomplished by stacking the wood in well-ventilated piles. In the cooler northern latitudes, however, the rate of drying may be so slow that even thin-barked pieces will be infested while still green. In such localities, both removal of bark and piling in well-ventilated piles are necessary to prevent infestation.

In the South, on the other hand, rapid sun-curing for the purpose of modifying the character of the food has been used successfully. Under this method, the pieces of wood are either piled in open piles or, in the case of logs, placed side by side on skids. In the latter case, it is necessary to rotate the logs every few days to ensure even curing on all sides. The rotation of the logs results not only in hastening the seasoning rate but also in killing directly by heat any insects that may attack the logs during treatment.

Seasoning wood for the purpose of producing unfavorable food conditions for insects is applicable not only to logs and bolts but also to sawed lumber. During the drying process, food modifications occur that prevent all future attacks of some insects, even though moisture conditions may later become favorable. Ambrosia beetles are representatives of this group. The more rapid the seasoning, the shorter the period during which attack by these pests will be possible. Careful piling of green lumber in such manner that there is free circulation of air through and around the piles hastens drying, but kiln drying is an even better procedure.

Still another means of protecting freshly cut wood from insect attack is by water treatment. The chief effect of this treatment is to change the moisture conditions unfavorably. This effect will be discussed in the next section, "Modification of Moisture." But water treatment also changes food conditions. It is a matter of general knowledge that after logs have been in water for some time, they cease to provide suitable food for many wood-boring species. Short periods of floating in water have comparatively little effect on food composition within the wood. For this reason, short periods of water storage followed by removal from the water have little or no effect either on the insects present in the logs or on the susceptibility of the wood to later infestation. To be effective in changing food conditions, water treatment must be continued over a period of several months to a year.

MODIFICATION OF MOISTURE

One of the most important factors determining the rate of insect development is moisture. Insect development is limited by the moisture requirements of each species to a definite moisture zone. Outside that zone of toleration, development is impossible; consequently, control of insects in logs may be accomplished either by lowering the moisture content of the logs below the limits of the insect's toleration or by raising the moisture content above that zone.

Reduction of Moisture

Piling, removal of bark, and sun-curing are methods that may be used to change the composition of the insects' food. The same methods may sometimes be used for the purpose of reducing the moisture content.

The eastern subterranean termite, *Reticulitermes flavipes,* can only exist in an atmosphere above 97 percent relative humidity, and the wood-destroying fungi associated with the termites do not develop in wood with a moisture content of less than 20 percent. Thus, serious termite problems can be prevented by a reduction of moisture below either the termite or the fungous zone of toleration.

Increasing Moisture

In many instances, it is much easier to raise the water content of freshly cut wood above the point of insect toleration than to lower it. Water treatment, like seasoning, can control insects by changing the composition of their food and can be used to change the moisture content to such an unfavorable condition that insect development will be checked and sometimes entirely stopped.

There are two different ways of applying the water treatment: (1) by sprinkling the wood or (2) by floating it in a pond. The method selected for any particular case will depend on existing conditions. Each has its advantages and disadvantages.

The sprinkler system has been used on pulpwood and with equal effect on saw logs. Under this system, the wood throughout a pile is kept continuously dripping wet during the season of insect activity by spraying water on it (Fig. 8-4).

Figure 8-4   Water sprinkling is an effective method for protecting stored logs in the South. *(U.S. Forest Service photo.)*

In treating wood by the sprinkler system, it should always be borne in mind that the effectiveness of the operation depends primarily on the thoroughness of the wetting. Unless all the wood in the pile is kept wet, the treatment will have little or no beneficial effect. This system is likely to give the best results when applied to irregularly piled wood of short length. On evenly stacked piles, the water is likely to run off rather than drip through the pile.

The second method of raising the water content of freshly cut wood above the point of insect toleration is by floating it in water. This method can be used to protect not only wood stored at the mill but also freshly cut material in the woods. There is no more economical way of preventing injury to wood that must be held in the forest during a season or more than to get it into water.

If the wood is free from infestation when it is floated, infestation will usually not occur, especially if the logs are being worked. If the wood has been infested by wood borers prior to floating, the borers will develop no further, and if the wood remains in the water long enough, the insect inhabitants will die. Like most generalizations, this one has certain exceptions; for instance, large logs floating in water may be attacked on the upper side by ambrosia beetles.

The chief objections to the floating method of protection are: (1) Adequate water space may not be available. (2) A considerable proportion of the floated wood will become waterlogged and will sink, resulting in a total loss or a heavy expense for salvage.

In connection with this discussion of water treatment, benefits other than the protection against insect injury should not be ignored. Wood treated by either sprinkling or floating is protected not only against insects but also against the growth of fungi. Both stain and decay are checked or prevented by water treatment. Another benefit is the reduced fire hazard. It is practically impossible to burn wood stored under either of the methods discussed. The danger of fire in and around mills using one of these systems is reduced. All these benefits combine to make water treatment an especially desirable means of protecting wood.

MODIFICATION OF TEMPERATURE

Insects have a definite zone of temperature in which they are active. Within this zone, the rate of development varies with the temperature. The higher the temperature, the more rapid their development. Conversely, the lower the temperature, the slower the rate of development. This reaction to temperature provides us with another useful weapon in insect control. Theoretically, we can either lower or raise the temperature of wood to a point at which insect activities are checked.

There are opportunities for controlled use of high temperatures. The old house borer, *Hylotrupes bajulus,* dies when temperatures within its galleries exceed 125°F (52°C). Forced heat treatments have been effective in infested buildings. Similar treatments will kill other wood borers infesting lumber or wood in use.

Fire has been used for centuries to destroy infested materials, and many nests of defoliators and wasps have been burned to eliminate their destructiveness or annoyance.

The use of low temperatures in insect control is mainly limited to insects infesting stored food products. Packaged foods, especially dried dog food, can be placed in the home freezer for a few days to kill insect pests that may have been discovered. A similar technique could be used on furniture infested with borers, though the practice would be limited.

The three topics discussed under the headings "Modification of Food Supply," "Modification of Moisture," and "Modification of Temperature" are generally not independent of each other. These are methods to be considered, and though they do not generally apply to outbreaks in the forest, the forester will use them in harvesting, in regeneration, and in utilization of forest products.

BIBLIOGRAPHY

Beirne, B. P. 1966. *Pest management.* Cleveland, Ohio: CRC Press.

Coppel, H. C., and Mertins, J. W. 1977. *Biological insect pest suppression.* Berlin: Springer Verlag.

DeBach, P. 1974. *Biological control by natural enemies.* London: Cambridge University Press.

Franz, J. M. 1976. Towards integrated control of forest pests in Europe. In *Perspectives in forest entomology,* ed. J. F. Anderson and H. K. Kaya, pp. 295–308. New York: Academic Press.

Geier, P. W. 1966. Management of insect pests. *Ann. Rev. Entomol.* 11:471–490.

Graham, S. A., and Knight, F. B. 1965. *Principles of forest entomology.* New York: McGraw-Hill Book Company.

Lanier, G. N., et al. 1976. Attractant Pheromone of the European elm bark beetle (*Scolytus multistriatus*): Isolation, identification, synthesis, and utilization studies. In *Perspectives in Forest Entomology,* ed. J. F. Anderson and H. K. Kaya, pp. 149–175. New York: Academic Press.

Metcalf, R. L., and Luckmann, W. H. 1975. *Introduction to insect pest management.* New York: John Wiley & Sons.

Nagel, R. H., et al. 1957. Trap tree method for controlling the Engelmann spruce beetle in Colorado. *J. Forestry* 55:894–898.

National Academy of Sciences. 1975. *Pest control: An assessment of present and alternative strategies.* Forest pest control, vol. 4. Washington, D.C.: National Academy of Sciences.

Popham, W. L., and Hall, D. G. 1958. Insect eradication program. *Ann. Rev. Entomol.* 3:335–354.

Richmond, H. A., and Nijholt, W. W. 1972. *Water misting for log protection.* Canada Forest Service, Pacific Forest Research Center, Vancouver, B.C.

Roelofs, W. L., et al. 1976. Pheromones of lepidopterous insects. In *Perspectives in forest entomology,* ed. J. F. Anderson and H. K. Kaya, pp. 111–125. New York: Academic Press.

Ruesink, W. G. 1976. Status of the systems approach to pest management. *Ann. Rev. Entomol.* 21:27–44.

Shorey, H. H. 1973. Behavioral responses to insect pheromones. *Ann. Rev. Entomol.* 18:349–380.

Smith, H. S. 1925. The commercial biological control in California. *J. Econ. Entomol.* 18:147–152.

————. 1926. Biological control work in California. *J. Econ. Entomol.* 19:294–302.

Smith, R. F. 1963. Principles of integrated pest control. *Proc. North Central Branch Entomol. Soc. Am.* 18:71–77.

————, et al. 1962. Integration of biological and chemical control. *Symp., Bull. Entomol. Soc. Am.* 8:188–201.

Stinner, R. E. 1977. Efficacy of inundative releases. *Ann. Rev. Entomol.* 22:515–532.

Strong, Lee A. 1938. Control legislation in the United States. *USDA, Bur. Entomol. P. Q.,* E-455.

U.S. Forest Service. 1972. *The outlook for timber in the United States.* USDA, Forest Service Resource Report 20.

Wagner, F. G., Jr. 1978. Preventing degrade in stored southern logs. *USDA Forest Service SE Area St. Pri. For., For. Prod. Util. Bull.* no. 1.

Warren, L. O. 1965. Controlling insect damage to young southern pine stands. In *Insects of southern forests,* ed. C. B. Marlin, pp. 88–102. Baton Rouge: Louisiana State University Press.

Chapter 9

Biological Control

Throughout the ages, whenever harmful insects have been observed, the attention of entomologists has been attracted by activities of parasites, predators, and pathogens. The possibility that these organisms might be used for control purposes was recognized early. They were especially attractive for use against forest insects because control of forest pests seemed, in early times, to present almost insurmountable difficulties. Unfortunately, we fully understand how to manipulate and encourage only a few of these organisms. Nevertheless, progress is being made in this direction.

APPROACHES TO BIOLOGICAL CONTROL

There are two distinct approaches to the use of biological factors for control: (1) the introduction of these organisms into areas where they do not occur and (2) the encouragement of those organisms already present. The introduction of parasites, predators, and pathogens has usually been directed against exotic pests without natural enemies in their new environment. Environmental manipulation designed to improve conditions for native parasites and predators applies equally well to both foreign and native pests.

Introduction of Organisms

The attempts to control forest insects by biological means have most often been directed toward introducing beneficial organisms to check the ravages of foreign pests. For over 70 years, much effort in the United States has been devoted to introducing natural enemies of the gypsy moth, *Lymantria dispar.* During this time, more than 50 species were introduced, and 10 parasites and 2 predators may have become useful in affecting outbreaks (Leonard, 1974).

In Canada, similar work has been done in introducing parasites. One of the earliest attempts was the introduction, in 1910, of parasites of the larch sawfly, *Pristiphora erichsonii* (McGugan and Coppel, 1962).During the 1930s, Canadian forest entomologists greatly expanded their work with parasites and introduced, among others, a number of parasites of the European spruce sawfly, *Gilpinia hercyniae.* Of these parasites, at least three have become widely established in both Canada and the United States and have proven to be valuable additions to our fauna (Reeks and Cameron, 1971).

A major program to introduce foreign predators for control of the balsam woolly adelgid, *Adelges piceae,* began in Canada in 1933. In 1957, the United States introduced 23 predaceous species from seven countries throughout the world in an attempt to control the balsam woolly adelgid. Releases were made in the fir forests of the Pacific Northwest, in the Appalachians, and in eastern Canada. Three dipterous flies and two beetles are now successfully established in the Pacific Northwest (Mitchell and Wright, 1967).

The introduction of biological control agents is not always restricted to species imported from other parts of the world. Sometimes, parasites are transferred from one locality to another within a country. For example, parasites were transported from Virginia to the Nebraska National Forest to control the Nantucket pine tip moth, *Rhyacionia frustrana* (Graham and Knight, 1965). Also, parasites of the spruce budworm, *Choristoneura fumiferana,* have been transported from British Columbia to eastern Canada.

Both in the United States and in Canada, the use of biological control has received considerable attention. Because most of the work has been done by entomologists, it is natural that parasitic and a few predaceous insects have held the center of attention. However, these animals are not the only organisms that are useful for control. Fungi, bacteria, protozoans, nematodes, and viruses all attack insects, thereby bringing about reduction of their rate of reproduction or death. Occasionally, microorganisms have been introduced, often accidentally, and have proved most valuable as control agents. Some of these will be discussed in the following sections.

Encouragement of Native Organisms

Although the introduction of exotic agents of control has received most attention, forest entomologists have by no means neglected the possibilities of encouraging the good work of resident organisms. These organisms, like the pests themselves, respond to their physical, nutritional, and biological environments. By manipulating the environmental factors, we can greatly increase the effectiveness of biological control. With this in mind, studies have been made of the ecological relations existing between the various beneficial organisms and their environment. Some of this work will be discussed in connection with silvicultural practices (see Chapter 10, "Preventive Control by Silvicultural Practices"). The encouragement of the biological factors must be closely tied in with forest management.

MICROORGANISMS

Insect populations are influenced by various kinds of microorganisms. Of these, bacteria, fungi and viruses are best known (Steinhaus, 1963; Burges and Hussey, 1971). Discussion will be limited to these, although we should remember that other groups also feed on insects.

Bacteria

References to bacterial diseases of insects are very common in entomological literature (Falcon, 1971; Faust, 1974).

Uncertainty of action under forest conditions seems to be characteristic of many bacteria, especially those that infect defoliators and other insects that live on the exposed surfaces of plants. In certain seasons, they may destroy almost every individual of a host species; in others, none of the hosts will be killed. Steinhaus (1949) contrasted the durability of bacteria that do not produce spores with those that do produce spores, pointing out that the spores are very resistant to drying and high temperature. Bacteria that do not produce spores are relatively sensitive to physical extremes. From this, we may logically assume the spore-producing bacteria offer greater possibilities for the effective control of forest insects.

This conclusion is supported by the fact that some of the most effective bacterial diseases belong to the spore-producing group. Prominent among these is the milky disease of the Japanese beetle, *Popillia japonica,* caused by *Bacillus popilliae* (Dutky, 1963). The endospores of this and similar species remain viable for several years in the soil, where they are in a position to infect any larva that may ingest them. The endospores produce vegetative cells that reproduce within the body of the host. When the host dies, its infected body remains in the soil. Endospores from the bodies of infected larvae are then eaten by other larvae. The endospores are spread locally through the soil by the movements of larvae and more widely by birds and mammals that have eaten the infected larvae and beetles. Not only the larvae but also the adult beetles may become infected. These infected beetles may also spread the inoculum from place to place.

The spores of the milky disease are commercially produced in vivo. Larvae of the Japanese beetle, *P. japonica,* are inoculated with the spores and incubated until they develop the disease. The infected larvae are dried, ground, and their spore content determined. The ground material is then diluted with talc, so that each gram contains 100 million spores (Fleming, 1968). Within a few seasons after inoculation, the soil becomes completely infectious and hence uninhabitable for the beetle larvae. An example of spore dust treatments is given by Steinhaus (1949) for a park area in the District of Columbia; between 1940 and 1943, the grub population dropped from 44 to 5 per ft².

The symptoms of bacterial diseases are variable but are commonly characterized by color changes and motor disturbances. Often, the larvae remain hanging head downward from a twig or leaf after death. If they are undisturbed, they usually dry and shrivel but still retain their typical form (Beard, 1945).

Recently, the highly pathogenic bacterium *Bacillus thuringiensis* has received considerable attention. It appears to hold more promise as an agent of biotic control than most bacteria. The virulence of *B. thuringiensis* is the result of the production of both endospores and proteinaceous crystals (Fig. 9-1). When ingested, the endospores germinate and multiply while the crystals dissolve into toxic compounds. An international standardization has now been established for the various strains of *B. thuringiensis* obtained from around the world. The standard prepared by the Pasteur Institute in France is *B. thuringiensis* var. *thuringiensis* (Berliner) and is designated E61. The potency is expressed in international units (IU)/mg. The IU is defined as 1000 times the amount of E61 needed to kill 50 percent of a standard insect. In the United States, a standard has been prepared from *B. thuringiensis* var. *Kurstaki,* designated HD-1, and assigned a potency of 18,000 IU/mg; the cabbage looper, *Trichoplusia ni,* was selected as the standard insect (Coppel and Mertins, 1977).

B. thuringiensis has been most successful against many insect pests in the Lepidoptera. The forest-insect defoliators have been grouped by Harper (1974) according to their response to *B.t.* For example, (1) easily controlled were the tent caterpillars, cankerworms, bagworms, and spanworms, (2) susceptible but not easily controlled were the Douglas-fir tussock moth, *Orgyia pseudotsugata,* gypsy moth, *L. dispar,* and spruce budworm, *C. fumiferana,* and (3) not susceptible were the saddled prominent, *Heterocampa guttivitta,* the western hemlock looper, *Lambdina fiscellaria lugubrosa,* and the lodgepole needle miner, *Coleotechnites milleri.*

The variability in the efficacy of *B.t.* can be attributed to inherent susceptibility among insect species; variations in *B.t.* strains, formulations, and dosages prior to standardization; delayed response of larval death; weathering of the spores; and timing of application of a microbial biological control with an insect's feeding habits.

(a)

Figure 9-1 An outbreak of the forest tent caterpillar, *Malacosoma disstria,* in this water tupelo pond was stopped by an experimental aerial application of *Bacillus thuringiensis. (a)* The *B. thuringiensis* cell components are the immature vegetative cell (v), mature cell with spore and crystal (m), protein crystal (c), and reproductive spore (s); as viewed under magnification (1300X) *(b)* and schematically *(c). (Auburn University.)*

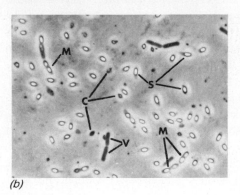

(b)

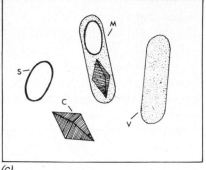

(c)

Fungi

Infections of insects by fungi are so common that they have come to the attention of everyone. The dead housefly on the window glass surrounded by a halolike circle of white spores and the mummified body of a dead aphid still attached to a leaf are both common examples of this sort of infection. At times, severe outbreaks of insects may suddenly be brought to an end by a fungal epizootic. An *epizootic* is a disease outbreak of unusually high incidence, in contrast with an *enzootic* disease of low but constant incidence in an area.

Although they are exceedingly numerous, fungal infections of insects have never been used successfully for controlling a forest insect. The reason for this is that whereas the parasitic fungi are able to check insect outbreaks only under relatively warm, moist conditions, insect outbreaks occur most frequently during periods of dry weather. Consequently, at the time when we need them the most, there is little or nothing that can be done to encourage fungal infections of insect pests. Nevertheless, we must recognize that during moist periods, the presence of fungous inoculum may aid materially in preventing the rapid multiplication of forest-insect pests.

Many fungi that attack insects may live as saprophytes for long periods. Thus, when conditions are favorable, the inoculum is present and ready to attack the living insects. At times, larch sawfly, *P. erichsonii,* prepupae within the cocoons are attacked and destroyed by one fungus of this sort, *Isaria farinosa.* Its effectiveness depends upon the amount of saprophytic growth in the moss, and that, in turn, depends chiefly upon moisture (Graham, 1956).

Viruses

Insect diseases that are caused by viruses are often responsible for checking outbreaks of both lepidopterous and hymenopterous defoliators. These viruses usually appear after 2 or 3 years of heavy defoliation and destroy so many larvae that the defoliator outbreak subsides.

Our knowledge of virus diseases had its beginning in Europe, where the *Wipfelkrankheit* of the nun moth, *Lymantria monacha,* has been studied intensively since its discovery in 1889. About 1907, a similar disease was observed in populations of the gypsy moth, *L. dispar,* in New England. It is commonly known as the *wilt disease,* a name suggested by the wilted appearance of the dead larvae. How the virus reached New England is not known, but according to Glaser (1927), it was not present in 1900. Presumably, it was introduced into America with the wholesale introduction of parasitic insects in 1905.

The wilt disease is infectious, gaining entrance into the body of the insect with ingested food. After becoming established, it usually kills the blood cells and certain tissues. The diseased larvae first become sluggish and then stop eating. Before they die, they have a tendency to climb high in the trees; after death, they remain hanging by their prolegs. Their tissues become darkened, decomposed, and liquified. Finally, they disintegrate and dry on the tree. These dried smears that once were insects remain infectious for a long time. However, if infection occurs late in the larval stage, the insect is often able to complete its development and reproduce without showing any symptoms of the disease. In such instances, the infection may be passed through the egg to the progeny. Because of the presence of chronic carriers, one can never be sure the disease does not exist in a locality, even though the insects appear healthy.

Another very striking case of a virus that has brought about the control of a forest insect is that of the European spruce sawfly, *G. hercyniae.* Between 1930 and 1938, this introduced sawfly threatened to destroy much of the spruce in eastern Canada and parts of New England. In 1938, a virus disease of the sawfly appeared, presumably with the importation of insect parasites. Once established, it spread rapidly until the sawfly was brought under control. According to Balch and Bird (1944), the disease built up in areas of very high population but, after attaining momentum, was able to invade areas where the sawfly was not so numerous. This disease has so reduced the numbers of the European spruce sawfly that this insect is no longer considered a threat to spruce stands (Neilson and Morris, 1964).

The use of viruses for control of forest insects in the United States was experimental until 1976, when the Environmental Protection Agency registered a nucleopolyhedrous virus (NPV) of the genus *Baculovirus* against larvae of the Douglas-fir tussock moth, *O. pseudotsugata*. [1] The registration of a virus indicates that techniques are now developed to determine ingredients, efficacy, safety to humans and the environment, the amount of residues, and methods of use. An NPV was registered for use against the gypsy moth, *L. dispar,* in 1978.[2]

INSECTIVOROUS VERTIBRATES

The insectivorous vertebrates, especially birds and mammals, have received much attention as agents of insect control. Undoubtedly they do have an influence upon some forest-insect populations. In Chapter 6 we discussed these animals in relation to their density-dependent influence upon their insect prey; some of the limitations of these organisms were brought out. These limitations do, however, influence the use of vertebrates for control.

Fish and Amphibians

The literature abounds with studies of the importance of insects as food for fish (Merritt and Cummins, 1978), frogs (Graham, 1956), toads, and salamanders. However, there are few studies on the influence of fish or amphibians on insect populations. Of importance to human health and outdoor recreation are the examples of mosquito suppression by the introduction of mosquito fish (*Gambusia* spp.). These fish are considered by Legner et al. (1974) to be the most widely used biological control agent. The fish have been successful in Virginia, California, Hawaii, Europe, Africa, and Malaysia. However, there are problems with these fish: the cost of rearing, mortality from cold water, their effect on the populations of native fish, and survival in ephemeral breeding places.

Birds

Of all the vertebrates, birds are probably the most important insect eaters. Many birds are insectivorous throughout their lives; the swallows, martins,

---

[1] Registration No. 27586–1, August 11, 1976, NPV *Baculovirus,* sub-group BV.
[2] Registration No. 27586–2, April 13, 1978.

and nighthawks are among these. Others are more or less omnivorous; these include the crows, blackbirds, thrushes, and gallinaceous birds. Almost all the perching birds, including even the seed-eating finches, are insectivorous in the nestling stage. Because these seed eaters are nesting during the period of greatest insect abundance, they are valuable aids in holding down insect populations. Therefore, the forester should give them every aid that is practicable.

The encouragement of birds can be accomplished in part by diversification in the forest. The more diversified the environment, the greater the variety of bird species; there the largest number of species will find food and favorable places to live. The forester must remember that in the carefully managed forest, favored nesting sites for certain birds, especially those that nest in the hollows or holes of tree trunks (Conner, 1978), are often incidentally eliminated. Without such nesting places, those birds will be eliminated from the forest, at least for the nesting season. In some European forests, the provision of nesting boxes for hole-inhabiting birds is common and is said to pay a high return on the investment (Bruns, 1960).

There have been few studies of the value of nesting boxes in America. In California, Dahlsten and Herman (1965) noted a 30 percent reduction in lodgepole needleminer, *C. milleri,* when nesting boxes increased breeding pairs of mountain chickadees. Foresters should remember that permanent nesting boxes should be cleaned out every year (Cole, 1964).

The introduction of birds is often urged as a means of controlling forest insects, but this procedure is attended with great risk. Flexibility of habitat and mobility make birds effective predators, but these characteristics may make them undesirable additions to the environment. Some introduced birds have become undesirable competitors with native species. For example, the English sparrow, although it may have accomplished some good, reduced the number of native birds in the areas where it became abundant. Similarly, the starling, although highly insectivorous, has become a pest wherever it roosts in great flocks.

The role of birds in controlling insects is a subject that promotes strong emotions among powerful interest groups. Foresters are advised to study in great detail the literature on this subject when the occasion arises. There are excellent studies of the role of birds on the following insects:

1 bark beetles (*Dendroctonus* spp.): Knight, 1958; Otvos, 1965; Shook and Baldwin, 1970
2 elm spanworm (*Ennomos subsignarius*): McAtee, 1925
3 fall webworm (*Hyphantria cunea*): Tothill, 1922
4 gypsy moth (*Lymantria dispar*): Campbell et al., 1977
5 larch casebearer (*Coleophora laricella*): Sloan and Coppel, 1968
6 larch sawfly (*Pristiphora erichsonii*): Buckner and Turnock, 1965
7 spruce budworm (*Choristoneura fumiferana*): Stewart and Aldrich, 1951; Mook, 1963; Morris et al., 1958
8 whitemarked tussock moth (*Orgyia leucostigma*): Dustan, 1923

Foresters interested in the interactions of insects and birds should consult excellent reviews by Bruns (1960), Franz (1961), Buckner (1966, 1967), and Coppel and Mertins (1977).

Mammals

Almost all carnivorous animals eat insects, and a few live almost entirely on insects and other arthropods. Some of the mammals that are not usually thought of as being insectivorous actually destroy many forest insects. For example, mice and voles, in some years and on certain sites, destroy more larch sawfly, *P. erichsonii,* cocoons than all the parasites combined (Graham, 1928).

Skunks, shrews, weasels, foxes, and many other mammals consume great quantities of insects on the ground. Their activities are especially beneficial during the dormant season, when many species of harmful forest insects are in hibernation either on or in the ground. Even during the summer, skunks are diligent predators on soil-inhabiting insects, digging them from beneath the surface. Numerous holes resembling small, inverted cones attest to the industry of these animals wherever white grubs, *Phyllophaga* spp., are common. These cones are considered a detraction in some recreational areas, such as golf courses.

Encouragement and protection of these valuable insect eaters should always be considered in planning the insect control program for any forest. Unfortunately, the opportunities to do this are usually overlooked. In general, desirable conditions for insectivorous mammals can be attained by diversification of both trees and ground cover.

When the use of mammals for the control of forest insects is considered, we naturally think of the introduction of new species. The danger of introducing predatory mammals is illustrated by experience with the mongoose. This animal was introduced into certain islands in the West Indies to control reptiles. This objective was accomplished reasonably well, but at the same time, the animal became a pest. Because of its taste for domestic birds, it jeopardized poultry raising.

In Canada one mammal has been introduced into the Province of Newfoundland. The masked shrew, *Sorex cinereus,* was introduced from the mainland to aid in control of larch sawfly, *P. erichsonii,* outbreaks (Buckner, 1966). The shrew is believed to have reduced cocoon populations of both larch and spruce sawflies (Warren, 1971).

PREDATORY ARTHROPODS

The predatory insects are found in a number of different families in several orders. Some of the most effective as control agents of forest insects are the lacewings, ladybird beetles, checkered beetles, ground beetles, and several families of sucking insects. In addition to the insects, there are other arthropods that prey on insects. Because they are closely related to the insects and because too little is known specifically about their effectiveness in the control of forest insects to justify separate treatment, they will be included in this section.

Spiders and Mites

Of the noninsect arthropod classes, spiders and mites are among the important predators in the forest. General observation indicates that they are of importance, but unfortunately, their effectiveness has never been adequately studied. During outbreaks of the spruce budworm, *C. fumiferana,* spider populations as high as 75,000 per acre were reported by Loughton et al. (1963). Coppel and Smythe (1963) found many species of spiders on eastern white pines infested with sawflies. A study by Clarke and Grant (1968) found predation by spiders to have an important effect on centipedes and collembola in the litter of a maple forest.

Spiders that spin webs capture not only flying insects but also small larvae, especially those that drop from trees on threads of silk. The hunting spiders that do not depend on web traps can move about actively in search of food and, therefore, are in a better position to concentrate where high populations of insect pests prevail. The exact part that spiders play in regulating forest-insect populations cannot be known until further study of their actions has been made.

Similarly, very little information can be found in the literature concerning the effects of predaceous mites on forest insects. Nevertheless, general observations indicate these organisms are active predators of scale insects. Tothill (1919) found that mites exercise an important controlling influence by eating the eggs of the oystershell scale, *Lepidosaphes ulmi.* The association of mites with bark beetles has been well summarized by McGraw and Farrier (1969). Doubtless many other predaceous mites are equally important, but unfortunately we have few studies.

Neuroptera: Chrysopidae

Members of the neuropteran family Chrysopidae are commonly named *lacewings.* They are predaceous on small insects, especially aphids, in both the larval and adult stages. They also eat scale insects, small larvae, and various other prey. These predators are especially valuable late in the season, at which time aphid outbreaks frequently occur. Unfortunately, there is apparently considerable mortality of lacewings during the winter. As a result, they are not usually abundant in early summer and therefore cannot exercise much control at that time of year. The rapid increase in number of the lacewings as the season advances is made possible by their short life cycle, some species passing through three or four generations during a season.

The adult lacewing is a slender, fragile-appearing insect about 18 mm in length, with delicate, many-veined, membranous wings. The wings, when at rest, are folded against the body and meet over the insect's back in a form resembling a tent. The adults present the appearance of anything but predators. The larvae, on the other hand, look quite ferocious. They are elongate, somewhat flattened, and have a pair of long, curved mandibles projecting forward from the mouth. These mandibles are hollow, and it is through them the insect sucks the fluids of its prey.

Coleoptera: Coccinellidae

The ladybird beetles are species of the family Coccinellidae in the order Coleoptera and are for the most part predaceous. Like the lacewings, they eat small insects in both the adult and the larval stages. Aphids or scale insects are their usual prey, but they will feed on small caterpillars, mites, insect eggs, or almost any small organism. In Michigan, Graham (1929) observed instances in which a comparatively large coccinellid, *Anatis 15-punctata,* was partly responsible for checking an outbreak of the jack pine budworm, *C. pinus.*

The life cycle of the ladybird beetles varies with the species. Many of them have two or more generations in northern latitudes; even more generations may be completed annually in southern lands. The insects pass the winter in the adult stage, hidden away in some protected location. Many of them may creep into buildings for the purpose of hibernating. When spring arrives, the beetles leave their winter quarters and seek suitable places for oviposition.

Some of our common native species deposit their eggs in groups on the surface of the foliage on which their prey feeds. The eggs are usually yellow. The larvae are elongated and flattened, with well-developed legs. Immediately after emergence, they begin to search actively for food. Their mandibles are long and sharp and well suited for capturing small insects. Many of them are mottled with contrasting colors and are covered with branched spines.

Coleoptera: Cleridae

Species within the Cleridae are commonly called the *checkered beetles* or *clerids* (Fig. 9-2). All are predaceous, mostly on bark beetles and other insects that bore into the bark of trees. One of the most common species in the United States is *Thanasimus undulatus* (= *dubius*), a large clerid nearly 13 mm in length. It is very brightly colored, with the bright red background marked with black and silver transverse bands. This type of marking is characteristic of most checkered beetles. These insects are predaceous in both the larval and the adult stages.

The eggs are deposited in the tunnels of bark beetles. The larvae, more or less grublike in form, with poorly developed legs but with powerful mandibles, are usually pinkish in color. When numerous, they materially reduce the number of bark beetles. After completing their development, the larvae cut

Figure 9-2 The redbellied clerid, *Enoclerus sphegeus,* is an important predator of western bark beetles and woodborers in both the adult *(a)* and larval *(b)* stages. *(U.S. Forest Service photo.)*

into the outer bark. There they hollow out cells in which they pass the winter in any stage of development. In the spring, they complete their development. By the time of the spring flight of the bark beetles, they have transformed to adults and are ready to attack their prey.

Coleoptera: Carabidae

The species within the Carabidae are commonly called the *ground beetles* or *carabids* (Fig. 9-3). They are the largest and the most conspicuous of all the ground beetles and are among the important predators of lepidopterous larvae. Their habits have been studied in great detail by Burgess and Collins (1915) in connection with investigations of the gypsy moth, *L. dispar,* in New England. The life cycles and habits of the species within this family are similar. The eggs are deposited in the ground at a depth of 100 to 150 mm. Upon emerging, the larvae make their way to the surface of the ground, where they actively search for food. At the end of the third instar, having completed their growth, they again penetrate the soil, where they form earthen cells in which to pupate. Some species, soon after transformation to the adult stage, emerge from their pupal cells and commence feeding; others remain in the ground until spring.

A few of these beetles even climb into the treetops, in contradiction to

Figure 9-3 A *Calosoma* beetle (Carabidae) preying upon larvae of the elm spanworm, *Ennomos subsignarius.* Note variation in larval color. *(U.S. Forest Service photo.)*

their common name. One of these is the fiery hunter, *Calosoma calidum,* which climbs trees not only while in the adult stage but also as larvae, a very unusual habit for a ground beetle. In the adult stage, the beetles of this genus are all unusually long-lived. Collins (Burgess and Collins, 1915) reported that unless they meet with some accident in the field, the beetles will live for 2 or 3 years. One fiery hunter lived for 4 years in captivity.

Hemiptera

The order Hemiptera, or "true" bugs, are mostly phytophagous, sucking sap from plants. Several families, however, are almost all predaceous, such as members of the Reduviidae, the assassin bugs. One of the well-known members of this family is the wheel bug, *Arilus cristatus,* common throughout the eastern and central United States. The wheel bugs lay their eggs on plants or other objects in groups of 50 or even more. The young nymphs, red in color, with long legs and four-jointed antennae, run about over plants in search of insect food. At first, they can attack only small insects, such as aphids, but the size of their prey increases with their growth. By the time they have reached their full growth of 18 mm in length, no insect seems too large for them.

Another family of the Hemiptera containing many species predaceous on forest insects is the Pentatomidae. They are called *stink bugs* because of their disagreeable odor or *soldier bugs* because of their predaceous habits. Some members of this family are entirely phytophagous, but many feed on the fluid of other insects during at least a part of their lives.

The spined soldier bug, *Podisus maculiventris,* is a very effective predator on tree-inhabiting insects, especially caterpillars. When fully grown, it is nearly 12 mm in length and yellow mottled with brown in color. Its life history is very similar to that of many other soldier bugs and, therefore, will serve to illustrate the group.

The winter is passed in the adult stage, usually buried in litter on the ground. On emergence from hibernation, the adults feed for a short time and then lay their yellow eggs in groups, often on the underside of a leaf. The young nymphs are either red or black in color and for a time feed on the sap of plants. After about 2 weeks, they adopt the predaceous habit. In the North, the soldier bugs complete two generations each season; in the South, they may have as many as four or five.

Several other families of the Hemiptera are also predaceous, but most of them are less important in the forest than those mentioned in this brief comment. Still other orders of insects are predaceous on forest insects and might well be considered if space permitted. Altogether, insect predators are important factors; they should not be overlooked in planning for the control of forest insects.

PARASITIC INSECTS

Of all the enemies of forest insects, the parasitic insects have received the most attention from entomologists. The parasites do not constitute a phylogenetic unit but are to be found in various orders and in widely divergent families. Adult parasites vary greatly in appearance, but the larval stages are surprisingly similar. They are practically all legless, maggotlike grubs. The mouthparts are usually much reduced or even vestigial. In short, they are specialized for a mode of life that is provided with an abundance of easily obtainable food, and they illustrate the condition called *specialization by reduction of parts.*

Parasitic insects are sometimes limited to a single species of host. These are called *monophagous* parasites. Others are able to attack a number of closely related species (*oligophagous* parasites), and still others are general parasites (*polyphagous* parasites).

Insects that are parasitic on other insects are often regarded as being beneficial to human interests inasmuch as they reduce the number of individual pests. Such a generalization is not entirely correct; there are parasites that attack other parasites, predaceous insects, or insects useful to humans. Thus, some insect parasites may be injurious. Those species that parasitize other parasites are called *hyperparasites.* Hyperparasitism is known to occur to the fourth or possibly the fifth degree, which reduces the effectiveness of parasitic control. Competition between parasites within a host is not uncommon. When dual parasitism occurs, one of the parasites usually destroys the other. Such competitive interactions complicate the problem of evaluating the effects of parasitism on pest populations.

The order Hymenoptera contains more species of insects that are parasitic than any other. Most of the species parasitizing forest insects are members of the two large superfamilies: Ichneumonoidea and Chalcidoidea. The former contains the species that are most familiar, the ichneumonids and braconid wasps; and the latter, the chalcid wasps. Next to the Hymenoptera in number of parasitic species is the order Diptera, the parasitic family Tachinidae being by far the most important.

Hymenoptera: Ichneumonoidea

The hymenopterous superfamily Ichneumonoidea contains thousands of species and hundreds that attack forest insects. The ichneumonids (Fig. 9-4) are all armed with slender ovipositors with which they are able to insert their eggs into their host. Some are large, wasplike insects; others are so minute that they can scarcely be distinguished with the naked eye. The habits and the life history of each species differ in detail, but nevertheless, there is considerable similarity among them.

One of the parasites introduced to control the gypsy moth, *L. dispar,* a braconid, *Apanteles melanoscelus,* will serve well for illustrative purposes. This parasite has proved to be one of the best thus far introduced in connection with the gypsy moth projects. It is a small, black insect, about 3 mm long. Its reproductive potential is tremendous, a single female being capable of laying 1000 eggs. It is synchronized with the phenology of the gypsy moth and, in addition, is able to parasitize a number of other defoliators, such as the satin moth, *Leucoma salicis,* the tent caterpillars, and several tussock moths.

The life history of *A. melanoscelus* is similar in most respects to that of other ichneumoids (Crossman, 1922) (Fig. 9-5). Perhaps the most unusual

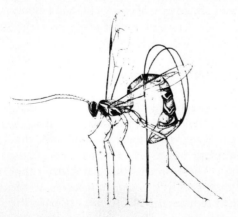

Figure 9-4 One of the largest ichneumons, *Megarhyssa lunator,* is able to insert her ovipositor deep into the wood and parasitize larvae of the pigeon tremex, *Tremex columba.* (U.S. Forest Service photo.)

feature is the fact that two generations of the parasite occur during a single generation of the host.

Unfortunately, this parasite suffers greatly from the attack of hyperparasites and ants, especially while in the overwintering stage. As a result, the parasite is so reduced during the winter that its first generation in the spring is able to parasitize only a relatively small proportion of the host. The second generation may parasitize a very high proportion.

Hymenoptera: Chalcidoideas

The chalcids comprise a superfamily that contains more species than the ichneumonids. A very large proportion of them are parasitic on other insects. The larvae of some of them, the ectoparasites, feed on the host externally; others, the endoparasites, live within the host. They are all small insects, some of them very minute. As a result of their small size and short ovipositor, they are unable to reach and parasitize most borers, although some of them do parasitize certain twig-inhabiting insects.

Some of the chalcids attack their host while it is within the cocoon. One of these, *Coelopisthis nematicida,* is a common parasite of the larch sawfly, *P. erichsonii.* It is distributed throughout the United States and Canada wherever the larch sawfly occurs. The winter is passed in the larval stage within the cocoon of the sawfly. The adults emerge in May or early June, while many of the sawflies are still in the prepupal stage. They immediately seek out cocoons and deposit their eggs. It is not at all uncommon to find from 50 to 75 of these parasites feeding within a single sawfly cocoon.

Figure 9-5 The braconid, *Coeloides dendroctoni,* is an important parasite of western bark beetles. She attaches her eggs to larvae within the phloem region. *(U.S. Forest Service photo.)*

Only about 3 weeks are required for completion of a generation of this parasite. Repeated generations are produced during the season, the number depending on temperature conditions. On first thought, it would appear that a parasite with a series of short generations would be limited in its effectiveness by the unavailability of suitable host material at certain seasons. However, this is not a limiting factor because sawfly cocoons are available at practically any time during the growing season.

Diptera: Tachinidae

Among the parasitic Diptera, the species of the family Tachinidae are the most valuable destroyers of forest pests (Fig. 9-6). The habits of the tachinid parasites vary considerably. Some of them are oviparous; others are larviparous. The oviparous species usually glue their eggs tightly to the body of the host to prevent them from being rubbed off before they hatch. The larvae in the larviparous species are either deposited on the body of the host or inserted into the body of the host by means of a larvipositor.

One of the imported tachnid parasites of the gypsy moth, *L. dispar,* is *Compsilura concinnata.* It attacks not only the gypsy moth but also the fall webworm, *H. cunea,* the whitemarked tussock moth, *O. leucostigma,* the forest tent caterpillar, *Malacosoma disstria,* and many other native insects. This parasite has been reared from more than 200 different host species (Dowden, 1962) and has played an important part in reducing the population of the gypsy moth even when that insect has been relatively scarce. At such times, as many as 50 percent of the larvae have been parasitized. *Compsilura* has also exercised excellent control of the introduced European pest, the satin moth, *L. salicis.*

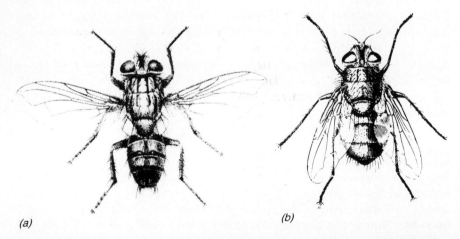

(a)   *(b)*

Figure 9-6   The introduced tachinid flies, *Compsilura concinnata (a)* and *Blepharia pracensis (equals scucellata) (b)*, are important parasites of the gypsy moth, *Lymantria dispar. (U.S. Forest Service photo.)*

Compsilura is a rather large, robust fly, about 7 mm in length. Its coloring is a combination of black and white, which gives the insect a gray appearance, somewhat resembling a housefly. It is one of those larviparous tachinids that place their larvae within the body of the host. In parasitizing the host, the adult female darts quickly at a nearby caterpillar, pierces the body wall with her chitinized piercing organ, inserts her larvipositor, and deposits a larva before the insect can escape. The larvae attach themselves by means of anal hooks to some part of the tracheal system, often near a spiracle. Thereafter, oxygen is obtained from the tracheal system of the host.

At the end of the third instar, the full-grown larvae leave the hosts to pupate. The pupal stage may be passed in bark crevices; in the webs of some host, such as the eastern tent caterpillar, *M. americanum;* or in the surface layers of the soil. As is the case with most of the other Diptera, *Compsilura* pupae are enclosed in puparia, not in cocoons. A *puparium* is the dry, brown larval skin of the last instar.

C. concinnata is active from the beginning of May to the last of October. During this period, there are three overlapping generations. The first adults that emerge in the spring depend for host material on species that overwinter in the larval stage. In May and June, caterpillars that have emerged from overwintering eggs become available. Among these are the gypsy moth, *L. dispar,* and the tent caterpillars. The second-generation adults appear in late June and July, when a great variety of hosts are available. The third-generation adults of *Compsilura* emerge in August and September, when comparatively few host species are available.

Diptera: Sarcophagidae

The family Sarcophagidae contains flies that are chiefly scavengers, as indicated by the common name *flesh flies.* Some of these, *Sarcophaga aldrichi,* for instance, are parasitic on certain caterpillars and their pupae. There is some doubt about the parasitic status of some members of this family. Nevertheless, reports by Hodson (1939) maintain that *S. aldrichi* is an important parasite of the forest tent caterpillar, *M. disstria.* Toward the end of a caterpillar outbreak, these flies frequently become so numerous that they are a nuisance.

CONCLUSIONS ABOUT BIOLOGICAL CONTROL

There is a popular idea that parasitic and predatory species can perform miracles in forest-insect control and that these organisms, properly handled, could eliminate the danger of all insect outbreaks. Valuable as these organisms undoubtedly are, they are only a part of environmental resistance. Furthermore, neither parasites nor predators will ordinarily eliminate their host or prey. The fact that all these forms of life have lived together for generations is prima facie evidence that the former do not usually destroy the latter.

Therefore, when we introduce one organism to control another, we have no right to expect better results than the beneficial species has been able to produce in its native land. Biological pressure can exert a powerful influence in holding down pest populations, but working alone, it can neither eliminate pests nor altogether prevent outbreaks.

Precautions

In the introduction of biological control agents, care must be exercised to avoid the release of undesirable species. Prominent among these are the hyperparasites. Some species are monophagous at one time or on one host but oligophagous at another time or on another host. Therefore, special care must be exercised in complying with quarantine regulations to avoid introduction of species of this sort. Once established in an environment, an insect species can seldom (if ever) be removed. Therefore, it behooves us to be sure of our choice of species before it is too late.

Ideally, the complete life history of the species, its reactions to the physical habitat, and its ecological relations with other organisms should be known before an introduction is attempted. Practical limitations of time and expense prevent the attainment of this ideal. But no parasite or predator should be introduced until at least the principal facts concerning its life history and habits have been determined.

In introducing an organism into a new environment, we should observe certain precautions. First, be sure that the organism is provided with satisfactory physical conditions. Introductions should be made only into new areas where the climatic conditions, in their essential characteristics, resemble the native home of the organism. This is an important consideration that has often been neglected and that explains many of the failures experienced in the past.

Introductions should be attempted only in locations where the host or prey is present in abundance. Furthermore, the host species must be in a stage and in a position suitable for attack; otherwise, the introduced animals will inevitably starve, and our efforts will be wasted.

Organisms should be introduced in large numbers. It is far better to release many individuals in one location than to release the same number scattered over a large area. Experience has demonstrated the truth of this over and over again, not only with insects but also with vertebrates.

The effectiveness of a parasite in controlling an insect pest depends to a considerable extent on certain characteristics of the parasite. The relative importance of these characteristics varies in individual cases; moreover, workers disagree among themselves in evaluating the various characteristics. But that certain characteristics are desirable is commonly accepted. Some of the more important of these are: (1) to reproduce rapidly, (2) to be synchronized with the host, (3) to be able to find and parasitize a large proportion of the host individuals even at low host density, and (4) to compete successfully with other parasites. Seldom (if ever) are all these qualities possessed in an equally high degree by any one parasite. However, we should always attempt to favor those parasites that most nearly approach this ideal.

Interpretation of Results

The results of parasite and predator activities are difficult to interpret. To do this accurately requires careful population studies, which may be impractical. However, we are often able to determine the proportion of a particular stage that has been destroyed by simple counts of parasitized and unparasitized insects in samples representative of the population. Sometimes these observations are expressed as if they represent the effect on the population of the insect as a whole, when actually it applies to only one stage. For example, a predator may destroy 75 percent of all the pupae. This obviously does not mean that the predator is responsible for a 75 percent reduction of the population; rather, it merely means that the pupae alone were destroyed to that extent. Care should always be used when quantitative statements are made regarding the effects of biological factors. The total effects of a resistance factor on the population of an insect species can be expressed only in terms of the potential population of that species. The effect observed on a single stage, therefore, is only a segment of the whole and should always be so indicated.

The effectiveness of biological control programs against forest-insect pests in Canada was reviewed by Reeks and Cameron (1971). Of the 12 species of forest insects considered, 8 were of European origin. Complete success was achieved with 3 species: the European spruce sawfly, *G. hercyniae,* the larch casebearer, *C. laricella,* and the winter moth, *Operophtera brumata.* The spectacular success of the control of the European spruce sawfly by parasites and a virus was estimated to have cost $300,000. The benefit can be theorized to have been $6 million.

When foresters are faced with an introduced insect pest, they should give serious consideration to all approaches to the solution of the problem. Regarding the duration of support, Clausen (1951:8) stated that "if establishment is not evident within three years after releases are begun, then there is little hope of success." Coppel and Mertins (1977:60) stated: "In general, after three years with no establishment, introduction efforts would be better spent on other species or perhaps on different strains of the unsuccessful species."

In conclusion, we repeat emphatically that the use of parasites and predators provides invaluable opportunities to suppress injurious forest-insect populations. These biological methods should not be overlooked in any integrated approach to forest-insect control if we are to manage our forests effectively and economically.

BIBLIOGRAPHY

Major Textbooks

Burges, H. D., and Hussey, N. W., eds. 1971. *Microbial control of insects and mites.* New York: Academic Press.

Coppel, H. C., and Mertins, J. M. 1977. *Biological insect pest suppression.* Adv. Series in Agric. Sci. 4, Berlin: Springer-Verlag.

DeBach, P. 1974. *Biological control by natural enemies.* New York: Cambridge University Press.

Huffaker, C. B., and Messenger, P. S. 1976. *Theory and practice of biological control.* New York: Academic Press.

Merritt, R. W., and Cummins, K. W. 1978. *An introduction to the aquatic insects of North America,* Dubuque, Iowa: Kendall/Hunt Publishing Company.

Steinhaus, E. A. 1949. *Principles of insect pathology.* New York: McGraw-Hill Book Company.

———, ed. 1963. *Insect pathology: An Advanced Treatise,* vols. I and II. New York: Academic Press.

Literature Cited

Balch, R. E., and Bird, F. T. 1944. Disease of the European spruce sawfly and its place in natural control. *Sci. Agr.* 25:65–80.

Beard, R. L. 1945. Milky disease of Japanese beetle larvae. *Conn. Agr. Expt. Sta. Bull.* no. 491.

Bruns, H. 1960. The economic importance of birds in forests. *Bird Study* 7:193–208.

Buckner, C. H. 1966. The role of vertebrate predators in the biological control of forest insects. *Ann. Rev. Entomol.* 11:449–470.

———. 1967. Avian and mammalian predators of forest insects. *Entomophaga* 12:491–501.

———, and Turnock, W. J. 1965. Avian predation of the larch sawfly in New England. *Ecol.* 46:224–236.

Burgess, A. F., and Collins, C. W. 1915. The calosoma beetle (*Calosoma sycophanea*). *USDA Bull.* no. 251.

Campbell, R. W., et al. 1977. Sources of mortality among late instar gypsy moth larvae in sparse populations. *Envir. Entomol.* 6:865–870.

Clarke, R. D., and Grant, P. R. 1968. An experimental study of the role of spiders as predators in a forest litter community. Part I. *Ecology* 49:1152–1154.

Clausen, C. P. 1951. The time factor in biological control. *J. Econ. Entomol.* 44:1–9.

Cole, B. L. 1964. Birds as forest protection agents. *Wash. St. Dept. Natural Res., Resource Management Reports,* no. 2.

Conner, R. N. 1978. Snag management for cavity nesting birds. In *Proc. Workshop, Management of southern forests for non-game birds. USDA For. Serv. Gen. Tech. Rept.* SE-14.

Coppel, H. C., and Smythe, R. V. 1963. Occurrence of spiders on eastern white pine trees infested with the introduced pine sawfly, *Diprion similis* (Htg.). *Univ. WI, For. Res. Notes* no. 102.

Crossman, S. S. 1922. *Apanteles melanoscelus. USDA Bull.* no. 1028.

Dahlsten, D. L., and Herman, S. G. 1965. Birds as predators of destructive forest insects. *Calif. Agric.* 19:8–10.

Deane, C. C. 1976. Ecology of pathogens of the gypsy moth. In Anderson, J. E., and Kaya, H. K. (eds.), *Perspectives in forest entomology,* New York: Academic Press.

Dowden, P. B. 1962. Parasites and predators of forest insects liberated in the United States through 1960. *USDA Handbook* no. 226.

Dustan, A. G. 1923. Natural control of the white-marked tussock moth. *Proc. Acadian Entomol. Soc.* 8:109–126.

Dutky, S. R. 1963. The milky diseases. In *Insect pathology,* vol. 2, ed. E. A. Steinhaus, pp. 75–115. New York: Academic Press.

Falcon, L. A. 1971. Use of bacteria for microbial control. In *Microbial control of insects and mites,* ed. H. D. Burges and N. W. Hussey, pp. 67–95. New York: Academic Press.

Faust, R. M. 1974. Bacterial diseases. In *Insect diseases,* vol. 1, ed. G. E. Cantwell, pp. 87–183. New York: Marcel Dekker.

Fleming, W. E. 1968. Biological control of the Japanese beetle. *USDA Tech. Bull.* no. 1383.

Franz, J. M. 1961. Biological control of pest insects in Europe. *Ann. Rev. Entomol.* 6:183–200

Glaser, R. W. 1927. Polyhedral diseases of insects. *Ann. Entomol. Soc. Am.* 20:319–342.

Graham, S. A. 1928. Small mammals and larch sawfly. *J. Econ. Entomol.* 21:301-310.

———. 1929. *Principles of forest entomology.* New York: McGraw-Hill Book Company.

———. 1956. The larch sawfly in the Lake States. *Forest Sci.* 2:137–160.

——— and F. B. Knight. 1965. *Principles of forest entomology.* New York: McGraw-Hill Book Company.

Harper, J. D. 1974. Forest insect control with *Bacillus thuringiensis,* survey of current knowledge. Auburn, Ala.: Auburn University.

Hodson, A. C. 1939. *Sarcophaga aldrichi,* a parasite of forest tent caterpillar. *J. Econ. Entomol.* 32:396–401.

Hopkins, A. D. 1899. Insect enemies in Northwest. *USDA Div. Entomol. Bull.* 21:13–15.

Knight, F. B. 1958. The effects of woodpeckers on populations of the Engelmann spruce beetle. *J. Econ. Entomol.* 51:603–607.

Legner, E. F., et al. 1974. The biological control of medically important arthropods. *Critical Reviews in Environmental Control* 4:85–113.

Leonard, D. E. 1974. Recent developments in ecology and control of the gypsy moth. *Ann. Rev. Entomol.* 19:197–229.

Loughton, B. G., et al. 1963. Spiders and the spruce budworm. In Morris, R. F. (ed.), The dynamics of epidemic spruce budworm populations. *Memoirs Entomol. Soc. Can.* 31:249–268.

McAtee, W. L. 1925. Vertebrate control of insect pests. *Smithsonian Inst. Annual Rept.*: 415–437.

McGraw, J. R., and Farrier, M. H. 1969. Mites of the superfamily Parasitoidea

(Acarina: Mesostiqmota) associated with *Dendroctonus* and *Ips* (Coleoptera: Scolytidae). *N.C. Agric. Exp. Sta. Tech. Bul.* no. 192.

McGugan, B. M., and Coppel, H. C. 1962. Biological control of forest insects, 1910–1958. In A review of the biological control attempts against insects and weeds in Canada. *Tech. Commun. Commonw. Inst. Biol. Contr.* 2:35–127.

Mitchell, R. G., and Wright, K. H. 1967. Foreign predator introductions for control of the balsam woolly aphid in the Pacific Northwest. *J. Econ. Entomol.* 60:140–147.

Mook, L. J. 1963. Birds and the spruce budworm. In Morris, R. F. (ed), The dynamics of epidemic spruce budworm populations. *Memoirs Entomol. Soc. Can.* 31:268–271.

Morris, R. F., et al. 1958. The numerical response of avian and mammalian predators during a gradation of the spruce budworm. *Ecology* 39:487–494.

Neilson, M. M., and Morris, R. F. 1964. The regulation of European spruce sawfly numbers in the Maritime Provinces of Canada from 1937 to 1963. *Can. Entomol.* 96:773–784.

Otvos, I. S. 1965. Studies on avian predators of *Dendroctonus brevicomis* Le Conte (Coleoptera: Scolytidae) with special reference to *Picidae. Can. Entomol.* 97:1184–1199.

Reeks, W. A., and Cameron, J. M. 1971. Biological control of forest insect pests in Canada, 1959–1968. In Biological control programmes against insects and weeds in Canada. Commonwealth Institute of Biological Control. *Tech. Communication* no. 4: 105–133.

Sabrosky, C. W., and Reardon, R. C. 1976. Tachinid parasites of the gypsy moth, *Lymantria dispar,* with keys to adults and puparia. *Entomol. Soc. Am. Misc. Publ.* no. 10 (2).

Shook, R. L., and Baldwin, P. H. 1970. Woodpecker predation on bark beetles in Engelmann spruce logs as related to stand density. *Can. Entomol.* 102:1345–1354.

Sloan, N. F., and Coppel, H. C. 1968. Ecological implications of bird predators on the larch casebearer in Wisconsin. *J. Econ. Entomol.* 61:1067–1070.

Stewart, R. E., and Aldrich, J. W. 1951. Removal and repopulation of breeding birds in a spruce-fir forest community. *Auk* 68:471–482.

Tothill, J. D. 1919. Natural control of the oyster-shell scale. *Bull. Entomol. Res.* 9:183–196.

———. 1922. Natural control of fall webworm. *Can. Dept. Agr. Bull.* 3 (new ser.).

Warren, G. L. 1971. Introduction of the masked shrew to improve control of forest insects in Newfoundland. *Proc. Tall Timbers Conf. Ecol. Animal Contr. Habitat Mgmt.* 2:185–202.

Chapter 10

Preventive Control by Silvicultural Practices

Every operation in the forest influences the environment and thereby creates conditions that are either favorable or unfavorable for harmful forest insects. Many human practices in the past have unconsciously encouraged insect outbreaks. These errors should be avoided in the future. Fortunately, it is possible not only to avoid creating conditions conducive to outbreaks but actually to produce safer conditions in the course of regular logging operations without any additional cost. In other instances, special operations may be required to correct past errors or to create desirable conditions. Usually, the cost of these operations should be less than the expense entailed by direct methods of control. However, we cannot assume blindly that this is true in every case. Without exception, the benefits from any control measure must be weighed against the cost.

BASIS FOR SILVICULTURAL CONTROL

Biological, Economic, and Social Relationships

Generally speaking, the control of insects by silvicultural[1] practices is accomplished through the adjustment of forest conditions and the avoidance of

---

[1] *Silviculture* is the theory and practice of controlling forest establishment, composition, and growth (Spurr, 1945).

practices that will be destructive to the general health of the stands under management. Much has been stated in the past about biotic balance. We are avoiding an emphasis on this because readers often expect more than is really meant by the term; instead, we will base this entire chapter on forest ecology, which is also often misunderstood but which is more readily accepted. Waters (1971:14) stated: "To manage forest insect populations by sound ecological means—to prevent rather than suppress outbreaks—is the continuing dream of forest entomologists and, generally, forest resource managers." Waters has clearly stated a goal that not all managers would fully accept as reasonable because this goal might not be economically or socially acceptable to some segments of our population. Such a goal might be attainable if we could base management purely on the ecological situation in the forest, without much attention to the economic and social aspects. But management must be based on all three. Marty (1965:39) said: "Timber owners ought not to invest in insect control except in situations where it is reasonably profitable for them to do so—that is, where there is a reasonable relation between cost and value saved."

The social relationships are also of great importance to all of us because we do enjoy the aesthetic view of an old-growth forest or of an individual old tree with its nest holes and/or dens. The ideal silviculture might suggest the elimination of all such overmature individuals from the forest, but such a recommendation does not take into account other needs of our citizens or the ecological requirements of other inhabitants of the forest. Thus, we must approach the use of silviculture as a preventive measure with logic and common sense. Above all, we cannot expect that the answers to all problems will be found through these procedures. We can greatly reduce losses from insects by the practice of intensive silviculture, but the actual practice must depend on the broad needs of our citizens for all the products and services from our forest.

Silvicultural Practices

The developmental sequence in the life of a forest is accompanied by a sequence of insect species characterizing each stage. As a result, different age-classes of a tree species are not ordinarily subject to injury by the same insects. This, as we shall see, is of importance silviculturally in handling tree species that grow naturally in pure stands. Experience has shown that whereas some forest types are highly subject to insect injury, others are relatively safe. Obviously, our objective should be to decrease the areas occupied by the former and increase those occupied by the latter. Unfortunately, we cannot always follow such suggestions because of the various constraints already discussed.

Many forest entomologists and foresters have made comments on the use of silviculture in the control of forest insects. The following generalizations may be made from their ideas and our experiences:

1   The vigor of a forest stand should be maintained as high as possible.
2   The abundance of various tree species should be controlled through management.
3   Management practices may be used to control the mixture of age-classes in the forest and in individual stands.
4   Plantations should be established only on suitable sites and within suitable climatic zones for the species selected.
5   Stands under intensive management should be managed to prevent stagnation and overmaturity.
6   The planting of exotic tree species should be based on more than an analysis of product usefulness and growth potentials.
7   If possible, maintenance of stand mixtures found in climax forests of the region and site should be a goal.

These guides to prevention of insect problems may seem like common sense to most readers. That is as it should be because most successes in prevention of losses have been based on application of common sense in forest management.

REGULATION OF FOREST COMPOSITION

Of all the silvicultural practices, the one that offers the widest opportunity for producing conditions that will discourage insect outbreaks is the regulation of forest composition. When such regulation is handled properly, we have at our command an invaluable means of insect control. On the other hand, if the possibilities for control are ignored, changes in forest composition as a result of human interaction may actually serve to bring on outbreaks.

Importance of Mixed Stands

Any study of our country's virgin forests will furnish proof that mixed stands are much safer from insect injury than are pure stands (Fig. 10-1). In fact, we may safely say that the greater the diversification of tree species, the less frequent insect outbreaks will be. This is an illustration of the general ecological principle that *other things being equal, the degree of environmental stability is in direct proportion to the number of species living together in an environment.* The truth of this becomes clear when we remember that in a diverse environment, important limitations on dangerous pests prevail. For example, the risk attendant on the search for food is far greater than in pure stands, where every tree provides a food supply. Furthermore, diversity, both of the insects living on the trees and of the parasites and predators feeding on them, results in a multiplicity of interactions that tend to limit the increase of all species in the

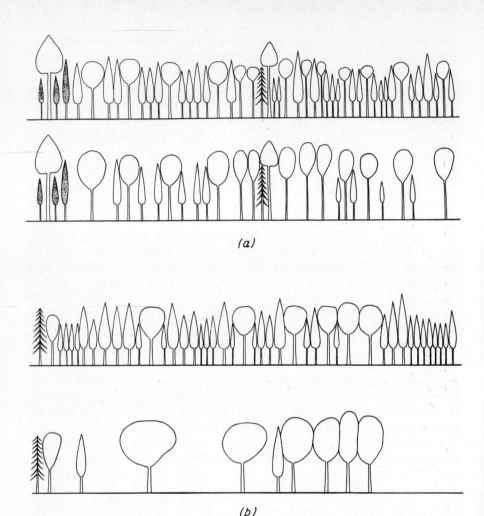

Figure 10-1 Diagrammatic cross section of two forest areas illustrating the effect of a spruce budworm, *Choristoneura fumiferana,* outbreak on forests of different compositions in the same locality. Balsam fir is represented by a narrow, unshaded cone, pine, by a broad cone; spruce, by the figure with drooping branches; other conifers, by the shaded cone; and hardwoods, by the rounded cone. The upper line of each pair represents conditions existing before the outbreak; the lower, conditions following the outbreak. *(a)* is a stand containing a considerable mixture of balsam fir with hardwoods, some pine, and other conifers. *(b)* represents a stand that was predominantly balsam fir. *(b)* was much more severely injured than *(a)*. It is doubtful if an outbreak could have built up in *(a)* if it had not been for a preponderance of adjacent stands similar to *(b)*. *(From Graham and Knight, Principles of Forest Entomology, 4th ed., Fig. 38, p. 214.)*

complex environment. As examples of mixed types in which disastrous out-breaks seldom or never occur, we can mention the so-called mixed mesophytic forests of the Ohio Valley, the mixed southern hardwoods, the mixed northern hardwoods, the mixed coniferous forests on the western slopes of the northern Rocky Mountains, and the mixed forests of the Douglas-fir region.

Each of these types is made up of a combination of tolerant and intolerant tree species. In them, the natural trend is toward the elimination of the intolerant trees and the perpetuation of the more tolerant, shade-loving kinds. In the northern hardwoods, for instance, only sugar maple, hemlock, and beech can persist indefinitely because they are the only species living in their climatic area that can reproduce and survive in shade. The more intolerant species, such as birch, elm, oak, and ash, find little opportunity to reproduce except in openings of at least 1/10 acre (0.16 ha). Under primeval conditions, the reproduction of these intolerant trees has been made possible by disasters—fire, fungi, and insects. Similarly, the trend in the northern Rocky Mountains is toward the tolerant true firs; the trend in the Douglas-fir region is toward hemlock.

Therefore, if we are to maintain the highly diversified insectproof condition characteristic of the true mixed forest, now that catastrophic events are largely eliminated, we must seriously consider this objective in planning logging operations. One of the most promising methods that will maintain the mixed condition is to cut in small groups. On the other hand, individual tree selection that does not tend toward groups is probably the worst practice from the standpoint of insect control. Individual tree selection tends to encourage pure stands of tolerant species and will ultimately lead to outbreaks.

Graham (1963) pointed out three very important steps in maintaining resistant conditions in northern hardwood forests.

1 Management practices should be aimed at creating maximum diversification.

2 Partial cutting should be relatively heavy, removing 50 percent or more of the crown cover to permit reproduction of all species.

3 Despite these efforts, when numerous deer are present, there is a greater tendency toward the development of the more hazardous pure maple forest. Therefore, steps should be taken to reduce the number of deer.

The three rules cited from Graham for northern hardwoods could be readily modified for any of the mixed forests showing similar characteristics.

There are stands within these same regions in which, for one reason or another, the diversity is not as great as might be desired. In such situations, logging should be conducted in such a manner that a maximum variety of tree species can develop. Even in the spruce-fir forests, the mixture following cutting can be improved by cutting in groups or strips. This system will create openings where spruce, aspen, birch, and balsam fir will grow up together, the hardwoods protecting the conifers from attack of insects until the first approach maturity. The forester must be able to view the forest condition and through professional knowledge and a developed understanding of forest ecology be able to provide the best-possible preventive silviculture within the economic constraints of the time. The manager thus is practicing forest entomology while applying basic silvicultural practices.

Under some situations, the application of pesticides may be utilized for silvicultural purposes. For example, if defoliation such as that caused by the gypsy moth, *Lymantria dispar,* is prevented by a treatment with an insecticide, the normal variety of plants and animals on the forest floor can be maintained. Without such treatment, defoliation and the resultant exposure to intense radiation will eliminate many light- or heat-sensitive plants and animals from the forest (Turner, 1963).

Ensuring Suitable Reproduction

Control of insects by regulation of composition depends for its success on the establishment, after logging, of an adequate amount of satisfactory reproduction of all the desired species. Many trees cannot reproduce successfully on soil covered with organic debris. Therefore, logging operations that do not scarify the ground, exposing mineral soil for a seedbed, may result in the elimination of certain species from a mixture. This is true of the fir-spruce-aspen-birch mixtures. If the soil is left undisturbed after logging, balsam fir will increase at the expense of the spruce and hardwoods, a most undesirable condition (Crosley, 1949). Scarification can be secured during summer logging by skidding. On areas logged in winter, scarification could be accomplished by harrowing subsequent to logging and prior to the ripening of a seed crop. When reproduction of the desired species does not occur naturally, it may be wise to resort to planting in order to maintain a safe mixture of species. Present-day mechanized logging provides some additional scarification, but there may still be reason for use of special equipment to assure reproduction of desirable quantities of all components. The common practice of planting following the harvest of southern pines usually requires site preparation utilizing specialized techniques.

The dangers that are attendant on changing mixed stands into pure stands are well illustrated by the history of jack pine in the Lake States. Originally, this species grew on comparatively small areas on poor sandy soils, interspersed with other forest types growing on better soils. After the white and red pine was logged off, the jack pine extended its domain over vast areas. The original forest had been notably free from insect injury, but since 1920, the jack pine, now growing over extensive continuous areas, has been more commonly attacked by pests such as the jack pine budworm, *Choristoneura pinus,* the pine tussock moth, *Dasychira pinicola,* the jack pine sawfly, *Neodiprion pratti bank-*

sianae, the pine tortoise scale, *Toumeyella parvicornis,* and others (Wilson, 1977).

Diversifying Single-species Forests

A question that may well be asked at this point is when can pure forests be grown without danger of major insect attack? Certain species are normally found in natural pure stands, for example, Douglas-fir and ponderosa pine in the West, loblolly and slash pine in the South, red and jack pines in the North, and aspen from coast to coast. Many of these natural stands were not excessively subject to devastation by pests. Therefore, the establishment of single-species stands that are relatively free from attack should be possible. To do this, it is necessary to understand the conditions of the natural pure forests that ensured survival and to be prepared to duplicate the essential conditions. The importance of these conditions seems to have been overlooked. Those natural pure stands that proved to be insect-resistant were composed of limited-sized groups of different age-classes. Managers have erred in not duplicating these features through silvicultural practices. A mixture of age-classes creates a safer condition than would prevail in an extensive single-age forest.

Mixed age-classes in pure forest types can be secured by either scattered cuttings or plantings. In either procedure, the aim will be to mix age-classes groupwise in small units so that no contiguous groups are in the same age-class. In this way, only a fraction of a large area can be subject to attack by a certain insect at any one time. This management practice must be done with a practical view to economic realities as well as long-term forest productivity.

Whenever this practice of diversification by age-classes has been ignored in pure stands, we have had to contend with insect outbreaks. An example of this is to be observed in the Lake States, where, from 1932 to 1938, thousands of acres of national and state forest lands were planted in solid blocks with red pine seedlings. Within 10 years, several serious insect pests became so numerous that direct control measures were necessary to save the plantations. Our intention is to be realistic in assessment of the problem because a combination of the single-species planting with lack of careful selection of planting sites (Fig. 10-2) and poor planting practices contributed to the problem. Problems with sawflies, pine root collar weevil, *Hylobius radicis,* and Saratoga spittlebug, *Aphrophora saratogensis,* were serious. The foresters of the South have had similar experiences on the many slash pine plantations established during the soil bank period. Sawflies, tip moths, and after plantations begin to stagnate, *Ips* spp. have all been problems.

Causes Leading to Hazardous Conditions

Pure even-aged stands have been created by destructive logging and fire as well as by planting; regardless of their origin, they are equally subject to insect attack. Examples of pure stands created by logging are common: for instance,

Figure10-2 Site is extremely important but not always easily recognized until after the results have been seen. This plantation in Ottawa County, Michigan, was attacked by the European pine shoot moth, *Rhyacionia buoliana.* The trees on the left were attacked similarly to those on the right but were on a better site for red pine. *(U.S. Forest Service photo.)*

aspen stands attacked by forest tent caterpillar, *Malacosoma disstria;* oak forest subject to attack by walkingsticks, *Diapheromera femorata,* gypsy moth, *L. dispar,* and oak leafroller, *Archips semiferanus;* stands of paper birch attacked by various defoliators; forests predominantly balsam fir attacked by the spruce budworm, *C. fumiferana;* and pure pine stands attacked by several species of sawflies.

Animals, both wild and domestic, have also been responsible for undesirable changes in forest composition—for example, the elimination of particularly palatable tree species, such as yellow birch, elm, and basswood, by deer browsing in localities where overzealous protection has permitted these animals to increase excessively—and also for the disruption of the normal distribution of age-classes in various forest types where overgrazing by domestic stock has been permitted. In each case, the trend toward pure stands and a single age-class has increased the danger of insect outbreaks.

REGULATION OF DENSITY

In addition to the regulation of forest composition, there are other silvicultural means by which the harmful activities of forest insects can be reduced. One of these is the regulation of forest density. One of the objectives of regulating density in the interest of insect control is to maintain thrifty growth. Thrifty, rapidly growing trees are generally far less subject to insect injury than slow-growing, unthrifty individuals are in a comparable situation.

Thinning to Stimulate Vigor

If young trees are standing too close together to permit satisfactory growth of dominant and codominant trees, removal of part of the stand by thinning or partial cutting will often ensure vigor and thereby lower the danger of insect devastation. Thinning is usually accomplished by use of the axe and the chain saw. The economic justification for such operations is apparent if the materials cut can be marketed for a price sufficient to cover the cost of the operation. If the products cut are not salable, thinning might entail an outright expenditure the justifiability of which might logically be questioned. Some authorities have recommended a less expensive method of thinning certain forest types, namely, controlled or regulated burning. Unrestricted burning cannot be condoned, but controlled burning properly planned and executed can be a useful silvicultural tool (Weaver, 1947).

Research on the mountain pine beetle, *Dendroctonus ponderosae,* on ponderosa pine reveals a very positive effect from thinning of stands (Sartwell and Stevens, 1975; Sartwell and Dolph, 1976) (Fig. 10-3). The authors have estimated 150 ft^2/acre (about 35 m^2/ha) as a maximum basal area at 90 years of age. Data were not exact and more research will be done, but the authors showed conclusively that thinning was effective in their studies (Table 10-1). The stand occupying the test site was about 55 years of age. Severe tree killing had occurred in the vicinity for several years.

The data gathered by Sartwell and Dolph show clearly that a basal area of 39.88 m^2/ha is excessive. The light thinning to a basal area of 26.87 m^2/ha was sufficient to greatly reduce losses from the mountain pine beetle.

The age at which a stand requires thinning varies with the species and with the amount of water available in the soil. Before the crowns of a young stand have closed, the evaporating leaf surface is so small that the trees can grow well in locations where the available moisture is relatively low. Later, when the trees have closed together, the evaporating leaf surface is greatly increased while at the same time interception of precipitation by the crowns reduces the amount of water reaching the soil surface. Therefore, a stand of large seedlings or small saplings may grow well during early years and suffer later from water deficiency. On dry sites this condition may develop long before the trees reach merchantable size, whereas on moister sites water shortage may never occur.

When severe competition for water develops, the growth rate and vigor of the trees are reduced, whereupon they become increasingly subject to attack and injury by pests. On the drier sites, this competition for water may develop at such an early age that the stand will fail to break up into the normal distribution of crown classes. As a result, it becomes stagnant. On better sites, individual trees will gain dominance over the others, so that ultimately they will eliminate their smaller competitors. In the latter case, thinning to regulate density is seldom necessary. In the former, it is essential if the trees are to be

Figure 10-3 Dense stands of ponderosa pine are a high risk. These trees were killed by the mountain pine beetle, *Dendroctonus ponderosae.* The loss might have been prevented by thinning. *(U.S. Forest Service photo)*

Table 10-1   Tree Mortality and Net Stand Growth During Five Years after Thinning of Ponderosa Pine Stands, 1967–1972

1967 basal area	Mountain pine beetle loss	Total mortality	Five-year net growth	1972 basal area
		Stem basal area, m^2/ha		
39.88*	2.70	4.89	−4.78	35.10
26.87	0.75	1.38	−0.63	26.24
19.74	0.06	0.12	0.75	20.49
14.22	0.00	0.17	0.69	14.91
8.06	0.00	0.17	0.52	8.58

Source: Adapted from Sartwell and Dolph 1976.
*Unthinned.

brought to commercial maturity. The age when thinning should be done and the intensity of thinning required will, as we have said, depend on the amount of available water.

From this discussion, it is evident that on sites with adequate water, trees can grow in dense stands and that on the drier sites, only scattered groups or individual trees separated from their neighbors reach maturity.

Encouraging Straight Growth

Although wide spacing may be desirable for trees above sapling size, seedlings and saplings are often benefited by growing in dense stands. This is especially useful in preventing injury from insects that damage the leading shoots. Obviously, young trees should never be grown so close together that satisfactory growth is impossible; on the other hand, they should be grown in sufficiently dense stands so that side shade will stimulate straight growth, thus maintaining the excurrent habit. This will cause trees injured by such insects as the white pine weevil or the Sitka spruce weevil, *Pissodes strobi,* the pine reproduction weevil, *Cylindrocopturus eatoni,* and the European pine shoot moth, *Rhyacionia buoliana,* to outgrow injury. Furthermore, the white pine weevil and most other shoot insects often decline in numbers soon after the crowns close in a dense stand.

Eliminating Alternate Hosts

Density also has an important influence on insects that require herbs or shrubs as alternate hosts. For example, the Saratoga spittlebug, *Aphrophora saratogensis,* is dependent on certain shrubs and herbs for food during the nymphal stage. Only in the adult stage does it attack and damage the young pines. The favored food plants for the nymphs are sweet fern and blackberries. When these intolerant shrubs are killed by shade, the source of infestation to the pines is eliminated. This does not happen until the trees have passed through the very susceptible seedling and sapling stages. Managers should be aware of the risk taken when plans call for plantations of susceptible pines in areas where the alternate hosts are abundant. Sometimes the sensible alternative is not to plant at all.

Influencing Physical Factors

Forest density influences the physical conditions found in the forest by modifying such factors as light, evaporation, air movement, and temperature. Although the effect of these factors is evident from general observations, specific information concerning their influence on insect life in the forest is very scarce. It is known that many of the sun-loving beetles are attracted only to those trees or parts of trees that are exposed to the direct rays of the sun; for instance, the work of the twolined chestnut borer, *Agrilus bilineatus,* is almost entirely confined to defoliated stands, to individual ornamental trees, to trees in open stands, or to the few exposed branches in fully stocked and stagnating stands. The same is more or less true of the bronze birch borer, *A. anxius,* the hemlock borer, *Melanophila fulvoguttata,* and many other species.

Site Conditions

From observations in Minnesota and in Michigan (Hall, 1933), it appears that thrifty forests are, as a rule, immune to the bronze birch borer, *A. anxius.* In a fully stocked birch forest, after the rate of growth has been reduced because of overmaturity or competition, especially under poor site conditions, this insect will attack branches and the upper parts of the bole of somewhat decadent trees. At first, the larvae may fail to survive, and their tunnels will be overgrown. But after a number of years, repeated injury appears to reduce the tree's vigor until it dies. Thus, openings in the stand are created. Then the exposed trees on the edge of the openings eventually become infested and die, thus enlarging the openings. Once this condition is reached, the borer population grows rapidly, and the infestation may change from a secondary to a primary status. Stands of white birch growing on poor sites are almost certain to deteriorate in this manner before they reach a diameter of 6 in (15 cm). On better sites, a similar deterioration does not occur until the trees reach a mature age of 75 to 100 years with diameters ranging over 10 in (25 cm) (Fowells, 1965). Much of the second-growth white birch in the northeastern forests is growing on poor soils, and for this reason, the bronze birch borer may prevent most of these stands from reaching full development.

Site, therefore, must be considered seriously in formulating plans for handling stands of white birch. White birch forests should always be utilized before they begin to deteriorate. On poor soils especially, provision should be made to convert temporary birch forests to other species better suited to the site.

Similarly, stands of aspen and other temporary forests of intolerant species growing on poor sites will deteriorate at an early age. Insect pests are often

associated with this deterioration. Therefore, all such species require silvicultural treatment similar to that suggested for birch. Site conditions should be
a primary consideration in formulating the silvicultural plans.

Site conditions have strikingly important influences on the susceptibility
of trees to insect injury. Trees growing under site conditions to which they are
adapted are thrifty and relatively resistant to insect injury. For example,
Craighead (1925) has shown that balsam fir is much more resistant to injury
by the spruce budworm, *C. fumiferana,* when it is growing on a favorable site
than when it is growing under poor conditions. He goes so far as to recommend
growing this species in pure stands, on sites where the tree grows rapidly, in
spite of the budworm menace. Whether or not such a recommendation should
be accepted has been questioned, but the fact remains that whereas vigorous
trees survived an outbreak when growing on favorable sites, less vigorous trees
in the same locality were almost all killed.

When unfavorable site conditions are due to water level, managers can
often make needed improvements. Sometimes, a site is poor because of an
excess of surface water. This may result in producing an acid soil and other
conditions that inhibit growth. In some cases, these conditions may be improved by the construction of a drainage system to carry off the surplus water.
Such operations, if they result in the stimulation of growth and the improvement of the tree's vigor, may justify their cost by the resulting reduced susceptibility to insect injury. However, caution is advised because under some
circumstances, the practice may cause other problems related to changes in
water levels in adjoining areas.

Site is of considerable importance in determining the susceptibility of trees
to bark beetle attack. Infestations are greatly influenced by tree density; stress
on the trees will result in a very strong response in the beetle numbers. Bark
beetles are generally considered secondary pests and thus become killers only
when trees are in poor health or have recently died. The data on the mountain
pine beetle, *D. ponderosae,* in Table 10-1 illustrate this effect and the very
dramatic difference when the stands were thinned. The dense stands were
under stress and were under attack by the beetles (Fig. 10–3). Thus, in stands
of the same physiological age and character, the number of trees attacked will
be higher on poor than on good sites.

Moisture is the most important site factor responsible for this condition;
therefore, the above statement may not apply at high elevations, where temperature is the chief limiting factor. Also, during seasons of unusually severe
drought conditions, trees on good sites are affected even more than those on
poor sites because those on poor sites are better adjusted to water deficiency
than stands usually more favorably located. Therefore, during periods of
drought, outbreaks of bark beetles may develop on the better sites, and the
forester must be prepared to control them by direct means.

We may conclude that adaptation to site conditions may be accomplished in a variety of ways: (1) by converting off-site temporary types to types better suited to the site, (2) by encouraging species suited to the site, (3) by adopting appropriate thinning and cutting practices designed to adjust the basal area of the residual stands to site conditions.

IMPROVEMENT OF PLANTING STOCK

Whenever reproduction is obtained by planting, it would be ideal if the forester could use stock that would produce a forest free from insect attack. Unfortunately, this cannot be accomplished. Only occasionally can stock that is resistant to even one insect be found. The desirability of breeding insect-resistant strains of trees has long been recognized, and the possibilities of doing this are evident from the resistant individuals occasionally observed in the field. Unfortunately, little breeding of this sort has been done. Hanover (1975:75), in his recent review of tree resistance, commented as follows on the lack of research and application: "Remarkably little use has been made of tree resistance for control of pest populations despite the fact that this is generally regarded as the ideal method."

Breeding Trees for Resistance

Breeding trees for resistance is both expensive and time-consuming. Unfortunately, from the viewpoint of forest entomology, almost all tree-breeding work has been for purposes other than insect control. Undoubtedly, breeding for insect resistance could produce valuable results.

Occasionally, breeding designed for other purposes produces strains having a high degree of insect resistance. For instance, a hybrid of Jeffrey and Coulter pines, developed at the U.S. Forest Service, Institute of Forest Genetics, proved to be very resistant to the highly injurious pine reproduction weevil, *C. eatoni.* The presence in the field of resistant individual ponderosa and Jeffrey pines indicates that resistant varieties of either species might be obtained by selection. The hybrid has not been utilized in planting programs.

In discussing this hybrid, Miller (1950) points out that resistance is apparently accomplished in part by walling off the attacked areas in the bark before the larvae reach the vital cambium and in part by the production of large quantities of resin. The former reaction is indicated by a thick layer of cork cells surrounding the necrotic areas; the latter is evidenced by a great concentration of resin ducts inside the cambium and also by the flow of the resin from these ducts into the bark. A similar condition is evident in resistant individuals of both Jeffrey and ponderosa pine. The production of hybrids on a commercial basis has not happened for various reasons. However, the principles have been developed; and when economic conditions permit, a practical application may result.

Selection from Desirable Parents

Although the possibilities of tree breeding are recognized, we cannot expect to obtain many resistant varieties in the near future. Therefore, at least for the present, the production of resistant planting stock must depend on the selection of seed from especially desirable parent trees. Selection from resistant individual trees has one glaring inadequacy. It is made on the basis of the desirability of only one parent. The characteristics of the pollen-bearing parent might be either good or bad. The chance of collecting seed from good parent stock on both sides could be improved by removing all undesirable individuals from seed-collection areas. The problems inherent in selection and breeding with a view to producing insect-resistant stock for planting purposes offer a real challenge to the enterprising forester.

Selection from Suitable Habitats

Although the development of tree varieties resistant to insect attack has not often been accomplished by human efforts, nature has been selecting through the ages those individuals best fitted for survival in every locality. Observations indicate clearly that trees are especially subject to insect attack when they have been taken from their native habitat and grown under different conditions. This rule applies not only to species but almost as much to varieties of species. Therefore, care should be exercised that planting stock is grown from seed collected in a climatic area similar to the place where it is to be planted. The importance of this from the viewpoint of forest entomology cannot be overemphasized.

MAINTAINING THRIFT BY LOGGING

Control of forest insects by silviculture is by no means limited to practices aimed at the regulation of composition and density, to the adaptation of forests to site, or to the development of desirable varieties. These practices are concerned chiefly with the developing forest. Other equally important practices can be applied to the mature forest to create safe conditions until the trees are utilized (Knight, 1976; Vité, 1971).

Use of Tree Classifications

As soon as a forest reaches merchantable size, it may be clear-cut; but usually, foresters find that the removal of the trees in a series of partial cuts is not only more desirable silviculturally but also more profitable. In planning for a partial cutting, the question of which trees should be cut and which left immediately arises. A number of tree classifications have been devised that make it possible for the forester to decide which trees should be removed first and which should be left for a later cut. Keen's classification (Fig. 10-4) of ponderosa pine on

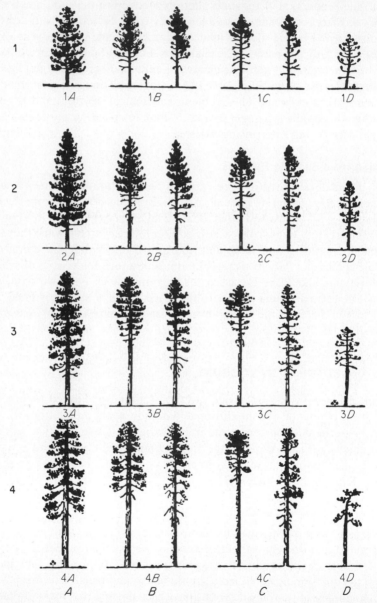

A PONDEROSA PINE TREE CLASSIFICATION
For comparison of barkbeetle susceptibility
Classes based on age and vigor

Figure 10-4 Keen's tree classification for ponderosa pine. *(U.S. Forest Service drawing.)*

the basis of age and vigor, as indicated by bark and foliage characteristics, is a useful tree classification for silvicultural purposes (Keen, 1943). By application of this classification, it is possible to select trees from those classes that possess the lowest growth capacity and the highest susceptibility to bark beetle infestation. The classification applies especially to the interior or east-side type of ponderosa pine. A similar classification has been devised for eastern hemlock based on age and crown characteristics (Graham, 1943). Generally speaking, the oldest trees with the poorest crowns are likely to be most subject to injury and, therefore, should be cut first. The location of the line between the best trees and the worst, above which the trees should be reserved and below which they should be cut, depends on the plan of management for the particular stand. The shorter the cutting cycle, the fewer the classes in which cutting is done.

Removing High-risk Trees

Within any class, some trees may be in better health than others (Keen and Salman, 1942). When a stand is marked for partial cutting, any sick trees should usually be cut regardless of their class because these subnormal individuals are most likely to succumb to the attack of insects (Salman and Bongberg, 1942). Therefore, it is desirable to take current health into consideration when using a tree classification.

The fundamental difference between the tree classification and the risk-rating systems must be kept in mind. The first is based on the actuarial probability that trees in a certain class will survive until the next cutting. The second rates each individual tree according to its current health. Applying the risk-rating system, sanitation-salvage cuts are being made in ponderosa pine for the purpose of removing trees immediately subject to bark beetle attack. Such an operation reduces the hazard of bark beetle attack to a point where it is hoped that losses will not occur for a period of from 5 to 10 years. After that, a second sanitation-salvage cut may be made if enough trees have reached a condition of high risk. Ordinarily, however, the second cut will have special silvicultural objectives leading to the establishment of reproduction for the next tree generation and the normal distribution of age-classes. The data on the sanitation-salvage method are documented in Miller and Keen (1960). The method is effective (Table 10-2).

In Montana, the risk-rating system is used in marking overmature stands of ponderosa pine. In these stands, a relatively heavy cut is made, and the period between cuttings is correspondingly extended, sometimes to 30 years. All the trees in the two highest-risk classes (Table 10-3) and a few of those in classes 1 and 2 are removed. Many workers feel that cuts based only on current health must be repeated more frequently than every 30 years in overmature ponderosa pine types. They point out that within a 5-year period, some trees may shift from one risk class to another and that within 10 years after the

Table 10-2 Reductions in Timber Killed After Sanitation-Salvage Operations in California, Ponderosa Pine

Location	High-risk volume cut, % of stand	Years after cut, no.	Reduction in losses, % of volume*	Volume killed	
				Cut stands, bd ft/acre	Uncut stands bd ft/acre
Blacks Mountain	15.7	10	82.4	106	601
Cayton Valley	8.5	5	84.7	81	529
Rail Glade	18.0	6	83.5	64	388
Swede Cabin	14.1	7	71.5	110	388

*Losses were direct comparisons of board feet per acre killed in adjacent uncut stands.

Table 10-3 Descriptions of Risk Ratings from Miller and Keen (1960)

Risk rating	Symptoms
Risk I, low risk	Full-foliaged, healthy-appearing crowns. Foliage of healthy appearance, needles usually long and coarse, color good dark green. Practically all twigs with normal foliage complement. No weakened parts of crown.
Risk II, moderate risk	Fair to moderately healthy crowns, imperfect in spots. Foliage mostly healthy, needle length average or better, color fair to good. Some twigs or branches may lack foliage, but such injury should not be localized to form definite weak spots in crown.
Risk III, high risk	Crowns of fair to poor health, somewhat ragged or thin in parts of crown. Foliage in parts of crown thin, bunchy, or unhealthy, needles average to shorter than average in length, color fair to poor. Some to many twigs or branches lacking foliage, some to many twigs or branches fading or dead. Small, localized weakened parts of crown usually present.
Risk IV, very high risk	Crowns in poor condition, ragged or thin, often showing evidence of active insect infestations in upper parts. Foliage thin or bunchy, needles short or sparse, color poor. Twigs and branches dead or dying, parts of crown definitely weakened. Active top-killing or partial infestations often present.

initial cutting, hazard will probably have increased to a degree requiring another cutting. On the other hand, it is possible that the heavier cutting made in the Montana stands will not only remove the trees currently in the high-risk classes but will also reduce competition in the residual stand to a greater degree than the light sanitation-salvage cuttings would. As a result, the health of the stand might improve rather than deteriorate.

Logging of Lodgepole Pines

Earlier, we discussed the use of thinning as a method for preventing losses of ponderosa pine because of attack by the mountain pine beetle, *D. ponderosae.*

This same insect is a severe pest in lodgepole pine, but the method of control is somewhat different. Cole and Cahill (1976) point out that in lodgepole pine stands, the mountain pine beetle generally infests the large-diameter individuals with thick phloem. The authors suggest that stands be managed so that trees do not reach 10 in (25 cm) diameter at breast height. Berryman (1976) reported on some of the theoretical aspects of beetle dynamics in lodgepole pine stands.

Safranyik et al. (1974) and Amman et al. (1977) have developed more specific guidelines for management in Canada and the United States. Beetles show definite preferences for trees of larger diameters and those over about 80 years of age. Losses are also greater in stands growing at low elevations than at higher altitudes, where beetle survival is reduced. The risk of attack is clearly defined as being made up of three components: (1) an elevation-latitude factor, (2) an age factor, and (3) a diameter relationship.

A sanitation-salvage method is suggested as a method of management. Harvesting should commence in outbreak areas containing more than one infested tree per acre. Areas with largest-diameter trees at the periphery of the infestation should be given cutting priority if outbreaks are too large to treat completely. Long-term management plans should take into account the beetle attack characteristics so that future losses may be reduced.

Logging and Spruce Beetles

The dynamics of a spruce beetle, *D. rufipennis,* outbreak are different from those for either the western pine beetle, *D. brevicomis,* or the mountain pine beetle, *D. ponderosae.* Logging can have a very important effect on what happens to residual stands of mature spruce. Schmid (1977) discusses some of the ways this problem should be treated. Spruce beetles are attracted to recently felled timber, which is the basis for the trap-tree method of control sometimes used in management of the problem.

Trap trees have been utilized very effectively in spruce beetle, *D. rufipennis,* control where the forest is accessible. Trees are felled prior to the beetle flight period and are left in the woods until all flight activity has ceased. Then the trees are removed from the forest and utilized. The spruce beetle does not infest standing green trees when trap trees or logs are available to them.

Another serious problem relates to logging operations in overmature stands with large quantities of cull materials. Such materials function exactly like trap trees, but because they are left in the forest, they may create problems because of the production of beetles that infest standing timber in the vicinity. Precautions can be taken to reduce such problems by requiring that trees be cut with very low stumps and that cull logs and tops be limbed and kept fully exposed to the sun. If the stand is heavily shaded, as might be the case in a selective logging operation, the cull material should be removed.

PRACTICES THAT FAVOR PESTS

The comments in this chapter have been designed to provide principles and examples of silvicultural methods that may be useful in preventing insect infestations. One must temper these ideas and successful applications with common sense and a broad understanding of forest ecology and economics. Our opportunity to practice silviculture is improving with the development of greater emphasis on intensive forestry. The development of markets for poor-quality materials to supply some of the energy needs of the country may provide a means for application of these principles, which for so long have been known mainly on an experimental or theoretical basis. Much that we have known to be practical from an ecological viewpoint has not been feasible because of economics.

We must be careful, however, because the same economic forces can lead to application of additional practices that may favor pests; many forests are still handled in ways that encourage dangerous infestations. The reasons given by foresters for these practices may or may not justify the risks inherent in them. But if a forester decides that the risks are worthwhile economically, after due consideration of the probable cost of controlling resultant outbreaks of pests, one can take no issue with the decision. If, on the other hand, the decision to use certain practices is based on immediate financial advantages without consideration of the risk of injury that such practices entail, then there should be no surprise if losses instead of profits are the end result of the operations.

The importance of balancing the *damage that pests may cause* against the *advantages of some silvicultural and logging practices designed to reduce immediate cost* justifies a discussion of practices that tend to encourage outbreaks of pests. The importance of balancing advantages against disadvantages should be emphasized. Also, we should emphasize that each case is an individual problem. A practice that in one case is undesirable because it favors pests may in another case be desirable when measured in economic terms.

Disregarding Site Quality

All too often, site quality has been disregarded in establishing forest plantations of pine and other trees. For example, in the North, red pine has all too frequently been planted on sites that can support only jack pine, or Scotch pine has been used on sites that are unsuitable for that species. The result has been outbreaks of spittlebugs, shoot moths, scale insects, and other pests that would not have occurred in stands growing on suitable sites.

In the Southeast, large areas of slash pine have been planted on sites unsuited to that species. Similarly, loblolly pine has been planted off site. Tip moths, scale insects, spittlebugs, and bark beetles may be expected in these off-site plantations.

Developing Single-species Forest

The single-species, even-aged forest is undoubtedly easier to manage and more profitable than the mixed forest. Therefore, foresters tend to favor large-scale, pure-stand silviculture over practices that grow trees in mixture, in rotations that alternate species, or in diversified age-classes and size classes. In the past, this has led to encouraging pests that under primitive forest conditions were unknown. There is no reason to believe this will change in the future.

Using Injurious Logging Practices

Some logging practices, such as high-lead skidding, full-length skidding, and the use of heavy equipment in cutting operations, have contributed much to the economy of logging. However, when similar heavy equipment is used in forests where partial cuttings are being made with the expectation of returning later for another crop, disastrous outbreaks of pests may occur. Full-length skidding has become popular in some localities and is resulting in an excessive amount of basal scarring of the trees. On pines, these scars attract several bark beetles, notably the black turpentine beetle, *D. terebrans,* in the Southeast, several pitch moths, ambrosia beetles, carpenter ants, and other insects that cause deterioration of the butt log or, in some instances, a serious amount of mortality.

The use of heavy tractors and trucks in partial cuttings may affect the growth of residual trees and make them more subject to pest attack by compacting the soil. Especially on soft ground, such equipment may break the roots of residual trees, thus creating infection courts for fungi and conditions favorable for root-eating weevils that later will damage nearby seedlings. Similar mechanical damage may lead to bark beetle outbreaks and to disastrous insect-fungous infestations that can completely destroy residual stands needed for a later cutting.

A point that is frequently overlooked is that practices known to favor pests that can be killed with pesticides also favor pests that cannot thus be controlled under forest conditions. For example, many insects can be killed only if a pesticide contacts their bodies. Therefore, because they cannot be contacted with pesticides effectively, control in the forest of scale insects or other sucking pests is virtually impossible with pesticides. The same is true for most shoot moths except where several heavy applications can be made from the ground. Systemic insecticides may be used effectively in the future, but have not yet been developed for use on the large scale of a forest operation. They may be effective against these pests on ornmental trees and on trees in special-use areas.

All these practices and others that might be mentioned under the heading of careless operations are conducive to insect injury. Unless these effects are understood and are carefully regulated, the practices can easily wipe out the expected profit from a forest operation.

In the preceding chapters, some general principles concerning the population dynamics of forest insects have been presented. Also, various forces that regulate these populations and some of the procedures that managers can use to protect forests against excessive depredations have been discussed. The division of this presentation into formal chapters and sections has been essential to orderly presentation. Nevertheless, the degree of categorization can be misleading unless we bear in mind that in nature the actions, reactions, and interactions between insects and their environment are all going on simultaneously and that the regulation of a forest-insect population is seldom (if ever) attributable to the operation of any single factor.

BIBLIOGRAPHY

Amman, G. D., et al. 1977. Guidelines for reducing losses of lodgepole pine to the mountain pine beetle in unmanaged stands in the Rocky Mountains. *USDA Forest Service,* General Technical Report Int-36.

Berryman, A. A. 1976. Theoretical explanation of mountain pine beetle dynamics in lodgepole pine forests. *Environ. Entomol.* 5:1225–1233.

Cole, W. E., and Cahill, D. B. 1976. Cutting strategies can reduce probabilities of mountain pine beetle epidemics in lodgepole pine. *J. Forestry* 74:294–297.

Craighead, F. C. 1925. Relation between the mortality of trees attacked by spruce budworm and previous growth. *J. Agr. Res.* 30:541–555.

Crosley, D. J. 1949. Reproduction of white spruce following disturbance of forest floor. *Can. Dept. Mines Res., Dominion Can. Forest Serv., Res. Note* no. 90.

Fowells, H. A. 1965. Silvics of forest trees of the United States. *USDA For. Ser. Agr. Handbk.* no. 271.

Graham, S. A. 1943. Causes of hemlock mortality. *The Univ. Mich. School Forestry and Conserv.* (now *Natural Resources*) *Bull.* no. 10.

————. 1963. Making hardwood forests safe from insects. *J. Forestry.* 61:356–359.

Hall, R. C. 1933. Post-logging decadence of northern hardwoods. *The Univ. Mich. School Forestry Conserv.* (now *Natural Resources*), *Bull.* no. 3.

Hanover, J. W. 1975. Physiology of tree resistance to insects. *Ann. Rev. Entomol.* 20:75–95.

Keen, F. P. 1943. Ponderosa pine tree classes redefined. *J. Forestry* 41:249–253.

————, and Salman, K. A. 1942. Progress in pine beetle control through tree selection. *J. Forestry* 40:854–858.

Knight, F. B. 1976. Management of the forest. In *Perspectives in forest entomology,* ed. J. A. Anderson and H. K. Kaya, pp. 41–60. New York: Academic Press.

Marty, R. J. 1965. How much can you afford to spend in controlling forest insects? In *Insects in Southern Forests,* ed. C. B. Marlin, pp. 38–50. *Proc. 14th Ann. For. Symposium,* Baton Rouge: Louisiana State University Press.

Miller, J. M. 1950. Resistance of hybrids to pine reproduction weevil. *USDA, Calif. Forest Range Expt. Sta., Forest Res. Note.* no. 68.

Miller, J. M., and Keen, F. P. 1960. Biology and control of the western pine beetle, *Forest Service, USDA Misc. Publ.* no. 800.

Safranyik, L., et al. 1974. Management of lodgepole pine to reduce losses from the mountain pine beetle. *Environment Canada, Forestry Tech. Report* no. 1.

Salman, K. A., and Bongberg, J. W. 1942. Logging high-risk trees to control insects in pine stands of northeastern California. *J. Forestry* 40:533–539.

Sartwell, C., and Dolph, R. E. 1976. Silvicultural and direct control of mountain pine beetle in second-growth ponderosa pine. *USDA Forest Service, Res. Note* PNW no. 268.

Sartwell, C., and Stevens, R. E. 1975. Mountain pine beetle in ponderosa pine.*J. Forestry* 73:136–140.

Schmid, J. M. 1977. Guidelines for minimizing spruce-beetle populations in logging residuals. *USDA Forest Service Research Paper* RM-185.

Spurr, S. H. 1945. A new definition of silviculture. *J. Forestry* 43:44.

Turner, N. 1963. The gypsy moth problem. *Conn. Agr. Expt. Sta. Bull.* no. 655.

Vité, J. P. 1971. Silviculture and the management of bark beetle pests. *Proc. Tall Timber Conf. Ecol. Animal Contr. Habitat Mgmt.* 3:155–168.

Waters, W. E. 1971. Ecological management of forest insect populations. *Proc. Tall Timber Confr. Ecol. Animal Contr. Habitat Mgmt.* 3:141–153.

Weaver, H. 1947. Fire as a thinning agent in ponderosa pine. *J. Forestry* 45:437–444.

Wilson, L. F. 1977. A guide to insect injury of conifers in the Lake States. *USDA Forest Service, Agric. Handbook* no. 501.

Chapter 11

Control by Chemical Methods

We have discussed many of the methods used to control insects through direct or preventive procedures and have detailed some of the biological and silvicultural applications. It would be fine if we could conclude at this time that these methods, with refinements, provide all the techniques needed. However, we still must depend on chemical methods for control of many pest problems. The forester should think of chemicals as the last resort and should continually strive to improve management practices to avoid the need for such applications. Chemical insecticides are still needed by the forester and should be included as options under integrated pest management.

Chemicals are extremely important as a means of controlling emergency situations. These materials should not be used unless needed; when they are needed, they should be utilized wisely. Thus, the continuing great need is for knowledge about the chemicals available and the techniques for using them efficiently. The use of chemicals remains vital to forest management. The usefulness of any insecticide is governed greatly by its application, the properties of the insecticide itself, the site on which it is used, and the nature of the pest involved.

It is not the purpose of this text to provide prescriptions for control or lists of registered materials; these are continually changing. Specific chemicals will be mentioned only as examples of applications utilized in a variety of situations.

REGULATIONS ON USE OF PESTICIDES

There are numerous publications dealing with the various regulations on the use of insecticides. General coverage of these is readily available to anyone in the current issue of *ENTOMA: Pesticide Handbook,* which is published by the Entomological Society of America. Any citizen can obtain information regarding specific regulations applicable to the area in which he or she resides by writing or calling the Cooperative Extension Service or the Environmental Protection Agency in his or her state.

The laws, policies, and regulations that influence pest control have been developed over many decades. There are four ways that we might view these laws as a benefit to people:

1   protection of the manufacturers of pesticides and innovators of pest control strategies
2   protection of the consumers[1] of pesticides
3   protection of all people and other organisms in the environment from pesticide contamination
4   protection of the individual from damage to property or person by uncontrolled populations of pests

These four purposes, much simplified here, are drawn from a report by the National Academy of Sciences (1975). One could develop a considerable discussion centered on each point as they should be interpreted in the broad sense.

Federal Insecticide, Fungicide, and Rodenticide Act

FIFRA was enacted in 1947. It was completely revised in 1972 in the Federal Environmental Pesticide Control Act (FEPCA); a further amendment was added in 1975. FIFRA as amended is supported by separate laws in each state, which must meet the minimum standards of the federal law.

Reorganization Plan no. 3 of 1970 established the Environmental Protection Agency (EPA), which carries out the provisions of the law. One of the major aspects relates to the registration of pesticides. The administrator of EPA is charged not only with the responsibility for registration but also with other aspects of regulation and monitoring. The EPA Office of Enforcement has issued statements on the policy of enforcements to assist those using

---

[1] *Consumer* in this sense would be the extension agent, the property owner, the retailer, the applicator, or other similar persons.

pesticides; these were first announced in the *Federal Register* of May 5, 1975. These Pesticide Enforcement Policy Statements (PEPS) have been very helpful to applicators who have had difficulty in accomplishing needed work under special circumstances.

Under the terms of the act, registration of a pesticide will be granted by the administrator if the administrator determines that

        A   its composition is such as to warrant the proposed claims for it;

        B   its labeling and other material required to be submitted comply with the requirements of this Act;

        C   it will perform its intended function without unreasonable adverse effects on the environment, and

        D   when used in accordance with widespread and commonly recognized practice it will not generally cause unreasonable adverse effects on the environment.

The provisions of the laws that are most important to the pesticide applicator relate to classification for use. *General-use* pesticides are those that are not likely to harm humans and the environment when used as directed on the label. *Restricted-use* pesticides may pose a threat, and only a certified applicator may purchase and apply them. Each state is responsible for certifying private and commercial applicators to apply restricted-use pesticides. All state programs must comply with the minimum standards set by EPA.

Foresters and forest scientists who may work with any type of pesticide should obtain certification as commercial applicators in either the forestry or the research category. Some may desire to be certified for both. The applicator must be certified if any restricted-use pesticides are to be used. The forest manager who hires someone to apply chemicals should require that the person hired is certified and that only registered materials are used. Foresters needing certification may contact the USDA Cooperative Extension Service or the state's department of environmental protection (or comparable agency) for information on the procedure for certification in the state of residence.

In addition to the commercial applicator certification, there is a separate category of certification as private applicator. This category allows property owners to apply restricted-use chemicals on their own property. Most foresters who need certification will probably need the broader certification as commercial applicators.

One other federal law is also of continuing significance. This is the Federal Food, Drug and Cosmetics Act (FFDCA) of 1938 and its amendments. The act allows for the setting of tolerances for various chemicals. The Miller Amendment of 1954 set up a procedure for setting tolerances of pesticidal residues on food, and the Delaney Amendment of 1958 declared that food additives found to induce cancer in humans or animals are unsafe.

Sometimes, the many bills under which an agency operates may seem confusing, and such is the case here. Actually, the 1972 amendment of FIFRA was an entirely new bill titled the Federal Environmental Pesticide Control Act (FEPCA) of 1972. The major part of the bill was the revision of FIFRA. It is recommended that all individuals with an interest in chemical use become familiar with the applicable laws and regulations.

DESIRABLE CHARACTERISTICS

The ideal forest insecticide has never been produced. In fact, we can never expect to have a single ideal product because special cases require different characteristics and some of these requirements are conflicting. Although we may never obtain the ideal material, we can set up ideal standards that will be useful in evaluating the insecticides with which we must work.

Table 11-1 lists some of the requirements and some of the desirable characteristics. The same qualities that are desired in an insecticide for spraying living trees may be of no value or even actually undesirable in treating forest products or nursery soils. Therefore, Table 11-1 indicates the suitability of each quality for each of the three classes. The required characteristics are in italics; those that are desirable but not absolutely essential are in roman type.

Toxicity

Insecticides must, of course, be toxic to the pests against which they are directed. At the same time, if they are used for spraying trees, they must be relatively nontoxic to the foliage. These two conflicting requirements limit the number of substances that are suitable for treating living plants.

Safety of application is accomplished in several ways. A substance that is toxic to insects can be safely applied to plants if it is a stable solid and is

Table 11-1 Desirable Qualities for Insecticides

Quality	For living trees	For dead trees or forest products	For nursery soils
High degree of toxicity for pest	*Yes*	*Yes*	*Yes*
Nontoxic to living plants	*Yes*	No	No
Nontoxic to humans or in safe form	*Yes*	*Yes*	Yes
Nontoxic, in dosages used, to beneficial organisms	Yes	No	Yes
Low cost	Yes	Yes	Yes
Simplicity of application	Yes	Yes	Yes
Short residual effect but not permanently stable	*Yes*	No	No
Long-time or permanent stability	No	Yes	Yes
Good storage qualities	Yes	Yes	Yes

insoluble in water. Such a substance cannot be absorbed by the plant tissues and, therefore, cannot cause injury.

Water-soluble material must be sufficiently toxic to kill the pest when it is diluted to a concentration point safe for the plants. Unless a substance meets these requirements, it should not be used as a spray. Oil solutions are capable of penetrating plant tissues, and most oils are toxic to plants. Therefore, all oil insecticides must be very thinly dispersed if injury to the plants is to be avoided. A thin dispersion can be accomplished for conventional spraying by emulsifying the oil and diluting the emulsion with water. In aerial spraying or in the use of the mist blower or the fog machine, the oil is separated into fine particles as it leaves the machine and is dispersed so widely that little or no plant injury will result.

Workers who handle insecticides day after day are inclined to become careless and, therefore, cannot be trusted to protect themselves from highly toxic materials. Almost all chemicals used for insecticides are toxic to humans if absorbed in sufficient quantities. Therefore, it is essential that all insecticide materials be prepared so that they can be handled safely. In order to meet this requirement, extremely toxic substances are usually placed on the market in partly diluted form. In spite of care on the part of the manufacturer, serious illness or even death may result from carelessness. *Many pesticides are deadly poisons, and every possible precaution should be used to avoid inhaling them, ingesting them, or absorbing them through the skin.*

The foresters in charge must be aware of their responsibility for those people supervised in field operations. Deaths directly caused by the use of insecticides are few in number and have almost always been directly related to carelessness and misuse.

In aerial spraying especially, both pilots and ground crews find it almost impossible to keep the spray materials off their faces and clothing. For this reason, some of the more toxic chemicals should not be applied from the air.

Stability

The quality of stability deserves special consideration. Some insecticides decompose slowly or not at all after application; others decompose or volatilize in a few minutes or hours. If the insecticide is placed in a location where the insects will eat or contact it later, enough stability to permit this contact is essential. If, on the other hand, the insecticide is applied directly to the insect, stability after treatment may not be of any consequence.

After an insecticide has done its work in the forest, the quicker it disappears from the environment the better. Unfortunately, many otherwise excellent insecticides have a higher degree of stability than is ideal. Insoluble arsenical salts of lead, copper, and calcium decompose slowly or not at all after they are applied. DDT and some other organic insecticides may remain active in the soil for indefinite periods. Repeated heavy application of these substances can result in undesirable accumulations. Many of these materials have been completely banned from use on forested areas. DDT, for example, is no longer registered for use on any forest insect. No insecticide should be used more often than is absolutely necessary or in heavier dosages than the minimum required to accomplish control. We must emphasize that use of insecticides without strict attention to label instructions is not only illegal but also wasteful and dangerous.

CLASSIFICATION OF INSECTICIDES

Before the advent of the various types of synthetic organic insecticides, the classification of insecticides was a simple matter. The relatively few materials in use could be separated with a minimum of confusion, according to either their mode of action on the insect or their physical characteristics. Today, it is difficult to devise a clear-cut and simple classification. This is because so many chemicals fall into more than one insecticidal subdivision, no matter how carefully these subdivisions are defined. The same insecticide may, in some cases, enter the insect's body through either the mouth or the sensory pores. It may be applied in the form of a dust, a suspension in water, an oil emulsion, or an oil solution. A logical classification could be based on the chemical characteristics of the various substances, but that method does not seem appropriate here. Therefore, there will inevitably be considerable overlapping in the following discussion.

Three classes of insecticides are generally recognized. These are (1) stomach poisons, (2) contact poisons, and (3) fumigants. The fumigants are the most effective but are limited in application because they must be used in confined locations. But when insects must be killed in the open, a spray or a dust is needed. For this purpose, we must use either a contact or a stomach poison.

The stomach poisons are sprayed or dusted on the materials to be protected or can be impregnated into the wood. The insects ingest the poison with their food. These insecticides are especially effective against insects with chewing mouthparts. Insecticides of this type may be used in four principal ways (Metcalf et al., 1962).

1 The natural food of the insect is covered with the poison, the insect consuming the poison when it eats.
2 The poison is mixed with a material that is attractive to the insect as a bait and is placed where the insect will find it.
3 Some poisons may be placed where insects travel so that the insects get the poison on their feet and antennae. They then consume the poison while cleaning their appendages.

4 In recent years, much attention has been given to systemic materials. These are toxic substances that are absorbed and distributed harmlessly throughout the tissues of living host organisms, either plant or animal, so that insects feeding on them are killed.

Contact poisons are of much greater significance than the other types. They enter the bloodstream directly through the chitin or the intersegmental membranes or into the respiratory system through the spiracles. Application may be directly to the body by spraying or dusting the insects or indirectly through a residue left on plant surfaces or other places visited by the insects. These insecticides are effective against either chewing or sucking insects.

Contact insecticides vary greatly in their stability. Some have a long residual life and are effective controls for long periods after application. Others are highly volatile and dissipate soon after application. These must be applied so that the spray or vapor from the spray will contact and penetrate the insect's body before dissipation. Most registered chemicals for present-day control of forest insects are from neither of these extremes. Most of the registered materials tend to break down a few days or at most a few weeks after application.

SOME COMMON INSECTICIDES

No attempt will be made here to provide the reader with a cookbook on registered materials for specific insects. Changes are taking place very rapidly, and the forester is advised to obtain current information when a control operation is required. Rather, this discussion will provide information on various types of insecticides and will use the names of specific materials only to illustrate the diversity that exists. Some of the insecticides are no longer in use, but the forester should be aware of the potential for future development of materials in these categories.

Botanicals

Among the commonly used materials that act as contact insecticides are many of botanical origin (Gould, 1966). These are not much used in forestry but should nevertheless be mentioned. The nicotine alkaloids are insecticides that are derived from tobacco combined chemically or mechanically with various substances. They constitute an important group. As early as 1763, tobacco extract was recommended as a control for plant lice (Metcalf et al., 1962). It has continued as an important control for piercing-sucking insects. Nicotine is extremely poisonous to warm-blooded animals.

The so-called fixed nicotine compounds act as stomach poisons. The commonest of these are nicotine tannate and nicotine bentonite. Nicotine sulfate is also stable where used in nonalkaline solutions; when mixed in a soap solution or with any other alkaline detergent, free nicotine is released and volatilizes promptly.

Nicotine is used not only as a plant spray but also as a dust. Nicotine dusts are prepared by mixing free nicotine with carriers of various kinds. When dusted on infested plants, the nicotine volatilizes and forms a gas that penetrates the bodies of the insects dusted. Nicotine may be combined with various other insecticides for special purposes. In nurseries and small forest plantations, nicotine sprays or combinations of nicotine and other substances are used to control aphids, tip moths, and other insects. Nicotine sprays are expensive and therefore are used only where cost is not a serious consideration.

Generally, modern synthetic insecticides are less expensive and often less dangerous to use. Insecticides containing the plant products rotenone and pyrethrin are excellent contact insecticides, but because of their cost they, as well as nicotine, have only limited possibilities for use on forests or forest products. Rotenone and the associated compounds (deguelin, tephrosin, toxicarol, sumatrol, and elliptone) are present in the roots of certain leguminous plants of the genera *Derris* and *Lonchocarpus* in tropical America and also in the roots of the genera *Tephrosia* and *Mundulea* in the tropics of the Eastern Hemisphere. The roots are dried, ground, and used to make insecticidal dusts. The pyrethrins are present in the flowers of certain plants belonging to the genus *Pyrethrum*. The flowers are ground into insecticidal dust, or the active pyrethrins are extracted and concentrated for use in insecticide formulation. All these plant products vary greatly in the amount of toxic materials that they contain. Therefore, in buying either the ground materials or the extracts, one should always know the proportion of active materials.

In recent years, *synthetic botanicals,* such as the synthetic pyrethroids, have been developed. These are being tested in forestry, and although none have been registered as of 1978, they are likely to be used in the future. There is also much interest in feeding deterrents derived from plants, but no recommendation appropriate to potential forestry uses has been developed.

Inorganic Compounds

Most of the important insecticides in this group are stomach poisons. Their use today is limited in forestry because they have been replaced by the cheaper and often more effective synthetic compounds.

Arsenical insecticides were formerly the most important materials for the treatment of forests and ornamental trees. Arsenicals that are applied to living plants are always insoluble salts. Copper arsenate, commonly called paris green, was developed earliest but was found to burn the foliage of many trees. Later, lead arsenate and calcium arsenate came into general use. In treating forests, calcium arsenate was the form usually preferred, chiefly because it is somewhat cheaper and yet equally as effective as lead arsenate. Calcium arsenate slowly decomposes in the presence of carbon dioxide and water vapor, releasing water-soluble forms of arsenic. This water-soluble arsenic will injure foliage, but in forest treatment, the amount of injury was never of any great consequence.

Calcium arsenate has been applied from the air as a dust, but it can be applied with equal effectiveness either in dilute suspension or in concentrated suspensions by means of the mist blower. For conventional spraying, the usual dilution was 3 lb (1.36 kg)/100 gal; in concentrated suspensions, from 3 to 6 lb (1.36 to 2.72 kg) were used. Calcium arsenate may be diluted with equal parts of hydrated lime for ground dusting or used undiluted when applied from the air. The amount needed per acre varied somewhat with the size and density of the trees but was usually about 20 lb/acre. The price of calcium arsenate varied greatly, depending on market conditions and the quantity purchased in a single order.

Fluorine compounds were developed as insecticides to provide substitutes that would not leave highly poisonous residues on edible crops. The two most commonly recognized are sodium fluoride (NaF) and cryolite. Cryolite has low toxicity to mammals and is effective against many chewing insects. These compounds have never been used intensively in forestry.

There are many other inorganic compounds that are toxic to insects and other organisms. Many have been used against agricultural pests, but few are of importance today.

Synthetic Organic Compounds

The word *synthetic* is used here only to distinguish the materials manufactured by the chemical industry from the organic compounds in the botanical group derived directly from plant materials. Many materials have been developed as illustrated in the current issue of *ENTOMA: Pesticide Handbook.* Table 11-2 shows some of the changes that have taken place. DDT was a widely accepted material, and from 1954 to 1967, it was used on all spray projects in Maine. Since 1970, a variety of registered materials have been developed, and testing of new materials was still going on in 1978. One very promising registered material (Zectran) was taken off the market a few years ago because the limited use in forest entomology could not support production. It is desirable to have more than one material registered for a particular problem because of the rapid changes that are continually taking place.

DDT, or *dichlorodiphenyltrichloroethane,* has been the most publicized of all the insecticides, and it still receives much attention even though its use is no longer permitted in the United States. This compound was first synthesized in 1874, but its insecticidal properties were not discovered until 1939. It is a remarkable chemical, but actually, it had no greater potentialities for insecticidal purposes than several other similar, but lesser known compounds.

DDT is one example of the group of insecticides called *chlorinated hydro-*

Table 11-2 Summary of Aerial Spraying for Spruce Budworm, *Choristoneura fumiferana*, in Maine, 1960 to 1978

Year	Insecticide	Acres
1960	DDT	217,000
1961	DDT	53,000
1963	DDT	479,000
1964	DDT	58,100
1967	DDT	92,162
1970	Fenitrothion	210,000
1972	Zectran	500,000
1973	Zectran	470,000
1974	Zectran	430,000
1975	Carbaryl	496,445
	Fenitrothion	1,499,260
	Zectran	238,000
1976	Carbaryl	3,460,000
	Dylox	40,000
1977	Carbaryl	808,400
	Dylox	55,400
	Orthene	58,400
1978	Carbaryl	966,216
	Dylox	53,657
	Orthene	96,487
	Bacillus thuringiensis	21,848
1979	Carbaryl	2,479,433
	Dylox	96,902
	Orthene	110,417
	B. thuringiensis	41,483

Source: L. C. Irland, Maine's spruce budworm program; Moving toward integrated management, *J. Forestry* 75 (1977): 774–777; H. Trial, Jr., and A. S. Thurston, Spruce budworm in Maine, The 1978 cooperative spruce budworm suppression project and expected infestation conditions for 1979, *Maine Department of Conservation, Entomology Division, Tech. Rept.* no. 8, 1978. 1979 data supplied by A. S. Thurston, *Maine Department of Conservation.*

carbons. Like DDT, many of these have been banned from use because of their extremely long residual activity. But not all these materials have the extreme problems common to DDT and the other materials that have been removed from the market.

The *organophosphates* are a second group of materials that have been developed by the industry. There are numerous examples of these compounds, and one listed in Table 11-2 (Fenitrothion) has been widely used in forest spraying. Some of the organophosphates are highly toxic materials; some are relatively safe to use. Some have a wide spectrum of activity; others are highly selective.

Malathion is one of the organophosphate chemicals that has been used effectively against forest insects. It is one of the safest of all insecticides for human handling and in addition is a persistent general-purpose insecticide.

Diazinon is somewhat more toxic to warm-blooded animals but is also an effective general-purpose insecticide.

Many of the organophosphates are systemic in their action; that is, they can be introduced into the bodies of animals or the sap streams of plants and will kill insects that feed on the treated organisms. Systox, Schradan, and Thimet have been used in this manner to protect plants from sucking insects, tip moths, and mites.

The *carbamate* insecticides are being used effectively in forest-insect control. The material carbaryl (Sevin) was used in Maine from 1975 to 1979 (Table 11-2). This material has been very effective in the control projects against the spruce budworm, *Choristoneura fumiferana,* and will likely remain as a registered chemical for that insect for some years to come.

The names *chlorinated hydrocarbons, organophosphates,* and *carbamates* are each representative of a number of commonly used insecticides. The very long persistence of the chlorinated hydrocarbons has resulted in a reduced use of these materials, but interestingly, the very volatile space fumigants such as methyl bromide, carbon tetrachloride, and ethylene dibromide are halogenated hydrocarbons (Anon., 1969). The impressive versatility of the organophosphorus and carbamate compounds is especially significant, as is their rapid degradation in the environment.

Among the fumigants, several are effective against forest insects. Methyl bromide is often used against termites, powderpost beetles, and other structural pests and for sterilizing soils and treating dormant nursery stock. Such fumigation is widely practiced using either tents of plastic films or fumigation chambers. Ethylene dibromide and orthodichlorobenzene have both been used successfully in bark beetle control but may also be used as soil fumigants (Fig. 11-1).

Insect-growth regulators are a new and rapidly developing group of chemical insecticides. Dimilin is an example that has been effective against several forest insects. These materials are potentially valuable additions to our arsenal of pest management tools.

The oils used on shade trees are derived from the lubricant fraction of petroleum. The less highly refined oils are commonly called *dormant oils.* The lighter *summer oils* are used on green trees. These materials are used for control of mites and of several species of scale insects. Oils have been judged to possess extremely low human health hazards and have been used for well over 75 years with no sign of the development of resistant insect races.

We have named a few of the many organic compounds for the purpose of illustration but are not recommending any of them. You will need to consult the current registration lists to determine those materials most suitable for solving your control problems. New materials are continually being developed.

Figure 11-1 A field mixing site of the 1950s where water was added to an ethylene dibromide concentrate for use in spruce beetle, *Dendroctonus rufipennis,* control. Such operations must be conducted with care to avoid pollution of the stream. *(U.S. Forest Service photo.)*

Therefore, the student must keep up with the literature in order to be sure of using the safest and most effective materials.

FORMULATION OF INSECTICIDES

Insecticides other than dusts are usually received from the manufacturer in concentrated form and hence must be diluted or otherwise formulated before being applied. The method used varies with the material and type of equipment.

Dilution for Conventional Spraying

Because hydraulic spraying machines can handle only liquid sprays of low viscosity, insecticides applied by them must be diluted with large quantities of water. Some insecticides, such as the nicotine compounds, are soluble in water and can be easily diluted to any desired concentration. Other substances that are oils or dissolved in oils can be dispersed in water only after being mixed with an emulsifying agent.

Some insecticides are insoluble powders and must be mixed with water; a few pounds of the powder with 50 to 100 gal. The suspensions thus formed have a tendency to separate. To avoid this, the larger conventional sprayers have tanks equipped with agitators that keep the solid particles of powder from settling to the bottom. In small sprayers, frequent shaking during spraying is necessary to avoid separation of the suspension.

Most insecticides must be diluted to make them easier to apply. They are seldom used full strength. Some of the common formulations are:[2]

1 *Dusts.* These are used dry. The toxic material is mixed with or impregnated on organic materials, such as walnut shell flour, or fine mineral particles, such as talc or bentonite.

2 *Granular Materials.* These formulations are similar to dusts but have a larger particle size. They are commonly used in dressings on or in soil and may be directly mixed with fertilizers before application. Under some circumstances they are applied from the air.

3 *Wettable Powders.* This material looks like a dust but is generally more concentrated. Wettable powders are meant to be diluted with water and used as sprays. To achieve suspension in water a dispersing and wetting agent is added to the formulation.

4 *Solutions.* Most of the synthetic organic compounds are insoluble in water but can be dissolved in certain organic solvents. Some of these solutions, especially oil solutions, are used directly for insect control, but seldom directly on plants by the use of ground equipment, because of phytotoxic reactions. Oil solutions are, however, applied from the air. In such treatments the phytotoxic action is prevented by the dispersion of the oil particles.

5 *Emulsifiable Concentrates.* This is a common and versatile formulation consisting of the insecticide, a solvent for the insecticide, and an emulsifying agent. This concentrate mixed with water forms an emulsion of the oil-in-water type. Since the quantity of solvent is small, these sprays may be used directly on plants from ground equipment. Emulsions are not stable and will eventually "break" into their component parts. Thoroughly shaking the container will return the formulation to its proper form.

6 *Insecticidal Aerosols.*[3] Aerosols are minute particles suspended in air as a fog or mist. These particules may be produced by burning, vaporizing with heat, atomizing mechanically, or, as in the case of the aerosol bomb, releasing with a liquefied gas.

7 *Fumigants.* Insecticides that are effective in their gaseous form are often formulated as liquids under pressure. Methyl bromide is one of these. When released into air the material rapidly volatilizes. Some fumigant gases are released chemically. Hydrogen cyanide, for example, is released by dropping a cyanide salt into dilute acid.

Addition of Spreaders and Adhesives

Because water is characterized by high surface tension, aqueous sprays tend to form droplets and run off the foliage. To avoid this, various kinds of spreading or wetting agents are added to aqueous solutions and suspensions.

[2]Adapted from R. E. Pfadt (ed.), *Fundamentals of applied entomology,* 1962, with permission of The Macmillan Company.
[3]Users of either aerosols or compressed liquefied gases should be warned against transporting these materials by air at high elevations unless they are in very strong containers. Lowered atmospheric pressure may cause containers to leak or explode with disastrous results.

These substances reduce the surface tension of the water and cause the liquid to spread over the surface, thus preventing excessive runoff. A few of the materials thus used also act as adhesives, causing the insecticides to stick to the foliage.

All water-soluble emulsifying agents also act as spreaders. Soap is readily available as an effective spreading agent. It has the objectionable characteristic of combining with calcium or magnesium salts to form insoluble solids. Therefore, the amount of soap needed will depend on the hardness of the water. Tests should be made to determine the quantity necessary for the water used. Enough soap to produce suds will be sufficient. The quantity most often required is 6 lb/100 gal (2.72 kg/3.78 L) of water. Other spreading agents are usually added to the spray solution at the rate of 1 lb/100 gal (0.454 kg/3.78 L). Many spreaders were developed for use as detergents in the textile industry. These organic substances are sold under various trade names too numerous to mention. (See the current issue of *ENTOMA.*)

Raw linseed oil, fish oil, cottonseed oil, soybean oil, and some petroleum oils are often added to sprays to increase the adherence of the insecticide. Calcium caseinate also acts as an adhesive when added to sprays.

Dilutions for Concentrated Sprays

When sprays are to be distributed by fog machines, mist blowers, or from the air, the materials are diluted as little as possible. Some of the synthetic organic wettable powders are diluted to similar heavy suspensions when distributed by air blast or as a mist. Wettable powders, for example, may be diluted for this purpose at the rate of 1 lb (0.454 kg) in enough water to make 1 gal (3.78 L) of finished solution.

Strong solutions of chemicals in oils are sprayed on the bark of trees or logs to kill bark beetles. A 12 percent solution of orthodichlorobenzene in diesel oil or fuel oil applied with conventional sprayers has been a mixture used for bark beetle control.

Oil solutions of some insecticides are effective against bark beetles and emulsions have also been used. For example, a recommended insecticide to control the spruce beetle, *Dendroctonus rufipennis,* consisted of an emulsifiable concentrate and water. The concentrate contained ethylene dibromide, emulsifiers, and fuel oil (Massey et al., 1953).

Dilutions for aerial spraying are usually concentrated oil solutions, but emulsions of a chemical may be applied (Fig. 11-2).

For Soil Poisoning

Insecticides for poisoning root-eating insects are mixed with the top few inches of the soil, where they kill the insects either by contact or by being eaten. Inert substances not only kill the insects present at the time of application but also

Figure 11-2 Helicopter spraying of a Douglas-fir tussock moth, *Orgyia pseudotsugata*, infestation. Spray deposits can be more precisely controlled by using this equipment. *(U.S. Forest Service photo.)*

prevent reinfestation by poisoning the soil for a year or more. Soil poisoning is too expensive to use in forest plantations but has a possible place in the control of insects in forest nurseries.

A specialized application has been developed for control of white grubs in plantations. The insecticide may be applied directly to the roots of the plants from an attachment on the planting machines. This procedure is very inexpensive and has given excellent results (Stoeckler and Jones, 1957).

Most of the tests of soil poisoning have been applied to lawns, golf greens, and similar situations, rather than to forest nurseries. Therefore, the recommendations that have been published cannot be accepted without reservations by the forest entomologist. Those tests that have been made in forest nurseries demonstrated that the effects on plants growing in the poisoned soil vary from place to place. Thus, in some nurseries, a treatment may be safe; whereas in others, injury to the young trees results.

In places where soil insecticides are safe, they provide the nursery manager with a powerful weapon against soil-inhabiting insects. On the other hand, if they injure the seedlings, the characteristics of stability may put a poisoned soil out of production for many years. Therefore, a nursery manager should be advised to test any soil insecticide that he or she may plan to use on a small plot before applying it extensively. By so doing, he or she can determine definitely whether or not the substance is suited to conditions in the nursery.

Formulation of Baits

Some insects, such as cutworms and grasshoppers, can be injurious to small trees in nurseries or newly established plantations. These insects and other organisms are attracted to and feed on various baits. By the addition of a poison to these attractive mixtures, the organisms can be killed. Many bait formulations are effective.

APPLICATION OF CHEMICALS

The use of chemicals for the control of insects has had a phenomenal development since 1868, when paris green was first used against the Colorado potato beetle, *Leptinotarsa decemlineata*. Since that time, insecticides have been improved and perfected to such an extent that chemical warfare against insects has come to be almost universal. For many years, its use in the control of tree insects was limited to ornamental and orchard trees. Because of the high costs involved in applying insecticides and because of the inaccessibility of forest lands, chemical control was little used in forests. Today, however, the development of new machines for applying concentrated spray materials has reduced costs greatly. Nevertheless, these new machines have by no means replaced the older types that spray the insecticides in greatly diluted solutions or suspensions.

The three general methods of applying insecticides arc (1) as sprays, (2) as dusts, and (3) as fumigants. These are discussed in terms of tree application in the following sections. The different types of equipment are numerous, and many new modifications of each type are continually appearing; therefore, this discussion will be limited to a consideration of general methods without describing specific models or designs of machinery.

To Ornamental Trees

Although the cost of treating forest trees with hydraulic sprayers is prohibitively high, it has always been practicable to spray ornamental trees with these machines. The chief advantage of these sprayers is the high degree of accuracy in control of the spray. It is also cheaper in original cost and may be safer when used by inexperienced operators.

However, the spraying of tall trees with the hydraulic sprayers requires the use of heavy and expensive equipment; large-capacity tanks, special nozzles and spray guns, and sufficient length of high-pressure hose to extend from a road to the infested trees. The spray liquid is discharged from the nozzle at pressures as high as 400 to 800 lb/in^2 and at rates up to 80 gal/min. Specially designed nozzles discharge a solid stream that breaks into a fine mist in the air. For tall-tree spraying, the conventional sprayers are mounted on a trailer or truck and are equipped with a tank holding at least 150 gal of spray material.

Some tanks hold as much as 500 gal if the machine is designed to be used on good roads. Because the long hose required to reach the trees causes a great reduction of pressure between the pump and the nozzle, the pump pressure must be at least 500 lb (325 kg) and sometimes as much as 1500 lb

(675 kg). Less elaborate equipment is adequate for small trees. Trees up to 30 ft (9 m) in height can be treated with any of the power machines used in orchards.

Current literature contains much detailed information on the numerous types of spraying equipment. References in *ENTOMA*, published biennially by the Entomological Society of America, serve as an excellent guide to this subject. Low vegetation and small trees can be treated with a poisonous dust instead of a liquid spray. The various dusting machines range in size from the small hand dusters that hold 1 lb (0.45 kg) or less of the insecticide to the elaborate power dusters used in field and orchard work. Even with the most powerful machinery, it is impractical to throw a column of dry dust to the tops of tall trees. Furthermore, dusting is wasteful of materials and cannot be accurately controlled. Therefore, dusting from the ground is not often used for the control of tree insects.

To Forest Trees from the Ground

Spraying equipment suitable for use in forests has been developed to a high point of efficiency. Small, accessible trees are treated from the ground with either dusts or liquid sprays. Dilute sprays are most often used for this purpose.

The knapsack sprayer or back pump is one of the most common types suitable for the hand treatment. In treating scattered infestations in forest plantations, hand treatment from the ground is sometimes the least expensive procedure. The back pumps, used for fire fighting, can be adapted for this work by equipping them with nozzles that deliver a spray instead of a stream of liquid. Another type of sprayer used frequently for hand spraying is the compressed-air sprayer. This type is more difficult to carry than the knapsack sprayer, but it is just as efficient.

The chief objection to both the knapsack and the compressed-air sprayers for use in forest plantations is the large quantity of water needed for dilution of the insecticide. The time involved in frequent trips to a dispensing point for a load is a major item of cost in a backpack operation. Even in using power sprayers, it is difficult to supply adequate water for dilution of spray materials. In order to reduce the amount of water needed for spraying, special atomizers have been designed to handle concentrated sprays. To atomize concentrated suspensions, higher pressure than is produced in a knapsack sprayer is required. Special atomizers have been designed and used effectively for applying these sprays to small trees, but this method has never become popular because overdosage with the insecticide is difficult to avoid.

The stirrup pump is commonly used in treating bark beetle–infested trees (Fig. 11-3). These pumps have been used in many large projects in remote areas inaccessible to power equipment. This type of pump is inserted into a can of insecticide and pumped to the tree at pressures up to 150 lb/in^2 through a neoprene hose and a 6-ft (1.83-m) wand. Surfaces up to 30 to 35 ft (9 to 10

Figure 11-3 Application of insecticide to an infested spruce during an outbreak of the spruce beetle, *Dendroctonus rufipennis*. This type of stirrup pump has been used on numerous projects in the West. *(U.S. Forest Service photo.)*

m) above ground can be thoroughly soaked with the insecticide, a height necessary in bark beetle control.

However, power machines for applying concentrated sprays have proved both practical and effective in places where the trees are accessible by truck or tractor. The mist blower is one of these modern developments. The insecticide is injected into a large-volume air stream moving at a rate of 150 to 250 mi/h. If the amount of material injected and the forward speed of the machine are regulated, the insecticide is blown onto the trees in any quantity desired. The smaller mist blowers are light enough so they may be transported readily (Fig. 1-1). The smallest backpack models have a smaller range but can be used in treatment of forest plantations. Some mist blowers (or air-blast sprayers, as some prefer to name them) can be used to spray tall trees because some models are capable of blowing the insecticide to heights of over 100 ft (30.5 m). When forest plantations are sprayed, the mist is discharged almost horizontally. In this manner, a swath about 300 ft (91.44 m) wide can be treated as the machine moves along.

Another modern development is the fog machine. A jet of concentrated insecticidal solution is directed against a heated surface, and thus an insecticidal fog is generated. The fog is then carried away from the machine by air currents. If advantage is taken of the temperature inversion that often occurs in the air at night, the fog can be held close to the ground. The particle size of the fog is important and can be controlled within certain limits by regulating the temperature or the generator.

Even the most effective fog has a killing range for many forest pests of only about 200 ft (60.96 m) from the machine, although it apparently spreads much farther. The failure of the fog to kill beyond a limited distance is the result of the small size of most of the particles. The particles of effective size are deposited mostly within 200 ft of the generator; the finer particles drift farther away. This is a serious problem associated with use of these machines.

Yeomans (1950) reports that "airblast machines" (mist blowers) can provide a more uniform deposit across a swath with larger spray particles than can be obtained with the wind-borne "aerosols" (fog machines). Only 25 to 50 percent of the particles less than 50 μ in diameter will be deposited within 2000 ft (609.60 m). The mist blower is, therefore, much better than the fog machine for treating trees with chemicals.

The comments on particle size lead into a discussion of nozzles and spray emission volumes as used in modern operations, whether on the ground or from aircraft. Nozzles that depend on hydraulic pressure can produce smaller droplets with increasing pressure and decreasing orifice size (Akesson and Yates, 1964).

Bifluid nozzles depend on the interaction of two fluids (air and liquid) to produce breakup of the spray liquid into the desired droplet sizes (Randall, 1975). Systems based on this principle are used in ultralow-volume spraying of the type now used against forest insects in the United States and Canada. Aerial spraying rates are now down to less than 1 qt/acre (2.32 L/ha) in oil solutions. With ultralow-volume spraying, the aim is to emit larger numbers of smaller droplets for better coverage and to avoid the volume wastage associated with emission of the larger droplets. Armstrong (1975) discusses some of the meteorological problems associated with spraying operations and the need for solutions with relatively low rates of evaporation so that the small droplets will reach the target.

To Forest Trees from the Air

Shortly after World War I, the airplane came into use for applying insecticides to fields, orchards, and forests. This method proved very popular. Its use made possible rapid and effective treatment of areas that could not have been reached otherwise.

Although the cost was relatively low, it was still too high for any but high-value forests. The low carrying capacity of most planes and the large quantity of insecticide required per acre were the chief factors that contributed to the expense. On the average, using arsenicals, it was necessary to apply about 20 lb of dust per acre in order to obtain satisfactory control of defoliating insects. Thus, with a pay load of 300 lb (91.44 kg), a plane could treat only 15 acres (6.07 ha) before returning to the landing field for reloading. Sometimes, difficulty was experienced in obtaining sufficiently rapid discharge of the insecticide to give adequate coverage. In such cases, a second flight had to be made over the same area.

In the early treatment of forests from the air, the limited carrying capacity of the planes precluded the possibility of using the common aqueous arsenical suspensions. Usually, therefore, the insecticide was a powder, most often calcium arsenate.

These limitations of carrying capacity were largely removed with the development of potent synthetic organic insecticides. Instead of 20 lb/acre (22.36 kg/ha) the synthetic organic compounds were effective against many forest insects at 1 lb/acre (1.12 kg/ha) or less (Fig. 11-4).

Helicopters, also used in spraying forests, have some advantages over other aircraft. The insecticide can be better controlled because these machines can operate at low speeds and can take off from a road or a small cleared space; the conventional plane, on the other hand, requires a landing field. As a result,

Figure 11-4   An early aerial spray project utilizing a synthetic organic insecticide. This Ford trimotor was used on a Douglas-fir tussock moth, *O. pseudotsugata*, project in 1947. *(U.S. Forest Service photo.)*

an operation using conventional planes often must load the insecticide at a point 20 mi (32 km) or more from the area to be sprayed; whereas the helicopter can be loaded nearby. Experiments comparing the cost of spraying forests from regular-type planes, ranging from single-engine agricultural spray planes to four-engine constellations, and helicopters indicate that the advantages of loading close to the spraying area may be offset by the high operating cost of the helicopter (Yuill and Eaton, 1949).

Because aerial spraying has become a standard practice for the control of many defoliating and some sucking insects, the operations have become the important concern of foresters. Operations are usually directed and controlled by administrative officers in charge of the infested area, but these individuals are usually aided and advised by professional forest entomologists. Sometimes, however, this professional advice is unavailable, especially on small projects. Therefore, every forester should understand aerial-spraying practices and how to plan, supervise, and execute a spraying job (Batch et al., 1956, and Prebble, 1975).

Laying out the Project

The spraying project is laid out on the basis of the survey that determined the extent and severity of the infestation. The area to be sprayed is calculated from previously prepared maps, and the infested portions of the forest must be divided into spraying units that can be recognized from the air. The boundaries of these units may be roads, ridges, edges of plantations, margins of cuttings, or other features that are easily identifiable from the air. The size of the units will, of course, vary with topography and the distribution of the infestation. They should be shaped or arranged so that they can be flown with a minimum number of turns. This is especially important if conventional planes are used. If the spraying is done with helicopters, this is less important.

Selecting the Loading Base

The next consideration is the loading field. This should be located as close to the area to be sprayed as possible and must be accessible by road. Frequently, a local airport is favorably situated, and arrangements can usually be made to use these established landing fields. On the other hand, the location of the spraying operation may be so far removed from an airport that the time consumed in flying back and forth will preclude its use. Then a temporary field must be provided.

Handling the Insecticide

For large spraying operations, ready-prepared spraying mixtures can often be obtained and are convenient and usually less expensive than materials prepared on the job. For smaller operations, however, it is usually necessary to mix the materials. This mixing must be done in advance of spraying so that the plane tanks can be loaded quickly. The economy of any spraying operation will be greatly influenced by the speed of loading the planes between flights. On small jobs, the insecticide may be mixed and carried in oil drums. By mixing in the tank of a power sprayer with a mechanical agitator, the process can be done more easily, and the materials for a small job can be pumped directly into the plane from the tank.

The amount of insecticide applied and the area covered must be recorded for each flight or group of flights so that dosage may be controlled. On small jobs, the amount of insecticide used on any run is estimated from the capacity of the tank in the plane or by gauging the amount taken from the oil drums or storage tanks. On larger operations, the prepared spray is usually hauled to the loading field in tank trailers or trucks and pumped from these directly into the planes by power pumps. The quantity of spray pumped into each plane is measured by a meter as it flows from the storage tank.

After the plane is loaded and ready to take off, the job is in the hands of the pilot, and the forester can do nothing to expedite the work. Careful planning prior to the takeoff will reduce delays, facilitate the operation, and reduce the cost.

Controlling the Spray

Controlling the application of the spray for both particle size and amount is essential. This is accomplished by nozzle adjustment (as discussed earlier) and by regulating the rate at which the insecticide is discharged from the plane, the speed of the plane while spraying, and the width of swath covered in each run. In spraying, care must be exercised by the pilot to lay down the swaths so that they do not overlap more than is necessary and that no gaps are left unsprayed. The suitability of the equipment to do a satisfactory job should be checked in ample time to permit adjustment. Even distribution of the proper amount of spray per acre is essential to good control. The amount of the insecticide applied per acre can be regulated not only by adjustment of the rate of discharge but also by modifying the swath width. During the operation, a constant check should be kept on the amount of insecticide applied and the area covered by each run or series of runs so that any errors that occur can be promptly corrected. Such checking will reduce careless tendencies to the minimum.

Prior to application of the spray, each pilot must be briefed concerning the area to be sprayed so that he or she can recognize it easily from the air. This is accomplished first by examination of a map showing direction, distance from the landing field, and general landmarks. Examination of the map will be followed by actual observation from the air. An aerial photograph mosaic is an important aid in briefing the pilots. A flier should never be permitted to spray a piece of forest without first flying over it in order to become familiar with the area.

It may be possible to provide corner markers with elevated flags or tethered balloons for specialized projects. Such methods are impractical on large projects. Flight line navigation by magnetic compass was found to be lacking, as was a system based on following the previously emitted cloud of spray. The next step in guidance was the use of spotter planes to provide information and guidance to the spray pilots. This was followed by the development of an inertial navigation system that accurately locates the start and stop points of each spray run (Randall, 1975). Present-day operations generally rely on both guide aircraft and electronic gear.

Securing Weather Information

Another responsibility that rests with the forester is the securing of weather information. Aerial spraying can be done effectively only when atmospheric conditions are favorable. The air must be clear, calm, and without appreciable convection currents. Such conditions commonly prevail for several hours in the early morning and for a shorter period in the evening. As soon as the air becomes turbulent or the wind rises above 5 or possibly 8 mi/h (8 to 13 km/h), spraying must cease.

Wind, fog, and rain are all serious handicaps to aerial spraying. They cause cessation of operations, and if they prevail for any length of time, they may prevent timely treatment. The full importance of timing in the control of various insects will be discussed at length in Chapters 12–20. Suffice it to say here that timing is of great importance in control work and that the atmospheric handicaps of aerial spraying are sometimes detrimental to proper timing.

Checking Distribution

In addition to other duties, the project manager must provide people to check to assure that a satisfactory job is being done. As has been noted, even distribution is essential to satisfactory control. Therefore, every practicable effort should be made to secure it. This checking for even distribution should be done while the operation is in progress. Pilots, like most other people, are more careful when they know the quality of their work is being evaluated.

A method utilizing chemically treated cards is perhaps the simplest way to determine spray distribution. Treated cards that are oil-sensitive are laid out at regular intervals over all or part of the area to be sprayed. After being sprayed, the cards are picked up and examined. If approximately the same number of insecticidal droplets are found on all the cards, the distribution is satisfactory. If more than the usual number are present on some or if there are no spots on some, uneven distribution is indicated. Serious misses should be corrected by another run over the skipped area, and any overdosage should be called to the attention of the pilot. Checking a large area in this manner is prohibitively time-consuming, but spot checks should be made on every unit.

Although these treated cards are very useful in indicating distribution of the spray, they do not indicate the amount of insect mortality that may be expected. This is because the amount of spray on the cards is not necessarily correlated with the amount of insecticide on the trees (Maksymiuk, 1963).

Determining the Effects

The general effect of the spraying treatment may be easily determined by the number of poisoned larvae that drop from the trees. But an accurate estimate of the number of insects surviving will require a more careful evaluation. An adequate number of sample counts should be made to determine the number of insects on a series of twigs or branches before spraying and again after spraying. The difference, of course, represents the number of larvae in the sample that were killed. This can then be expressed in either percent of the population killed or percent of the number surviving. A sample count of this sort should be made with care so that the samples examined before the spraying will be correctly comparable with the postspraying count. If the insects cannot be counted directly, the number of twigs infested before and after spraying or the amount of frass dropped before and after will indicate the effects of the spray.

Many projects are planned with foliage protection as their major objective. Estimates of numbers of larvae killed may be less important than the appearance of the foliage on the infested trees; in such cases, systematic visual or photographic evaluations are essential.

Estimating Costs

The cost of control varies with every operation. For small areas or units of moderate size, the cost per acre of flying is usually far greater than it is for large operations. Similarly, when materials can be purchased ready mixed in tank cars or tank trucks, they are often much cheaper than when purchased in small lots and mixed by hand on the job.

The three chief items that must be included in an estimate of spraying costs are (1) application, (2) materials, and (3) overhead. The first two will depend largely on the level of wages, the quantity of materials used, and market conditions. The third will depend chiefly on the size of the project.

Nearly as many supervisory people are required on a project of moderate size consisting of a few hundred or a thousand acres as would be needed for 100,000 acres. Thus, overhead on a very large project might be only a few cents per acre; whereas overhead on a small project could easily rise to $1 per acre if accidents or bad weather caused delays. From the realistic viewpoint, therefore, the forester is often compelled to cut the service and supervisory personnel to the minimum number required to handle the insecticide, load the plane, and establish any ground control in the field that may be needed. Weather observations are left to the pilot, and the checking of spray distribution and the making of mortality estimates are often done visually and almost casually. Inadequate supervision inevitably leads to carelessness and will account for some unsatisfactory results.

OTHER TECHNIQUES OF CHEMICAL CONTROL

The use of chemicals against insects is not confined to the dusting and spraying of trees in the forest. Other techniques for applying chemicals are in common use, especially where the insect pests are operating not in the forest, but in nurseries or in forest products.

Dipping

Nursery stock that has been dug for shipment and forest products of various kinds may be treated with dips and washes to control insect infestations. The insecticides are often the same as those used in the treatment of standing trees. In dipping nursery stock, the chemicals must be used in diluted form to avoid injury to the trees. Suspensions, aqueous solutions, or oil emulsions, diluted as for conventional spraying, are the common dipping materials. On wood products, more concentrated materials are suitable. Chemicals in various solutions that are too toxic for treating plants may be applied to processed wood without causing injury.

Insecticidal dips and washes must be selected carefully to suit conditions. For example, wood that is to be painted or varnished should never be treated with creosote or Carbolineum, even though these substances are both excellent insecticides. These same materials should not be used in cases where their color is objectionable. In contrast, oil solutions of such substances as pentachlorophenol will not affect either the paintability or the color of the wood. Dips for the treatment of living plants must be selected with equal care and used in dilutions that will kill the insects without harming the plants.

Poisoning the Soil

Some soil-inhabiting pests in nurseries, such as white grubs, wireworms, and other root eaters, may be killed by poisons mixed with the soil. The substances used for this purpose are usually insect powders or granules intended to make the soil uninhabitable for the pests for several years.

Usually, the poisons are applied to nursery soils in the course of preparation for seed or transplant beds; but occasionally, they may be drilled between the seedling rows in established beds. The concentrated types of soil poisons should be diluted before application by thoroughly mixing the required quantity with a bulky fertilizer. This will increase the total volume of material to be applied and will make the even spreading of the insecticide easier. After the poison has been spread evenly or drilled into the prepared soil, a light cultivation will mix it into the top few inches of soil.

Fumigating

Fumigation is usually a technique that is applied in enclosed spaces; occasionally, it may be used to kill insects in wood. Because a fumigant must vaporize readily at room temperatures, the number of useful insecticides of this type are limited. Generally, compounds that boil at or below room temperature are most useful. Examples are methyl bromide and ethylene dioxide. Many of these chemicals are odorless and should be used with an additive to warn people that the gas is present. In soil fumigation, some of the other compounds that boil at higher temperatures have also proved effective. Examples are lindane, trichlorobenzene, and orthodichlorobenzene. Many of the contact insecticides exhibit a limited amount of fumigant activity (Mallis, 1969).

Fumigation is used for many purposes, but perhaps the greatest use in forestry is in nursery practice. Fumigation of the soil with methyl bromide under a polyethylene covering is standard practice in many forest-tree nurseries as a preliminary to sowing seedbeds. The treatment controls both insects and fungi infesting the soil. A fumigation chamber may be used to treat nursery stock prior to shipment to the field. The method is effective, simple, and inexpensive, and it assures the land manager that the seedlings purchased are free from insects.

Treating Seasoned Wood

Wood preservative treatments are generally designed to prevent fungous attack and the resultant decay (Gjovik and Baechler, 1977). The same substances will also protect the wood against insects by killing any that are present and by keeping the wood free from subsequent attack. Until recently, most wood preservation was intended to prevent rot in wood contacting the ground, but the addition of synthetic organic compounds includes protection from insects.

Wood in buildings can be made permanently immune to insects by treatment with organic compounds Furthermore, the wood thus treated can be painted or varnished without danger of peeling or discoloration of the finish.

Treatment of wood with chemicals is accomplished by methods that vary from simple surface application by brushing or spraying to a complex system of impregnation by application of vacuum and pressure. The latter are available only in wood preservation plants. Insect control is usually accomplished either by surface treatment or by dipping. To ensure success, all surfaces must be treated; and if only the surface layers of the wood are penetrated, any cuts made through the treated surface in shaping and fitting must be treated before the piece is placed in position.

Sterilization

One of the techniques that has been successful for control of a few very specific problems is the sterilization method. This has not been used specifically for control of forest insects, but it deserves comment here because of its potential (Knipling, 1959, 1960; Hall et al., 1963). The discussion of sterilization may seem out of place in this chapter on chemical control, especially when one learns that the sterilization is often achieved through radiation of the insects. However, insects may also be sterilized by chemicals; thus, we have chosen to include this short discussion here, rather than in Chapter 9, "Biological Control."

Probably the greatest success story concerns the use of the sterile male technique in eradication of the screwworm, *Cochliomyia hominivorax,* from the island of Curaçao and from portions of the southeastern United States. The method, in brief, involves the release of large numbers of sterile males into the native population. In the successful projects against the screwworm, the males were sterilized by radiation. However, for some purposes, such sterilization is not practical; and for some of these, the use of chemical sterilants shows promise (Metcalf and Luckmann, 1975).

SIDE EFFECTS OF INSECTICIDES

No insecticide can be applied in the forest so that its effects will be entirely restricted to the pest against which it is directed. Inevitably, some associated organisms will be injured. When we use poisons against noxious insects, our objective should be to minimize injury to desirable forms of life as far as possible.

Most people were completely unaware of the serious problems associated with the widespread use of pesticides until the publication of Rachel Carson's book *Silent Spring* in 1962. This very readable book on an important subject was a best seller in the United States. Its effect generally has been good because it has resulted in an increased awareness of the need for research on chemical effects and for care in the handling and use of pesticides. All Americans can be thankful that the book has not resulted in widespread banning of *essential chemicals.* Fortunately, many of the evils referred to in *Silent Spring* have been corrected, and other safeguards are being considered. This is not the place to discuss the many facets of the insecticidal use problem, but the reader is advised to study this problem. Rudd (1964) published *Pesticides in the Living Landscape,* which was widely accepted by scientists as a more objective statement. Many of the early writings that followed discuss the insecticide problem in relation to DDT. This unfortunately detracts from our current work because we do not use DDT in the forest and our attention should focus on today's materials.

Monitoring

Monitoring and research on pesticide effects have become an important aspect of all insecticide development and a continuing part of control efforts. The

National Academy of Sciences (Anon., 1975) clearly expresses the need for efforts in this area. There are difficulties, and the costs are significant; no one expects that all organisms can be monitored. The report suggests the use of seven guidelines for selection of organisms for observation and measurement.

1   Humans are an obvious candidate and are continually monitored.
2   Species from a variety of taxonomic groups should be monitored.
3   Species should be from a variety of trophic levels: primary producers, herbivores, carnivores, omnivores, and decomposers.
4   Individuals should be from a variety of geographic locations representing a variety of land and water uses by humans and by other organisms.
5   The selected individuals should represent a variety of ages and physical conditions and both sexes.
6   The distribution of residues within individuals of a species should be kept clearly identified to avoid loss of information.
7   Endangered species should not be killed for monitoring purposes.

Monitoring efforts contribute significantly to the overall research effort. The present system of monitoring is designed to protect our environment by providing answers to many questions that relate to pesticide usage. Normal research efforts cannot provide the broad coverage of organisms and locations that is achieved through monitoring.

The Forest Situation

Side effects should be understood and evaluated in relationship to the overall benefits and losses in the forest environment. The development of pest management using all our best efforts to keep damage to the forest as low as possible by silvicultural and biotic means will be our best means of preventing serious side effects caused by excessive usage of chemicals. The only positive way to assure that the chemical usage we will need has a minimal detrimental effect to the air, water, and soil in and close to our forests is through research and careful monitoring. We know there will be some losses of beneficial insects when chemicals are used, but we must be concerned with effects on other organisms. Information obtained from research required for registration of chemicals and from monitoring of organisms is essential for the manager to make decisions on control and to balance benefits of alternative procedures in pest management.

LIMITATIONS OF CHEMICAL CONTROL

Control of forest insects by means of chemicals has a strong popular appeal, partly because some of its fundamental limitations are not immediately apparent. Nevertheless, these limitations must be recognized if the forester is to approach insect problems intelligently.

Temporary Character of Treatment

One of the more conspicuous limitations of chemical control is the temporary character of its effects. The treatment does not improve forest conditions in respect to insect susceptibility. A few weeks after treatment, the trees are as subject to new infestations as they were before. Therefore, we have no guarantee that the application of chemicals will not be required again the next year or within a few years; the same causal conditions that gave rise to the original outbreak are apt to be still present. If we are to rely on chemical control, we must recognize the probability that frequent treatments will be required during a tree rotation.

The exception will be a situation in which a certain age-class of trees is attacked by a pest immediately prior to the time when such trees will grow out of a susceptible condition. For instance, serious injury by the Saratoga spittlebug, *Aphrophora saratogensis,* is limited to that period after the pines have reached a height of 3 to 4 ft (1 m) but before the crowns close in and kill the ground cover. Spraying immediately prior to closure, therefore, would be likely to carry the trees through the few remaining years of high susceptibility.

A favorite criticism leveled against chemical control projects is that the effort may save trees but prolong the outbreak. Such an accusation cannot be discounted, but managers are usually upset by the inference because of the value of the resource protected. The manager must always make a decision regarding the resource, and risk of further loss is a part of the process. The question of how much one should invest in a forest that is susceptible has not been satisfactorily answered. Insecticides are only one tactic, and we must not become overly dependent on them (Luck et al., 1977).

Resistance of Certain Insect Races

Repeated treatment of an insect population with an insecticidal material is likely to result in developing resistant strains that cannot be controlled by the poison. For example, the San Jose scale, *Quadraspidiotus perniciosus,* a serious pest of fruit trees, developed a marked resistance to lime-sulfur sprays that formerly had produced good control. Similarly the codling moth, *C. pomonella,* has developed great resistance to arsenical sprays, so that new insecticides for this species have been required in some localities. This phenomenon produced in some strains of the house fly, *Musca domestica,* a great resistance to DDT, a characteristic which continued through many generations. The same is true of mosquitoes and other insect pests that receive repeated treatments. A variety of insects have developed similar resistance to many other chemicals.

Presumably, this building up of resistance is the result of selection by the insecticide, susceptible individuals being killed and nonsusceptible individuals surviving. The progeny of the survivors are also likely to be nonsusceptible, so that after several repeated applications of the same insecticide, resistant races appear. This phenomenon has been observed repeatedly, and entomologists recommend that different insecticides be used in successive treatments in order to avoid developing resistant strains. Unfortunately, the development of resistance to one of a group of chemicals may result in resistance to other related chemicals. Insecticides are seldom applied in the same place year after year in forests. Thus, resistance is much less likely among forest insects than in other situations.

A Realistic Viewpoint

Insecticides provide the forester with a powerful weapon against forest insects. If intelligently and conservatively used, they can save, for human purposes, millions of trees that otherwise might be destroyed by insects. On the other hand, when carelessly or unwisely used, insecticides have the potentialities for great injury to valuable organisms and the serious disruption of the biotic balance.

For these reasons, *insecticides must be used with caution and only when and where they are required to save losses that otherwise could not be avoided.* The promiscuous application of insecticides to forests cannot be justified.

BIBLIOGRAPHY

Akesson, N. B., and Yates, W. E. 1964. Problems relating to the application of agricultural chemicals and resulting drift residues. *Ann. Review Ent.* 9:285–318.

Anon., 1969. Insect-pest management and control. In *Principles of plant and animal pest control.* vol. 3. Washington, D.C.: National Academy of Sciences.

Anon., 1975. Pest control: An assessment of present and alternative technologies. Vol. I, *Contemporary pest control practice and prospects.* Washington, D.C.: National Academy of Sciences.

Anon., 1976. The federal insecticide, fungicide, and rodenticide act, as amended. Reprint. Washington, D.C.: Environmental Protection Agency.

Armstrong, J. A. 1975. Meteorological influences. In *Aerial control of forest insects in Canada,* ed. M. L. Prebble, pp. 56–58. Ottawa: Department of Environment.

Balch, R. E., et al. 1956. The use of aircraft in forest insect control, *Forestry Abstr.* Leading Article Series 23, *Forestry Abstr.* 16, no. 4, 1955; 17, nos. 1 and 2, 1956.

Carson, Rachel, 1962. *Silent Spring.* Boston: Houghton Mifflin Company.

Caswell, R. L., ed. 1977. *ENTOMA: Pesticide Handbook,* 27th ed. Entomological Society of America, College Park, Md.

Gjovik, L. R., and Baechler, R. H. 1977. Selection, production, procurement and use of preservative treated wood, supplementing Federal specification TT-W-571. *USDA, For. Serv., General Tech. Rept.* FPL-15.

Gould, R. F. (ed). 1966. Natural pest control agents. *Advan. Chem. Ser.* no. 53. American Chemical Society, Washington, D.C.

Hall, S. A., et al. 1963. New approaches to pest control and eradication. *Advan. Chem. Ser.* no. 41. American Chemical Society, Washington, D.C.

Irland, L. C. 1977. Maine's spruce budworm program: Moving toward integrated management. *J. Forestry* 75:774–777.

Knipling, E. F. 1959. Sterile-male method of population control. *Science* 130:902–904.

———. 1960. Use of insects for their own destruction. *J. Econ. Entomol.* 53:415–420.

Luck, R. R., et al. 1977. Chemical insect control: A troubled pest management strategy. *Bio-Science* 27:606–611.

Maksymiuk, B. 1963. Spray deposit on oil-sensitive cards and spruce budworm mortality. *J. Econ. Entomol.* 56:465–467.

Mallis, A. 1969. *Handbook of pest control.* New York: MacNair-Dorland Company.

Massey, C. L., et al. 1953. Chemical control of the Engelmann spruce beetle in Colorado. *J. Econ. Entomol.* 46:952–955.

Metcalf, C. L., et al. 1962. *Destructive and useful insects,* 4th ed. New York: McGraw-Hill Book Company.

Metcalf, R. L., and Luckmann, W. H., eds. 1975. *Introduction to insect pest management.* New York: John Wiley & Sons.

Pfadt, R. E., 1962. *Fundamentals of applied entomology.* New York: The Macmillan Company.

Prebble, M. L., ed. 1975. *Aerial control of forest insects in Canada.* Ottawa: Department of Environment.

Randall, A. P. 1975. Application technology. In *Aerial control of forest insects in Canada.* ed. M. L. Prebble. pp. 34–55. Ottawa: Department of Environment.

Rudd, R. L. 1964. *Pesticides and the living landscape.* Madison: University of Wisconsin.

Stoeckeler, J. H., and Jones, G. W. 1957. Forest nursery practice in the Lake States *USDA, Forest Serv., Agr. Handbook* no. 110.

Trial, H., Jr., and Thurston, A. S. 1978. Spruce budworm in Maine, The 1978 cooperative budworm suppression project and expected infestation conditions for 1979. *Maine Department of Conservation, Entomology Division, Tech. Rept.* no. 8.

Yeomans, A. H. 1950. Wind-borne aerosols. *U.S. Dept. Agr. Bur. Entomol. P. Q.* ET-282.

Yuill, J. S., and Eaton, C. B. 1949. The airplane in forest-pest control. *USDA Yearbook,* pp. 471–476.

Chapter 12

Leaf-eating Insects (Lepidoptera)

The foliage of trees furnishes food for a host of insect species, many of them dangerous forest pests, but the majority of the leaf-eating insects usually occur in comparatively small numbers. When they are not numerous, they scarcely affect the welfare of the trees because thrifty trees always have a greater amount of foliage than is actually required for their maintenance. However, when a leaf-eating species multiplies so rapidly that it outstrips environmental resistance, then it may attain tremendous numbers and so create a menace. The resultant defoliation may seriously reduce growth or even kill the trees attacked.

This is the first of several chapters in which various forest insects will be discussed. There is no intention to provide a listing of the many insects that may be found on trees nor of all those that may at times cause problems. There are manuals and references that may be of help to the forester who wishes to identify specific problems, and forest entomologists are available for consultation. There are many references to problems of particular regions, insect groups, damage classifications, and so on; these are too numerous to list. Three of significance that relate rather broadly are *Insects that Feed on Trees and*

Shrubs (Johnson and Lyon, 1976), "Eastern Forest Insects" (Baker, 1972) and "Western Forest Insects" (Furniss and Carolin, 1977).

DEFOLIATION

Sometimes outbreaks of defoliators arise with amazing suddenness. For instance, epidemics of the spruce budworm, *Choristoneura fumiferana,* the larch sawfly, *Pristiphora erichsonii,* the Douglas-fir tussock moth, *Orgyia pseudotsugata,* the forest tent caterpillar, *Malacosoma disstria,* and the pine sawflies, *Neodiprion* spp., have appeared to erupt over wide areas in the same year. Actually, in most instances, the insect numbers had been building up unnoticed over several seasons. Failure to observe increasing injury caused by growing numbers of insects produces the impression that the outbreak has developed within one season.

Failure to observe defoliation is not surprising because in moderate amounts it is inconspicuous. If it were evenly distributed over the crown of a tree, defoliation would have to be from 50 to 75 percent before the tree would appear abnormal. Even the concentrated defoliation of gregarious leaf eaters escapes the attention of the casual observer until the injury is severe.

The only way to be prepared to control defoliator outbreaks is to anticipate them at least a season in advance. This can be done by annual observations on the status of potentially dangerous pests. If these observations are made correctly, they will sound a warning of approaching danger before they begin to record the first evidences of actual injury. They will reveal disturbances favoring the increase of defoliators. They will also record any rare flights from localities that hold potentialities of danger into areas in which highly favorable conditions prevail.

Effects of Defoliation

Defoliation (Fig. 12-1) injures trees by reducing photosynthesis, by interfering with transpiration, and by interfering with the processes of translocation of food within the tree. A combination of these effects is reflected in the rate of growth. Defoliation has such a profound effect on growth that it is possible to trace the history of past outbreaks of leaf-eating insects by studying the annual rings of the surviving trees.

Kulman (1971) reviewed the various studies and reports on effects of defoliation on growth and mortality of trees. The general effects of defoliation result in variable responses in different species of trees and trees growing under varying physical conditions. Defoliation may result in incomplete or even missing growth rings, a condition that leads to additional problems in evaluating losses. These will vary greatly both individually and by species in reaction to defoliation. Conifers will often recover gradually as a full crown develops over a period of several years. Deciduous trees may recover very rapidly, so that a year after defoliation, the tree may exhibit normal growth.

Figure 12-1 Ponderosa pine defoliated by the pine butterfly, *Neophasia menapia,* Boise National Forest, Idaho. *(U.S. Forest Service photo by L. W. Orr.)*

In an evaluation of the effects of defoliation on the basis of the pattern of annual rings, care must be exercised to ensure correct interpretation. For example, on conifers such as firs, spruces, and hemlocks, which hold their needles for a number of years, defoliators that prefer the new growth may destroy nearly all the foliage of the current year without materially affecting the growth pattern at the base of the tree. Not until one or two sets of young leaves have been lost will the loss of photosynthetic tissue severely affect the trees. Thus, in the case of a spruce budworm, *C. fumiferana,* outbreak, the wider basal ring is the one laid down during the second or third year after severe feeding on the new needles begins.

Another point that should not be overlooked is the fact that the growth rate in a stand of trees is affected by many factors. Competition among trees, their relative position in reference to microtopography or macrotopography, the genetic constitution of the individual tree, the soil nutrients available, and the precipitation during the year are among the many influences on the growth rate of each individual tree or stand. As a result, the growth pattern is relative rather than absolute. Therefore, attempts to express the growth pattern as the average of a series of absolute measurements can lead to confusion. Much more remains to be learned about how and why trees grow as they do. Nevertheless, the relative growth and the resultant patterns can be observed and defoliation effects evaluated approximately.

All species of trees are not equally susceptible to injury from defoliation. Hardwoods are relatively resistant and ordinarily will successfully withstand 3 or more years of defoliation. This is a consequence of their relatively large supply of stored food and their ability to replace the destroyed foliage following defoliation. However, two defoliations during the same season may be disastrous (Giese et al., 1964). Also, dieback may occur several years after a major defoliation as secondary agents become established and flourish on weakened trees. To facilitate timely salvage, hardwood stands should be watched for signs of deterioration.

Great resistance to injury from defoliation is characteristic of deciduous conifers, as exemplified by the larch. Evergreen conifers, such as balsam fir or spruce, are more easily killed; for instance, white spruce, hemlock, and probably many others will die as a result of a single complete defoliation.

The time of year when defoliation occurs also has a material influence on the effects of defoliation. For example, if complete defoliation of a conifer, such as hemlock, fir, or spruce, occurs before midsummer, the trees will not have formed buds for the following year Then a single defoliation of 100 percent can kill the trees. On the other hand, if complete defoliation occurs late in the season, after buds for the next year have been formed, a new crop of leaves will be produced the following spring, and even conifers will survive. However, we should not develop an attitude which suggests that the late-season defoliators are harmless. Allen[1] stated that substantial dieback has occurred in some northern hardwood stands attacked several years earlier by the saddled prominent, *Heterocampa guttivitta.*

Resistance to defoliation injury varies not only with the species but also with different individuals of the same species. Dominant trees and those growing in the open without competition are less affected by defoliation than are those growing under less favorable conditions. Those trees that dominate their location have larger reserves of stored food than those growing in an inferior position. Many times, secondary pests such as bark beetles, borers, or fungi actually kill the trees. The dominant individuals may have more resistance to these secondary problems, and those growing in the open may be unattractive for some other reason.

Direct cause of death is difficult to assess because of the complex of secondary problems and the general condition of the host tree prior to defoliation. The determination of the causal agent may be important to the researcher, but to the forester, the cause is less important than the effect. If the tree were eventually killed by secondary pests, the defoliator would still remain

[1]Personal communication from Dr. D. C. Allen, State University of New York, November 7, 1978.

the primary problem for the forest manager. The management of the forest requires close attention to maintaining vigorous trees that will not attract the numerous secondary pest problems.

A defoliator may also be a secondary pest that becomes a serious problem when trees are under stress from other causes. This statement does not apply to all defoliators, but there is clear evidence for some species (Kulman, 1971). An example might be one of the leafrollers associated with maple mortality (Giese et al., 1964).

Types of Insect Defoliators

Generally speaking, defoliating insects may be separated into groups, according to their habits.

1 Leafminers   These feed between the epidermal layers and consume the chlorophyll-bearing tissues, thus leaving the epidermis (Fig. 12-2a).

2 Skeletonizers   These insects remove the chlorophyll-bearing tissues plus the epidermis from one side of the leaf, leaving one epidermal layer and the veins intact (Fig. 12-2b). Some species remove both the upper and lower epidermis, which results in a true skeletonlike appearance.

3 Whole-leaf Feeders   These insects feed on the margins of the leaves and consume or remove all the leaf tissues.

Figure 12-2a   Blotch-type leaf mine on sugar maple. *(Photo by D. C. Allen.)*

Figure 12-2b Skeletonized leaf of yellow birch. *(Photo by D.C. Allen.)*

4 Shot-hole Feeders Larvae consume all leaf tissues but do not feed from the leaf margins. Instead, they chew holes in the leaves (Fig. 12-3).

Some defoliators are miners during a part of their developmental period and skeletonizers at a later time. Others may be skeletonizers during their early stages and leaf feeders during later stages. But there are many species that belong to only one of these classes. Regardless of the manner of their work,

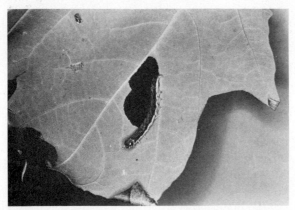

Figure 12-3 Shot-hole–type feeding on maple leaf. *(Photo by D. C. Allen.)*

all the defoliators have essentially the same effect on the life processes of the tree. The severity of the injury is directly proportional to the amount of chlorophyll-bearing tissues destroyed.

The manner of feeding is only one of several characteristics that may distinguish insects on trees. One can also compare those that feed in a solitary manner with those that feed as a colony. Some species may also feed in groups in the early instars and become more independent in the late instars. An example is the orangehumped mapleworm, *Symerista leucitys* (Fig. 12-4).

Space will not permit even a brief discussion of all the destructive defoliators. In fact, merely to mention them all would fill a volume. Therefore, only a few typical examples of the different groups will be discussed here. The species chosen for discussion will serve to illustrate the most serious types of injury and the ways in which the forester may tackle defoliator problems.

CONIFEROUS LEAF EATERS

Some of the more serious defoliating insects are pests of coniferous trees. The most costly of these in terms of tree mortality and expense of aerial spray projects has been the spruce budworm, *C. fumiferana.*

Budworms

Several species of coniferous-feeding budworms were described in early years as the spruce budworm. These have more recently been separated into distinct species by Canadian scientists (Freeman et al., 1967). Some of the forms

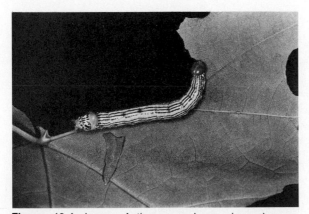

Figure 12-4 Larva of the orangehumped mapleworm, *Symerista leucitys*. The orangehumped mapleworm is a whole-leaf feeder in its late instars. *(Photo by D. C. Allen.)*

described by Freeman as the spruce-fir–Douglas-fir feeders and the pine feeders are listed in Table 12-1.

The spruce budworms have been selected for a large cooperative international research project between the governments of the United States and Canada known as the *CANUSA* program. A cooperative agreement to do research on both the western and eastern forms was signed in 1977. The work will continue for at least six years, commencing in the 1978 field season. Research already accomplished on the spruce budworm represents far more than that on any other native pest of forests in North America. The need for better methods for management of these pests has resulted in the decision to begin a large-scale effort at this time.

Spruce Budworm, *Choristoneura fumiferana*

One of the outstanding defoliators of fir and spruce in our northern forests is the spruce budworm, *C. fumiferana* (Fig. 12-5). This is a native insect,

Table 12-1   Some Budworm Species of North America (Lepidoptera: Tortricidae), Genus *Choristoneura*

Species	Major hosts	Range	Comments
Spruce budworm, *Choristoneura fumiferana*	Balsam fir, spruces	Virginia to Labrador west to Prairies and the Yukon	Sympatric with *C. pinus pinus*
Western spruce budworm, *C. occidentalis*	Douglas-fir, white fir, silver fir	New Mexico to southern British Columbia	At lower elevations than *C. biennis*
C. biennis	Subalpine fir, white spruce, Englemann spruce	British Columbia Alberta, and the Yukon	Two-year life cycle
C. orae	Silver fir, Sitka spruce	British Columbia, Pacific Coast	Coastal distribution
C. viridis	White fir, grand fir	Oregon and northern California	Body of larva is green
Jack pine budworm, *C. pinus pinus*	*Pinus* spp. esp. jack pine, red pine	Lake States, Saskatchewan to Nova Scotia	Sympatric with *C. fumiferana*
C. pinus maritima	Virginia pine, pitch pine	Massachusetts to Kentucky	Larger than *C. pinus pinus*
C. lambertiana	Ponderosa pine, limber pine, lodgepole pine	Interior mountain regions of the West into Canada	Species separations not fully completed
C. subretiniana	White fir, lodgepole pine, Jeffrey pine	California and Nevada, full range not described	Generally larger than *C. lambertiana*

Figure 12-5 *(a)* Egg mass of the spruce budworm, *Choris-toneura fumiferana,* on needle of balsam fir; *(b)* larva of the spruce budworm feeding on foliage of balsam fir. *(Photo by D. C. Allen.)*

distributed throughout the range of its host trees. For years it may remain innocuous, a rare and inconspicuous resident of the forest; but when conditions are right, it multiplies prolifically and spreads rapidly through the range of its host trees.

The life history of the spruce budworm, *C. fumiferana,* exhibits a number of interesting features. The eggs are deposited during late summer in elongate clusters on the needles of the host trees. These clusters are green and contain from 10 to 30 or more flattened eggs, overlapping one another like shingles. The young larvae that soon hatch from these eggs seek suitable places of concealment on the tree, spin a light covering of silk about themselves, molt once, and go into hibernation. The silken hibernating cases are called *hibernacula.* In the spring, about the time that the buds of fir are expanding, the larvae emerge from hibernation and begin to feed. At first, they mine the old needles or, in the case of the black spruce, the unopened buds. If staminate flowers are available, the young larvae prefer them to other food. Later, the larvae are leaf chewers, eating the foliage of the current year. As they work, they web the needles together to form a crude shelter. The larvae develop rapidly and, under most favorable conditions, are full grown in the course of 3 weeks.

During the early instars, the larvae are a pale yellowish green with black heads and thoracic shields. Later, they become much darker until the general color is brown with black markings. They then transform to the pupal stage on the trees. No cocoon is spun; instead, the pupal stage is passed within the web spun about the twig tip by the larva while feeding. The adult moths emerge in July and early August and soon thereafter mate and deposit their eggs. The moths have a wingspread of nearly ¾ in (18 mm) and vary in color from gray to copper.

The effects caused by the spruce budworm, *C. fumiferana,* in eastern North America on the forests of balsam fir and spruce depend on the duration and intensity of defoliation and vary greatly from forest type to forest type and from outbreak to outbreak. For example, during the outbreak that occurred between 1909 and 1919 in eastern Canada, the injury in many localities was much more severe than that reported by Macdonald (1962–1963) for an outbreak in New Brunswick. In central and western Quebec, where balsam fir was the dominant species in 1909, the mortality of balsam fir was well over 95 percent, whereas in other localities, where fir was in a more subsidiary position in mixture with hardwoods, the mortality was less than 50 percent.

Macdonald presented data on balsam fir mortality in New Brunswick between 1952 and 1961 on representative areas selected for intensive study. In an unsprayed area reserved for comparison with a surrounding sprayed forest, the mortality in large- and small-diameter classes was "very severe," but in the diameter classes ranging between 4 and 7 in (10–17.5 cm) mortality was only about 50 percent. In neither outbreak did excessive mortality in spruce occur directly as a result of defoliation, although windthrow following defoliation was locally high.

Direct control of the spruce budworm, *C. fumiferana,* may be accomplished by treatment of vast areas in aerial spray projects. The objective has been to prevent tree mortality by treatment of trees when budworm damage is severe. The treatments have generally been successful in reaching this objective, but the conditions leading to the outbreaks have generally not been changed. It has been suggested that spraying may have prolonged outbreaks. The correct ecological interpretation relates to the survival of the trees. The prevention of mortality may provide green foliage that allows the budworm to continue to exist in large numbers without starvation. The decision to spray must be based on the quantity of timber that would be lost without treatment and the effect those losses might have on the general economy of the region, rather than on an idea that spraying will eliminate the budworm.

Discussions such as these are not based on precise data, and managers need more information on which to base decisions. This has resulted in several studies in which the computer has been utilized in efforts to simulate populations (Stedinger, 1977) so that long-range effects of decisions may be better understood. There are other techniques that show considerable promise for the future. Sanders and Weatherston (1976) have continued to refine their work on the sex pheromone of the eastern spruce budworm. This information will be of immediate value in the monitoring of spruce budworm populations.

Forest entomologists have studied the spruce budworm, *C. fumiferana,* intensively, and recommendations for its control by silvicultural methods have been made. If in eastern Canada or the northeastern part of the United States the host trees of the budworm were to grow in pure stands, it is probable that the budworm would never become epidemic on any species other than balsam fir. This association of budworm with balsam fir is the result of close synchronization of budworm habits with the phenology of its favored host. The emergence of the overwintering larvae is synchronized with the expansion of balsam fir buds. The balsam, then, provides suitable food at the right times; whereas the later-developing black and red spruces furnish less food of poorer quality. In them, the larva must mine the needles or unopened buds until fresh expanding shoots are available. White spruce, like balsam fir, is synchronized well with emergence of the larvae, but the needles harden quickly and therefore become undesirable food before the budworm completes development.

This close connection between budworm activities and balsam fir is reflected in the effects of an outbreak on a mixed forest.

Table 12-2 shows that in outbreaks in Canada, the balsam fir died within 4 years after defoliation became severe; whereas red spruce died much later. This situation reflects the fact that defoliation on spruce is not so severe as on

Table 12-2   Tree Mortality from a Budworm Outbreak

Years after excessive defoliation	Percent dying each year	
	Red spruce	Balsam fir
1	0	9
2	0	32
3	0	29
4	0	18
5	0	0
6	8	0
7	12	0
8	15	0
9	25	0
10	10	0
Total	70	88

balsam fir and also that the spruce continues to die from secondary effects after the outbreak ends. These data are the result of studies by a number of Canadian workers in New Brunswick and Quebec and indicate what may be expected from a very severe outbreak in a mixed stand of balsam fir and red spruce.

Young, vigorous stands of balsam fir are almost never severely damaged by budworms. This is partly due to the vigor of the trees but may also be due to the less favorable conditions for the budworms in young stands. One of the probable conditions influencing the susceptibility of different ages is the presence or absence of staminate flowers. Bess et al. (1947) have observed that early-stage larvae are larger than usual when they feed on pollen-bearing buds. These flowers are most abundant on overmature, dominant trees and those that are growing under unfavorable conditions. In contrast, they are least numerous on vigorous, fast-growing individuals. Undoubtedly, outbreaks of the spruce budworm, *C. fumiferana,* occur most frequently in stands that contain a large proportion of stamen-bearing trees.

The control of the budworm by silviculture is based on a knowledge of its life history and the correlation of its habits with the phenological and other characteristics of the host trees. The principles involved were covered in Chapter 10 and need not be repeated. If we know the conditions that are conducive to outbreaks and those that are not, we can adopt those practices that will avoid dangerous conditions and encourage safe ones. Foresters must recognize, of course, that only those practices that can be justified from an economic standpoint will be practical on a large scale.

From an examination of this summary, the following practices discussed in Chapter 10 suggest themselves:

1 Conducting logging operations on small, scattered units of not more than 40 acres (16 ha).

Conditions Favorable and Unfavorable for Spruce Budworm Outbreaks

Conducive to outbreaks	Not conducive to outbreaks
1. Slow-growing or mature stands of pure balsam fir	1. Thrifty sapling and seedling stands of pure balsam fir without balsam over-story
2. Mature or nearly mature stands of mixed fir and spruce	2. Thrifty sapling or small-pole stands of mixed fir and spruce
3. Mature balsam fir mixed with smaller spruce. The higher the percentage of balsam, the greater the hazard	3. Fir mixed with spruce and overtopping hardwoods
4. Dominant fir, predominant in number, mixed with hardwoods, with or without spruce	4. Dominant fir, subsidiary in number, mixed with spruce and hardwoods
5. Large contiguous areas of any of above especially conducive to outbreaks	5. Areas containing balsam broken into small units so arranged that those in the same age-classes are separated from one another
6. Very dense, stagnated stands of pole-size trees	

2 Favoring spruce over balsam fir and encouraging mixture with intolerant hardwoods by summer logging or scarification.

3 Marketing balsam fir as soon as it reaches commercial maturity and marketing spruce on a longer rotation.

4 Planting spruce following logging where natural regeneration of that species fails.

Eradication of balsam fir to control budworm has frequently been suggested. Even if this undertaking could be accomplished, it would be an economically unsound procedure. Balsam fir is a valuable tree that can be grown on a short rotation. It almost always finds a ready market. It reproduces well and grows rapidly. Therefore, efforts should be directed toward its continued production in reasonable quantities, but under conditions that are not conducive to outbreaks.

Western Spruce Budworm, *Choristoneura occidentalis*

In the West, there is a complex of species that feed on spruces, firs, and Douglas-fir. There have been outbreaks in the Rocky Mountains in which the budworm has been active on many occasions. The general biology of the budworm in the West is similar to that of the eastern species, though there are differences.

Freeman (1967) pointed out many of the similarities and differences between the budworms. Sanders et al. (1974) found that the sex pheromone for *C. fumiferana, C. occidentalis,* and *C. biennis* are the same or very similar.

Control of the western spruce budworm, *C. occidentalis,* may be accomplished using methods and materials similar to those used in the East. Large-scale aerial spraying has been done, but not in recent years. Information on the biology and control of this insect was summarized by Carolin and Honing (1972). McKnight (1968) reviewed the literature to that date on the eastern, western, and two-year cycle budworms (*C. fumiferana, C. occidentalis,* and *C. biennis*).

Jack Pine Budworm, *Choristoneura pinus pinus*

The jack pine budworm, *C. pinus,* is the most well known of the several species that attack pines. Like most of the budworm species, this insect is similar morphologically to the others. The eastern spruce budworm is sympatric with this species, and most people would have difficulty distinguishing one from the other unless the host was known. The average adult specimen is more coppery in color than the grayish spruce budworm, *C. fumiferana.*

The life histories are nearly identical, though emergence of the various stages is later in the jack pine budworm, *C. pinus.* The species is apparently more dependent on pollen production for rapid increase in numbers. An abundance of staminate flowers frequently occurs on overmature, round-topped trees; on scattered orchard-type trees; or on suppressed trees. Thrifty, fully stocked stands seldom produce excessive quantities of staminate flowers.

Top-kill is common after jack pine budworm outbreaks, and tree mortality has been as high as one-third of the merchantable volume (Batzer and Millers, 1970). Because jack pine stands are partially killed, the remaining stands are understocked. The most severe damage often occurs in such stands. Thus, partially killed stands should be evaluated carefully because it may be wise to remove the remaining trees.

Silvicultural control of the jack pine budworm, therefore, calls for the following practices:

1 Growing hard pines in fully stocked stands or groups
2 Eliminating large-crowned wolf trees
3 Utilizing the trees before they become overmature
4 Encouraging species suited to the site

Douglas-fir Tussock Moth, *Orgyia pseudotsugata*

The Douglas-fir tussock moth, *O. pseudotsugata* (Fig. 12-6), is a destructive defoliator of Douglas-fir and true fir in the northern Rocky Mountains. Since 1918, when it was first discovered, it has periodically defoliated extensive areas. Outbreaks occur only in areas where Douglas-fir or true firs predominate. An especially extensive infestation occurred in 1946 and 1947 in Idaho and eastern Oregon, where it threatened to destroy valuable timber on more than 400,000 acres (around 162,000 ha). In 1947, DDT was applied from the air with outstanding success. A saving from this operation of $4,280,000 in timber at a cost of less than one-fifth that amount was recorded.

A more recent serious outbreak was detected in 1971 (Fig. 12-7). By 1972, this outbreak covered nearly 200,000 acres (81,000 ha) in eastern Oregon and

Figure 12-6 Larvae and adults of the Douglas-fir tussock moth, *Orgyia pseudotsugata: (a)* adult male; *(b)* adult female; *(c)* larva. *(U.S. Forest Service photo.)*

(a)

(b)

(c)

Figure 12-7 Defoliation by the Douglas-fir tussock moth, *O. pseudotsugata,* on *Abies concolor,* Modoc National Forest, 1965. *(U.S. Forest Service photo.)*

Washington. A special request to use DDT was denied by the Environmental Protection Agency (EPA), and in 1973, the infestation expanded to 800,000 acres (324,000 ha) and an estimated 850 million bd ft of timber had been killed. Special permission was granted by EPA to use DDT in 1974, when a total of 427,000 acres (around 173,000 ha) was treated (Mounts, 1976).

This experience with DDT was dramatic, but it made people realize that research was needed to provide other ways to control the pest. Thus, the Douglas-fir tussock moth, *O. pseudotsugata,* was selected for study in the USDA expanded research and development program (Brooks et al., 1979). This program had two objectives: (1) to implement methods that are presently available to reduce damage by tussock moth outbreaks and (2) to develop the new knowledge necessary to prevent or suppress future outbreaks (Anon., 1976; Campbell et al., 1978).

The moths are drab-colored, gray buff insects with a body length of ½ in (12 mm) or less; males are winged; females are smaller and wingless. Eggs are deposited in a mass, usually on the cocoon from which the female emerged, and pass the winter without hatching. The larvae feed through June and July, spinning their cocoons in early August and emerging as adults late in that month.

The larvae are strikingly marked insects, with red spots and tufts of cream-colored hair on the dorsum, two hornlike pencils of hair behind the head, and two longer, similar black pencils near the posterior end. The sides are marked with somewhat broken, narrow orange stripes. Additional details on the life cycle and habits may be read in Wickman et al. (1971). The expanded program has been successful in many ways, and there are now several materials registered for control of this insect including a virus material.

Hemlock Looper, *Lambdina fiscellaria fiscellaria*

The hemlock looper, *Lambdina fiscellaria fiscellaria* (Fig 12-8), and closely related species called by the same common name are probably the most serious forest insects belonging to the family Geometridae. This looper is capable of killing hemlocks in a single season by completely defoliating them in midsum

Figure 12-8 The hemlock looper, *Lambdina fiscellaria fiscellaria,* on foliage of eastern hemlock. *(Photo by D. C. Allen.)*

mer. In localities where hemlocks are of major economic value, the direct control of the looper becomes imperative.

A severe outbreak in the state of Washington in the early 1960s threatened to destroy a tremendous acreage of valuable hemlock but was controlled by aerial application of pesticides. Careful planning, supervision, and evaluation of results in terms both of pest control and of possible deleterious side effects have demonstrated that with due precautions, pesticides can be used to control forest insects without serious deleterious effects.

Other Lepidoptera on Conifers

Numerous other species of Lepidoptera feed on coniferous trees (Fig. 12-9). Some may be as important at times as the ones selected for examples in the preceding descriptions. The forester will find many more listed in the manuals mentioned earlier and will become familiar with some of them. A few are:

1 bagworm (Wollerman, 1971)
2 blackheaded budworms (Schmiege and Crosby, 1970; Miller, 1966)
3 larch casebearer (Denton and Tunnock, 1972; Baker, 1972)
4 lodgepole needleminer (Koerber and Struble, 1971; Struble, 1972)
5 pandora moth (Wygant, 1941)
6 pine butterfly (Cole, 1971)

(a)   *(b)*

Figure 12-9 Adults of two western coniferous defoliators; *(a)* the pine butterfly, *Neophasia menapia; (b)* the pandora moth, *Coloradia Pandora. (U.S. Forest Service photo.)*

    7   pine looper (Dewey, 1975)
    8   white fir needleminer (Washburn and McGregor, 1974)

DEFOLIATORS OF BROADLEAF TREES

Every hardwood tree has several species of leaf-feeding insects that usually are of very minor significance to the health of the tree. Publications such as Baker (1972), Furniss and Carolin (1977), and MacAloney and Ewan (1964) provide details on many of them. We will confine our comments to a few notable examples. The first is an introduced species that has been a serious problem since soon after its arrival in the United States.

Gypsy Moth

The gypsy moth, *Lymantria dispar,* is one of the many destructive insect pests that have come to America from foreign lands. It was accidentally introduced at Medford, Massachusetts, in 1869, and since that time has spread into all of the northeastern states. Scattered infestations have been located as far away as Wisconsin and California. It is particularly injurious to broadleaf trees, although conifers in mixture with hardwoods are by no means immune.

    The eggs are laid during July and August in clusters of 400 or more. Because the heavy-bodied female moth is unable to fly, the eggs are usually deposited near the site of pupation. The eggs hatch the following spring. The larvae in the first stage feed on the foliage of susceptible trees such as oak, basswoods, birches, and aspens; later, they may also feed on hemlocks, pines, and spruces. If the infestation is severe, the trees may be completely defoliated by the end of June. When full grown, the larvae pupate, either on the tree or in other convenient places; and after a pupal period of about 10 to 14 days, the adults emerge. The male moth is brown in color; whereas the female is white with black markings. Unlike the female, the male is a good flier.

    At first, it might seem that a species whose adult females were unable to

Figure 12-10 Lodgepole pine defoliated by the pandora moth, *C. pandora*, Ashley, N. F. Utah. *(U.S. Forest Service photo.)*

fly would have little chance of spreading from one locality to another and would, consequently, be confined to a restricted area and there very easily be controlled. Unfortunately, this has not proved to be the case. One of the important means of distribution for the gypsy moth, *L. dispar,* is wind. The first-stage larvae are clothed with long hairs that greatly increase the surface area of the body in proportion to its weight. The larvae at this stage are so light that they may be carried long distances by air currents. How far they can be blown in this way can only be a matter of conjecture, but it is known that they have been carried for a distance of more than 20 mi (32 km). However, such long-distance movement is not common; most individuals are carried only short distances. Other means of dispersal are of a more or less accidental character. For example, egg masses or larvae may be carried on persons or vehicles leaving the infested area, logs, mobile homes being moved, or quarry stones.

Campbell (1967) has provided an analysis of the causes of numerical changes in gypsy moth, *L. dispar,* populations. There is a complex of factors associated with the problem. He found that variation in the survival rate of large female larvae and female pupae was the largest sources of variation when population densities were high. Both small and large female larvae were important sources of variation in sparse populations. Disease, parasites, dispersion rates, late frosts, bird and mammal predation, weather, and site conditions all may have effects (Campbell, 1975).

Godwin (1972) has summarized information on the gypsy moth, *L. dispar*. He points out that the host tree complex is important and that management of stand composition can be used to reduce damage. Campbell et al. (1976) point out that artificial objects along the forest edges may be a factor in population increase because of the provision of larval resting and pupation locations. They found that artificial objects left along the forest edge in suburban areas contained about one-half of the egg masses at low densities.

Because of the differences in susceptibility of various tree species, the regulation of forest composition by the application of silvicultural principles promises to be a most effective method of protecting forests from the gypsy moth, *L. dispar*. If the most susceptible species are eliminated, a forest may be made comparatively safe. Much further work along this line is necessary before completely satisfactory methods can be developed.

Unfortunately, many forest lands in the northeast and central states are especially suited to oak, the favored food of the gypsy moth, *L. dispar*. Such lands cannot always be converted to other forest types. Studies of these oak types have shown that all of them are not equally subject to injury. On abused woodlands that have been repeatedly burned, heavily grazed, or trampled, and on areas where the stands are poorly stocked, outbreaks are frequent. On areas where desirable forest practices have resulted in good stocking, normal distribution of age-classes, and the development of a ground cover of young trees, outbreaks may be less frequent. On the lands where good forest practices have been applied, the predatory and parasitic insects are also far more effective than in the abused types (Bess et al., 1947). Houston and Valentine (1977) made comparisons of susceptibility of forest stands and developed methods for predicting where problems might arise.

The gypsy moth, *L. dispar,* has been one of those insects under study in the USDA's expanded research program (USDA, 1979). This program has resulted in further information on population regulation and has provided needed registered materials for use in direct control (Ketchum and Shea, 1977).

Elm Spanworm, *Ennomos subsignarius*

The elm spanworm, *Ennomos subsignarius,* sometimes called the snow-white linden moth, is a serious defoliator of forest hardwoods (Fig. 12-11). The species is widely distributed from Canada and the Atlantic states to Colorado. A severe outbreak of this species started about 1954 in the Great Smoky Mountain region and has spread eastward, involving more than 2 million acres in three states. Interestingly enough, this same area was involved in a similar outbreak of the same insects just prior to 1880, and since then the population

Figure 12-11 Eggs, larvae, pupae, and adults of the elm spanworm, *Ennomos subsignarius*. *(U.S. Forest Service photo.)*

had been very low until the 1950s. Although a vast area was involved, only about 7 percent of the infestation was truly serious.

Elm spanworms, *E. subsignarius,* feed on the foliage of a great variety of trees and shrubs (Fedde, 1971). Highly favored species are ash, hickory, and walnut, but nearly all hardwoods may be attacked. Mortality of host trees is usually caused by infestations of secondary insects in the weakened trees. The twolined chestnut borer, *Agrilus bilineatus,* is one of the insects commonly associated with outbreaks.

Tent Caterpillars

The tent caterpillars include several species of direct concern to the forester (Fig. 12-12). The best-known forest species is the forest tent caterpillar, *M. disstria;* this insect has been found throughout most of the United States and Canada wherever hardwoods may be found (Batzer and Morris, 1978). Three tent caterpillar species of importance are outlined in Table 12-3 (Stelzer, 1971; Anon., 1970; Furniss and Carolin, 1977).

The life cycles of tent caterpillars are similar, and the moths and larvae are also similar in size and appearance. Adults (wingspan 20–35 mm) appear in midsummer, fly at night, and are readily found around lights. Females deposit eggs in groups of 150 or more in bands around twigs or in flattened masses on limbs or boles of trees. Young larvae emerge at about the time new leaves appear in the spring. Young larvae are gregarious, and many species form substantial tents, where they rest when not feeding. Mature larvae are about 50 mm long (Figure 12-13).

Figure 12-12 Aspen defoliated by the Great Basin tent caterpillar, *Malacosoma californicum fragile.* This is one of several subspecies of the western tent caterpillar, *M. californicum.* Carson National Forest. *(U.S. Forest Service photo.)*

Table 12-3   Three Species of Tent Caterpillars, Genus *Malacosoma*

Species	Range	Hosts	Characteristics
Forest tent caterpillar, *Malacosoma disstria*	Throughout the United States and Canada	Sugar maple, aspens, oaks, sweetgum, cottonwood, elm, and others	Caterpillars do not spin a tent; larvae have pale bluish line on sides and series of keyhole shaped white spots on back.
Eastern tent caterpillar, *M. americanum*	East of Rocky Mountains into eastern Canada	Wild cherry and apple trees favored	Caterpillar has black, white, and blue markings, with white stripe along back.
Western tent caterpillar *M. californicum*	Rocky Mountains to the Cascade Range	Quaking aspen, cottonwoods, and willows favored	Caterpillar has pale blue stripe down middle of back bordered by black stripes interlaced with traces of orange.

Figure 12-13   Mature larva of the forest tent caterpillar, *M. disstria*. Note the characteristic keyhole-shaped spots along the back. *(Applied Forestry Research Inst. photo.)*

Other Lepidoptera on Broadleaf Trees

The number of occasionally destructive insect species on hardwood trees is very large. Some of them that the forester may encounter are as follows:

1   aspen blotchminer (Martin, 1956)
2   California oakworm (Wickman, 1971)
3   cankerworms (Jones, 1953; Miller and Allen, 1973)
4   fall webworm (Baker, 1972)
5   greenstriped mapleworm (Wilson, 1971a)
6   large aspen tortrix (Beckwith, 1973)
7   leaf rollers (Giese et al., 1964)
8   redhumped oakworm (Millers and Wallner, 1975)
9   saddled prominent (Miller et al., 1969)
10   satin moth (Reeks and Smith, 1956)
11   spearmarked black moth (Werner and Baker, 1977)
12   variable oakleaf caterpillar (Wilson, 1971b)
13   whitemarked tussock moth (Baker, 1971)
14   winter moth (Cuming, 1961)
15   oak leafroller, *Archips semiferanus* (Baker, 1972)

BIBLIOGRAPHY

Anon. 1970. Controlling the eastern tent caterpillar. *USDA, Home and Garden Bulletin* no. 178.

Anon. 1976. The tussock moth program. USDA Forest Service, Portland, Ore.

Baker, W. L. 1972. Eastern forest insects. *USDA Forest Service, Misc. Publ.* no. 1175.

Batzer, H. O., and Millers, I. 1970. Jack-pine budworm. *USDA Forest Service, Forest Pest Leaflet* no. 7.

————, and Morris, R. C. 1978. Forest tent caterpillar. *USDA Forest Service, Forest Pest Leaflet* no. 9.

Beckwith, R. C. 1973. The large aspen tortrix. *USDA Forest Service, Forest Pest Leaflet* no. 139.

Bess, H. A., Spurr, S. H., and Littlefield, E. W. 1947. Forest site conditions and the gypsy moth. *Harvard Forest Bull.* no. 22.

Brooks, M., Start, R., and Campbell, R. (eds). 1979. Douglas-fir tussock moth: a synthesis. *USDA, Forest Service Tech. Bull.* no. 1585.

Campbell, R. W. 1967. The analysis of numerical change in gypsy moth populations. *Forest Science,* Monograph 15.

————. 1975. The gypsy moth and its natural enemies. *USDA Forest Service, Agr. Info. Bull.* no. 381.

————, et al. 1976. Man's activities and subsequent gypsy moth egg-mass density along the forest edge. *Environ. Entomol.* 5:273–276.

————, et al. 1978. USDA Douglas fir tussock moth research and development program. *Journ. For.* 76:37–40.

Carolin, V. M., Jr., and Honing, F. W. 1972. Western spruce budworm. *USDA Forest Service, Forest Pest Leaflet* no. 53.

Cole, W. E. 1971. Pine butterfly. *USDA Forest Service, Forest Pest Leaflet* no. 66.

Cuming, F. G. 1961. The distribution, life history and economic importance of the winter moth, *Operophtera brumata* (L.) (Lepidoptera: Geometridae) in Nova Scotia. *Can. Entomol.* 43:135–142.

Denton, R. E., and Tunnock, S. 1972. Larch casebearer in western larch forests. *USDA Forest Service, Forest Pest Leaflet* no. 96.

Dewey, J. E. 1975. Pine looper. *USDA Forest Service, Forest Pest Leaflet* no. 151.

Fedde, G. F. 1971. Elm spanworm. *USDA Forest Service, Forest Pest Leaflet* no. 81.

Freeman, T. N. 1967. On coniferophagous species of *Choristoneura* (Lepidoptera:Tortricidae) in North America. I. Some new forms of *Choristoneura* allied to *C. fumiferana. Can. Entomol.* 99:449–455.

Freeman, T. N., et al. 1967. On coniferophagous species of *Choristoneura* (Lepidoptera:Tortricidae) in North America. *Can. Entomol.* 99:449–506.

Furniss, R. L., and Carolin, V. M., Jr. 1977. Western forest insects. *USDA Forest Service Misc. Publ.* no. 1339.

Giese, R. L., et al. 1964. Studies of maple blight. *Univ. of Wisconsin, Res. Bull.* no. 250.

Godwin, P. A. 1972. Gypsy moth. *USDA Forest Service, Forest Pest Leaflet* no. 41.

Houston, D. R., and Valentine, H. T. 1977. Comparing and predicting forest stand susceptibility to the gypsy moth. *Can. Journ. For. Res.* 7:447–461.

Johnson, W. T., and Lyon, H. H. 1976. *Insects that feed on trees and shrubs.* Ithaca: Cornell University Press.

Jones, T. H. 1953. Cankerworms. *USDA Leaflet* no. 183.

Ketcham, D. E., and Shea, K. R. 1977. USDA combined forest pest research and development program. *J. For.* 75:404–407.

Koerber, T. W., and Struble, G. R. 1971. Lodgepole needle miner. *USDA Forest Service, Forest Pest Leaflet* no. 22.

Kulman, H. M. 1971. Effects of insect defoliation on growth and mortality of trees. *Annual Rev. of Entomology* 16:289–324.

MacAloney, H. J., and Ewan, H. G. 1964. Identification of hardwood insects by type of tree injury. *North Central Region, USDA Forest Service, Res. Paper* LS-11.

Macdonald, D. R., et al. 1962–1963. *Canada Department of Agriculture, Forest Entomology and Pathology Branch Information Reports XVII and XVIII.*

Martin, J. L. 1956. The bionomics of the aspen blotch miner, *Lithocolletis salicitoliella* Cham. (Lepidoptera:Gracillariidae). *Can. Entomol.* 88:155–168.

McKnight, M. E. 1968. A literature review of the spruce, western and 2 year cycle budworms. *USDA Forest Service, Res. Paper* RM-44.

Miller, C. A. 1966. The black-headed budworm in eastern Canada. *Can. Entomol.* 98:592–613.

Miller, H. C., and Allen, D. C. 1973. Cankerworms. *State Univ. of New York, New York State Pest Leaflet* no. 8.

————, Krall, J. L., and Grimble, D. 1969. Saddled prominent. *State Univ. of New York, New York State Pest Leaflet* no. F.24.

Millers, I., and Wallner, W. E. 1975. The redhumped oakworm. *USDA Forest Service, Forest Pest Leaflet* no. 153.

Mounts, Jack. 1976. 1974 Douglas-fir tussock moth control project. *J. For.* 74:81–86.

Reeks, W. A., and Smith, C. C. 1956. The satin moth, *Stilpnotia salicis* (L.) in the Maritime Provinces and observations on its control by parasites and spraying. *Can. Entomol.* 88:565–579.

Sanders, C. J., et al. 1974. Sex attractants for two species of western spruce budworm, *Choristoneura biennis* and *C. viridis* (Lepidoptera:Tortricidae). *Can. Entomol.* 106:157–159.

———, and Weatherston, J. 1976. Sex pheromone of the eastern spruce budworm (Lepidoptera:Tortricidae) optimum blend of *trans*-and- *cis*-11-tetradecenal. *Can. Entomol.* 108:1285–1290.

Schmiege, D. C., and Crosby, D. 1970. Black-headed budworm in western United States. *USDA Forest Service, Forest Pest Leaflet* no. 45.

Stedinger, J. R. 1977. Spruce budworm management models. Ph.D. Thesis, Harvard University.

Stelzer, M. J. 1971. Western tent caterpillar. *USDA Forest Service, Forest Pest Leaflet* no. 119.

Struble, G. R. 1972. Biology, ecology and control of the lodgepole needle miner. *USDA Forest Service, Tech. Bull.* no. 1458.

USDA. 1979. The gypsy moth—research toward integrated pest management. *USDA, Forest Service Tech. Bull.* no. 1584.

Washburn, R. I., and McGregor, M. D. 1974. White fir needle miner. *USDA Forest Service, Forest Pest Leaflet* no. 144.

Werner, R. A., and Baker, B. H. 1977. Spear-marked black moth. *USDA Forest Service, F. I. & Disease Leaflet* no. 156.

Wickman, B. E. 1971. California oakworm. *USDA Forest Service, Forest Pest Leaflet* no. 72.

———, Trostle, G. C. and Buffam, P. E. 1971. Douglas-fir tussock moth. *USDA Forest Service, Forest Pest Leaflet* no. 86.

Wilson, L. F. 1971*a*. The green-striped mapleworm. *USDA Forest Service, Forest Pest Leaflet* no. 77.

———. 1971*b*. Variable oak leaf caterpillar. *USDA Forest Service, Forest Pest Leaflet* no. 67.

Wollerman, E. H. 1971. Bagworm. *USDA Forest Service, Forest Pest Leaflet* no. 97.

Wygant, N. D. 1941. An infestation of the pandora moth, *Coloradia pandora* Blake, in Lodgepole pine in Colorado. *Journ. Econ. Entomol.* 34:697–702.

Leaf-eating Insects (Hymenoptera, Coleoptera, and Other Orders)

Serious pests among the leaf-eating insects are identified from several orders other than the Lepidoptera, though the greatest numbers and most serious pests are in that order. There are well-known forest-pest problems identified with some of these insects, especially in the hymenopterous group commonly called the *sawflies*.

HYMENOPTEROUS LEAF EATERS

By far the greater proportion of all the leaf-eating Hymenoptera belong to the families Tenthredinidae and Diprionidae. The adults of these families are called sawflies because of the sawlike ovipositor of the female. The larvae of most species are very similar to caterpillars of the Lepidoptera and are called *false caterpillars*. Most of the common species can be distinguished from the true caterpillars by the presence of at least six pairs of prolegs on the abdomen (Fig.13-1), as opposed to the two to five pairs characteristic of most lepidopterous larvae.

Figure 13-1 A colony of sawflies on pine. Note the number of prolegs, an identification character for sawflies. *(U.S. Forest Service photo.)*

The sawflies attack both coniferous and broadleaf trees. Some are leafminers, some are skeletonizers, but the most important are leaf feeders. From the economic viewpoint, those that attack conifers are of paramount interest.

The taxonomy of these insects has been confused and is constantly changing, but fairly comprehensive statements can be obtained in articles and books such as Becker et al. (1966), Coppel and Benjamin (1965), Rose and Lindquist (1973), Knerer and Atwood (1973), Lyons (1977), and Wilson (1977). Sawflies infesting conifers may often be divided into two groups according to their typical life cycles. The more destructive are those referred to as *summer* sawflies; they generally have more than one generation per year and consume both the needles of the current year and those produced earlier. The *spring* sawflies are less destructive; they feed early in the spring on mostly the older foliage and have only one generation each year.

Many of the larvae are similar in appearance, but some are far more injurious than others. Foresters are often puzzled about the identity of the sawfly larvae observed in the field. In order to help them in recognizing the harmful species, Tables 13-1 and 13-2 have been prepared. The distinctive characteristics presented in these tables apply to the larger larvae. In many instances, the distinctive markings do not appear until the larvae reach the third or fourth instar; therefore, the smaller sizes are not always easy to

Table 13-1   Recognition of Sawfly Larvae on Conifers Other than Pine

Species	Host	Larval season	Distinctive characteristics
European spruce sawfly, *Gilpinia hercyniae*	Spruces	May–Sept.	Dark green with five white lines; lines disappear in last instar; solitary
Balsam fir sawfly, *Neodiprion abietis*	Fir	June–Aug.	Head black; body dull green, lighter beneath; six dark longitudinal stripes
Larch sawfly, *Pristiphora erichsonii*	Larch	June–July	Head black; body grayish green, darker above; no stripes
Twolined larch sawfly, *Anoplonyx laricivorus*	Western larch	June–Aug.	Body brownish green with two narrow dark green stripes on the sides
Western larch sawfly, *A. occidens*	Western larch	June–Aug.	Body brownish green with single green dorsal stripe
Yellowheaded spruce sawfly, *Pikonema alaskensis* (Wilson, 1971*d*)	Black spruce preferred	June–July	Yellow head; body dark yellowish green above, lighter beneath; marked with gray green stripes
Greenheaded spruce sawfly, *P. dimmockii*	Spruce	June–July	Head and body green
Eastern cedar sawfly, *Monoctenus milliceps*	Cedar and juniper	June–July	Head light brown with black eyespots; body dull green with three dark longitudinal stripes
Arborvitae sawfly, *Monoctenus juniperinus*	Cedar	June–July	Head reddish brown; body yellow with light green longitudinal stripes
Hemlock sawfly, *Neodiprion tsugae* (Hard et al., 1976)	Western hemlock	July–Aug.	Head black; body green, about 1 in (2.5 cm) long; longitudinal stripes along each side

Table 13-2   Recognition of Sawfly Larvae on Pine

Species	Host	Larval season	Distinctive characteristics
Redheaded pine sawfly, *Neodiprion lecontei*	Hard pines	May–Oct.	Head reddish brown; body yellowish with six rows of irregular black spots
White pine sawfly, *N. pinetum*	Eastern white pine	May–Oct.	Head black; body yellowish white with four rows of irregular black spots
Loblolly pine sawfly, *N. taedae linearis*	Loblolly pine, shortleaf pine	May–June	Head reddish brown with black eyespots; body greenish white with dull longitudinal stripe and row of black spots on each side
Jack pine sawfly, *N. pratti banksianae*	Jack pine	May–June	Head and legs black; body yellowish green with two pale green stripes and a third that tends to be in broken spots
Redheaded jack pine sawfly, *N. rugifrons*	Jack pine	June–Sept.	Head reddish brown; body white with black spots and stripes
Nursery pine sawfly, *Diprion frutetorum*	Red and Scotch pine	May–Sept.	Head reddish brown with black blotch; body light green with narrow stripes on each side of dark longitudinal stripe
Brownheaded jack pine sawfly, *N. dubiosus*	Jack pine	May–July	Head brown; body yellow with black spots and stripes
Swaine jack pine sawfly, *N. swainei*	Jack pine	July–Aug.	Head brown; body yellow green and faintly striped
Pitch pine sawfly, *N. pinirigidae*	Pitch pine	May–Oct.	Head reddish brown; body dull green with double longitudinal dorsal line and broken lateral line, below which a double row of spots may occur
Red pine sawfly, *N. nanulus nanulus*	Red pine	May–June	Head black until last instar, then brown; body dark above, light below; three light green stripes on dorsum and dark stripe at base of legs
European pine sawfly, *N. sertifer* (Wilson, 1971*a*)	Red and Scotch pine	May–June	Head black; body grayish green with light dorsal stripe and dark lateral stripe bordered by narrow lighter stripes
Introduced pine sawfly, *Diprion similis* (Wilson, 1966)	White pines	Apr.–Oct.	Head black; body yellowish green with black stripe above yellow and white spots
Lodgepole sawfly, *N. burkei*	Lodgepole pine	June–Sept.	Head brown with black eyspots; body greenish gray with double dorsal stripe and heavy lateral stripe

Table 13-2 Continued

Species	Host	Larval season	Distinctive characteristics
Monterey pine sawfly, *Acantholyda burkei*	Monterey pine	March– Sept.	Head black; body green or brown; web needles together as they feed
Pine false webworm, *A. erythrocephala*	Red and white pines	July–Aug.	Head yellow dotted with brown; body green with dorsal, ventral, and lateral stripes; gregarious web spinner
Nesting-pine sawfly, *A. zappei*	Hard pines	July–Aug.	Head brown; body green with darker dorsal stripe; solitary; webs needles together
Pinon sawfly, *N. edulicolus*	Pinon	May–July	Pale green with dark green stripe on each side and greenish white dorsal stripe

recognize, but neither are they likely to be observed. The descriptions apply to those stages that are most conspicuous.

Larch Sawfly, *Pristiphora erichsonii*

One of the most serious defoliators of larch in both Europe and America is the larch sawfly, *Pristiphora erichsonii* (Drooz, 1971). In America, this pest was first reported in the 1880s in New England, where it was responsible for much damage to the native larch. Since this early outbreak, the species has played a prominent part in defoliating larch throughout the range of eastern larch.

Hewitt (1912) assumed that it was one of our uninvited guests and that it was introduced into America prior to 1880. Graham (1956), on the other hand, maintained that it is a native of America. The first historical record of larch sawfly injury in Michigan was in the year 1906; in Minnesota, it was during 1909. A study of the annual rings of living tamaracks, however, shows that reduction from defoliation has occurred from time to time throughout the life of the oldest trees. In addition to the period of reduced growth around 1910 to 1918, there have been two other periods of heavy defoliation: one starting just prior to 1880 and another older one about 1840. Other minor defoliations occurred between 1855 and 1860, one about 1870, and another in the late 1890s. This evidence shows that a defoliator, probably the larch sawfly, *P. erichsonii,* has repeatedly been epidemic on tamarack and supports the contention of Graham that the larch sawfly is native.

The life cycle of the larch sawfly, *P. erichsonii,* is quite similar to that of many other sawflies. The eggs are deposited alternately in a double row of slits cut along one side of a rapidly expanding young shoot. When first deposited, they are translucent and very small. They soon swell, however, as a result of water absorption until they protrude from the slit in which they are placed. The oviposition injury usually kills one side of the shoot; the other side continues to grow. This type of injury causes shoots to twist, sometimes forming a complete loop. The number of these twisted twigs can be used as an index of sawfly abundance.

The eggs hatch in about 1 week, and the larvae work gregariously, completely defoliating one branch before moving to another. Full growth is reached by midsummer, when the larvae drop to the ground and spin their tough, brown, oval cocoons in the moss or litter beneath the trees. The larvae remain in the prepupal stage within these cocoons until the following spring, at which time the majority of them transform to the pupal stage and emerge as adults. The remainder hold over in the prepupal stage until the second spring after cocooning, when they, too, transform to the adult stage. The adult sawfly is a handsome black insect somewhat over ⅜ in (10 mm) in length with a bright orange band about the base of the abdomen.

The larch sawfly, *P. erichsonii,* being very sensitive to moisture conditions, is adversely affected at time of cocooning by either high or low water. Under drought conditions, the larvae burrow deeply into the accumulations of sphagnum beneath the trees, sometimes 1 ft (30 cm) or more beneath the surface. There they are certain to be drowned within their cocoons when the water rises to a normal level. Droughts during the summers of 1928 and 1929 resulted in the almost complete elimination of sawfly throughout the Lake States. Regulation of water level, where it can be accomplished practicably, offers a means of control for the larch sawfly.

Pine Sawflies

A large number of sawflies (Fig. 13-2) are injurious defoliators of pine. Some, such as the jack pine sawfly, *Neodiprion pratti banksianae* (Wilson, 1971*b*), attack the older trees; others, such as the redheaded pine sawfly, *N. lecontei* (Baker, 1972), attack the younger trees.

All the members of this group deposit their eggs in slits cut in the edge of living pine needles. Some species pass the winter in the egg stage, with the young larvae hatching out in the spring (spring feeders); others pass the winter in the prepupal stage within the cocoon (summer feeders). Most species have a divided emergence similar to that of the larch sawfly, *P. erichsonii;* that is, some of the adults emerge in the fall or spring following cocooning, and others remain in diapause, or dormancy, for a year or longer. This habit protects the species against seasonal catastrophes that may arise directly or indirectly from adverse weather conditions.

The genus *Neodiprion,* to which most of the pine sawflies belong, contains species that are very similar in both appearance and habits. Furthermore, the larval markings vary from instar to instar. Therefore, some of these species are easily confused in the field. Table 13-2 provides information that will be helpful in distinguishing some that are frequently encountered.

The jack pine sawfly, *N. pratti banksianae,* is one of this group that is a

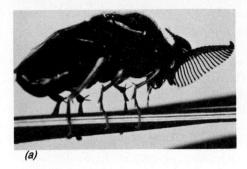

Figure 13-2 The introduced pine sawfly, *Diprion similis* (Hymenoptera: Diprionidae): *(a)* male, *(b)* female. The photographs suggest why these insects are called *flies,* even though they have two pairs of membranous wings rather that one pair, as is case in the dipterous flies. Actual length of female is approximately 8 mm. *(Photos by A. T. Drooz.)*

serious defoliator of jack pine, often in conjunction with the jack pine budworm, *Choristoneura pinus.* When the two work simultaneously, the sawfly feeding on the old needles and the budworm on the new growth, the result is serious mortality of the jack pine. The greatest damage is to mature or nearly mature stands. The adults appear in the autumn. The sexes are so different in appearance that they might easily be mistakenly identified as different species. The males are only about ¼ in (7 mm) in length, yellowish brown in color, and with threadlike antennae.

The winter is passed in the egg stage in the jack pine needles. Hatching in May, the larvae feed gregariously, completing their growth in 4 to 5 weeks. The cocoons are spun in the litter beneath the infested trees, where the larvae remain in the prepupal stage until late August or September. Then a part of them transform to pupae and later to adults; the others remain in diapause until the following autumn.

In contrast with the jack pine sawfly, *N. pratti banksianae,* which injures large trees, several pine sawflies concentrate on trees in smaller size classes. These species usually disappear after the stand becomes closed. One of the most important is the redheaded pine sawfly, *N. lecontei* (Fig 13-3). The adults of the overwintering generation of this sawfly begin to appear in May and continue to emerge through June. They deposit their eggs in slits cut in the edge of the needles, as do the other pine sawflies. The larvae feed gregariously, stripping one branch at a time, until they reach full development in about 5 weeks. Then they drop to the ground and spin their cocoons.

In the northern part of its range, the redheaded pine sawfly, *N. lecontei,* passes through a single generation per season; whereas in the South, two or more are completed. Many larvae that mature prior to July 15 are able to pupate and emerge during late summer and produce another generation. In contrast, those that develop after that time do not emerge until the following spring. Apparently, some controlling force such as day length prevents the transformation to adults when lateness of season would endanger completion of their progeny's development.

The cocoons (Fig. 13-4) are spun in the litter beneath the trees and are similar in appearance to those of other sawflies. Within the cocoon, the prepupae pass the winter, transforming to pupae and then to adults the following spring.

European Spruce Sawfly

The European spruce sawfly, *Gilpinia hercyniae,* was discovered near Ottawa, Ontario, in 1922 (Baker, 1972). Later, in 1929, it was found in New Hampshire. How or when it reached this side of the Atlantic from its native home in Europe is not known. This sawfly has slowly spread westward, being found for the first time in Michigan and Minnesota in 1950.

Figure 13-3 Larvae of the redheaded pine sawfly, *Neodiprion lecontei,* on pine needles. *(Photo by D. C. Allen.)*

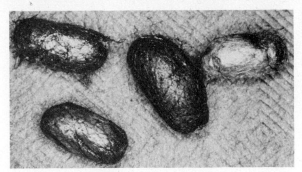

Figure 13-4 Cocoons of the redheaded pine sawfly, *N. lecontei. (Photo by D. C. Allen.)*

Serious outbreaks have occurred in the Gaspé Peninsula, New Brunswick, and the northern New England states (Neilson and Morris, 1964). They were brought to an end by a virus disease that appeared in New England in 1940 and spread throughout the entire area. During the initial epidemic in the Gaspé Peninsula and in scattered localities in other parts of Quebec, this sawfly killed more than one-half of the merchantable spruce. In most other localities, the damage was less severe.

Whether or not an outbreak of equal severity will occur at a later time, no one can say. This outbreak may have been the result of the insect's reaction to a new environment in which it met greatly reduced environmental resistance. If that is the case, we may never have another serious outbreak of the pest. On the other hand, we know that this sawfly causes injury from time to time in its native land. Therefore, we may logically expect it to cause trouble occasionally in the future. However, a disastrous outbreak of the intensity of the Gaspé epidemic would be surprising, considering the fact that the insect will be subject to the controlling influence of the parasites that are now well established in America. Morris in New Brunswick and other workers in New England (Morris, 1949) found that small mammals, especially mice, voles, and shrews, were active predators on the sawfly in the cocoon stage. These animals are most abundant in mixed and diversified forests and can be encouraged by silvicultural practices designed to maintain or produce these desirable conditions.

The adult of the European spruce sawfly, *G. hercyniae,* is about ⅜ in (9 mm) in length, thick-waisted, and like the other members of this family, provided with a sawlike ovipositor. In color, it is black with yellow markings on the abdomen, thorax, and head. It is a parthenogenetic species, and males are seldom found. This, of course, adds materially to its potentialities for reproduction.

According to Balch (1939), the adult females emerge in the spring from the oval, brown cocoons that were spun by the larvae in the duff layer beneath the trees. They lay their eggs in slits cut in the needles with the ovipositor. The larvae that hatch from these eggs pass through six instars; during the first three they are a pale green in color. During the fourth and fifth instars, they become striped with five longitudinal white stripes that again disappear in the sixth and final instar.

The eggs are laid in the old needles of all species of spruce. White spruce is especially favored. On hatching, the larvae feed on the old needles until the current year's needles are mature, after which some feeding may occur on them. Unlike so many sawflies, they are not gregarious but feed singly, at first chewing notches in the needles and later consuming the entire needle. This feeding habit makes their injury inconspicuous until it has reached serious proportions. Their presence is indicated by a thin appearance of the foliage and the presence of frass dropped by the caterpillars under the trees.

In the Gaspé Peninsula, only one generation occurs; but in parts of New England, as many as three may be completed. When full grown, the larvae drop to the ground and spin cocoons in the duff above the mineral soil. Some of the larvae may lie dormant within the cocoon for several years, but most of them will transform to pupae and adults the spring following cocooning.

Sawflies on Hardwoods

The damage to hardwoods by leaf-eating insects of the sawfly group is not so common, and outbreaks in the forest are rare, though not unknown. Many species have been troublesome on ornamental plantings.

The birch leaf-mining sawflies are among those that may reach outbreak proportions in the forest. Although these seldom cause significant numbers of trees to die, growth is affected, and trees are weakened. The birch leafminer, *Fenusa pusilla,* was first recorded in North America in 1923 (Baker, 1972). Infestations spread rapidly during the summer months, when up to three or four generations may develop. This species and the birch leaf-mining sawfly, *Heterarthrus nemoratus,* which has one generation per year, have caused considerable concern to homeowners. Species causing similar damage to other tree species are the elm leafminer, *Fenusa ulmi,* the European alder leafminer, *F. dohrnii,* and *Profenusa collaris* on sour cherry and *Crataegus* (Johnson and Lyon, 1976).

Two examples of sawflies that skeletonize or consume the entire leaves of broadleaf trees are the pear sawfly, *Caliroa cerasi,* and the mountain-ash sawfly, *Pristiphora geniculata.* The pear sawfly is an introduced species that is found coast to coast in the United States and in southern Canada. The larvae are sluglike and about 12 mm long and are covered with a shiny, olive green material. They feed on the surface of the leaf, creating a scorched appearance on the trees. There are two generations each year.

The mountain-ash sawfly, *P. geniculata,* is a common pest on both American and European ornamental mountain-ash trees. This insect feeds in colonies on trees in much the same fashion as its close relative, the larch sawfly, *P.*

erichsonii. Infestations are persistent, but homeowners can prevent the excessive loss of foliage that is common by inspecting these trees frequently and removing the small larvae when they first appear on the leaves.

A species of special interest because of its habits is the maple petiole borer, *Caulocampus acericaulis.* This introduced species feeds by boring into the petioles of maple leaves. This results in breakage and leaf fall. People will often question why so many fully developed and seemingly perfect leaves have fallen from their trees. The damage to the trees is generally insignificant. This is another type of defoliation that was not defined in Chapter 12. In this case, the leaf is not consumed by the insect.

Texas Leafcutting Ant, *Atta texana*

The Texas leafcutting ant is often referred to as the *town ant.* Most ant species living in temperate localities are not defoliators of trees. Some feed on other insects, the honeydew excreted by aphids and scale insects, fungi, fruit, or seeds, and some feed on low-growing vegetation. All of them live in colonial nests constructed, according to the species of ants, in the soil, beneath the bark of fallen trees, or in many other situations.

The town ant is a soil-inhabiting species that constructs nests with a ramification of galleries and many chambers with hundreds of openings to the surface. The galleries penetrate deeply into the soil, sometimes to a depth of 15 ft (4.5 m) or more. The numerous openings distributed over a considerable area of ground have given rise to the name *town ant.*

These ants defoliate trees and carry the leaves into the nest, where the leaves are chewed into bits and formed into fungous culture beds in special chambers excavated for this purpose. The fungus grown in these beds provides the only known food of the insects. It is grown in virtually pure culture.

In their harvesting activities, the ants march in a column to and from the tree or other vegetation that they are defoliating. On the return trip, each ant carries above it a bit of leaf. This habit has earned them the local name *umbrella ant.*

Although a large colony of these ants is capable of defoliating a 4-in (10-cm) tree in a single day, the greatest damage is to forest reproduction during the fall and early spring when other green vegetation is scarce. In eastern Louisiana and eastern Texas, they are serious pests of young pine plantations. Pines with relatively small buds, such as slash pine, are killed outright when both the needles and buds are cut off and the bark of the stem gnawed. Because of their much larger buds, the longleaf pines are seldom killed outright but may be severely weakened.

In winter, the ants tend to congregate together in chambers near the center of the nest. At that time, they can be controlled by fumigation. Methyl bromide has been recommended for this purpose, the gas being injected through a tube inserted to a depth of 2 ft (60 cm) into the central part of the colony through a principal tunnel. A pound of the fumigant is said to be enough to kill all or almost all the ants in an average-size colony.

COLEOPTEROUS LEAF EATERS

The leaf-eating insects that belong to the order Coleoptera are often members of a single family, the Chrysomelidae, although the genus *Phyllophaga* and several other genera of the family Scarabaeidae are defoliators in the adult stage. The Japanese beetle, *Popillia japonica,* is an introduced pest that defoliates trees in the beetle stage; also the pine chafer, *Anomala oblivia,* sometimes called the *anomala beetle,* seriously defoliates jack pine in the Lake States. The larvae of these Scarabaeidae are root eaters and will be discussed in Chapter 15, "Meristematic Insects."

Leaf Beetles (Chrysomelidae)

The leaf beetles are capable of causing severe defoliation, and at times, their work is conspicuous.

The elm leaf beetle, *Pyrrhalta luteola,* has attracted more attention than any other leaf beetle. It is an introduced pest that first appeared in Baltimore, Maryland, about 1834. Since then, it has spread throughout the United States, reaching almost every area where elms are grown. Occasionally, pure stands of elm are attacked severely; but for the most part, the elm leaf beetle is an ornamental-tree pest. Many municipalities are compelled to spray their trees regularly in order to control this beetle.

Like most chrysomelids, the elm leaf beetle, *P. luteola,* passes the winter as an adult, emerging from hibernation in the spring to feed on and to oviposit on elm foliage. The yellow eggs are laid in groups on the underside of the leaves. The larvae are flattened dark grubs with well-developed legs. Working gregariously, they first skeletonize and later chew holes in the leaves. After completing their development in 2 or 3 weeks, the larvae drop to the ground, where they pupate, unprotected by cocoons. In this stage, they are very vulnerable to small mammalian predators. In a week or 10 days, the adults emerge to repeat the life cycle, producing several generations during a single season in the South but only one generation in the North.

The cottonwood leaf beetle, *Chrysomela scripta,* defoliates any species of poplar or willow and at times becomes excessively numerous. The greatest damage from this insect is to young willow grown for basket production, where defoliation seriously affects the growth of willow wands. The life history is similar to that of the elm leaf beetle, *P. luteola.* This beetle is widely distributed in North America wherever its host trees grow. When necessary, it can be controlled by spraying.

Many of the leaf beetles are pests on ornamental trees and shrubs. One of the tiny members of this group is the locust leafminer, *Odontota dorsalis,* which may become a serious pest of the black locust. The larval stage of this beetle forms a mine in the inner layers of leaf tissue (Johnson and Lyon, 1976).

Two additional leafminers are of interest as examples of beetles from other families. The willow flea weevil, *Rhynchaenus rufipes* (Curculionidae), is tiny (2 mm long) and black; its larvae make small blotch mines in leaves of willows. *Brachys aeruginosus* (Buprestidae) mines the leaves of *Fagus* species. Damage is minor, but the insect and two related species of the same genus are fairly common on their host species.

OTHER LEAF FEEDERS

The great majority of the leaf feeders have been covered by the examples discussed from the orders Lepidoptera, Hymenoptera, and Coleoptera, but we would not want the reader to assume that there are no representatives from other orders. A few examples will illustrate this great diversity.

The flies, order Diptera, have a number of species that may defoliate trees. Many are especially significant on ornamentals. The leafminer flies of the genus *Phytomyza* include a species, the native holly leafminer, *Phytomyza ilicicola* (Agromyzidae), that is a serious pest in commercial orchards of Christmas holly. Damage is done by both the mining of the larvae and the wounds created by the sharp ovipositors of the adult females.

The dipterous gall midges (Cecidomyiidae) may also be of importance on ornamentals and on plants grown by specialized industries such as the Christmas tree growers. Some examples are the balsam gall midge, *Paradiplosis tumifex* (Fig. 13-5), the boxwood leafminer, *Monarthropalpus buxi,* the juniper tip midge, *Oligotrophus betheli,* the Monterey pine midge, *Thecodiplosis piniradiatae,* and the red-pine needle midge, *T. piniresinosae* (Baker, 1972; Furniss and Carolin, 1977).

The premier example of a defoliator in the order Orthoptera is the walkingstick, *Diapheromera femorata* (Phasmatidae), (Wilson, 1971c). This pest is found in southern Canada and the eastern United States west to the Plains and Texas, New Mexico, and Arizona. The insect feeds on a variety of species of deciduous trees and may defoliate trees over large areas. Adults are long, slender animals reaching over 80 mm in length; while motionless, they closely resemble the twigs of their hosts. Some species of grasshoppers may also be found feeding on the leaves of trees. One of these, the post-oak locust, *Dendrotettix quercus* (Acrididae), is widely distributed and has been recorded in outbreak numbers. Oaks are its preferred hosts, though it has been recorded on other species.

Other grasshoppers have been defoliators of forest trees where forests or windbreaks border the western ranges. Furniss and Carolin (1977) report that *Bradynotes obesa opima* in 1968 damaged 4200 acres of pine plantations in northern California, killing nearly all trees on 800 acres. Chemicals were used

Figure 13-5   Damage to balsam fir foliage caused by the balsam gall midge, *Paradiplosis tumifex. (Photo by E. Osgood.)*

to prevent further losses. Such losses may be rare, but when they happen, an individual manager may have to take prompt remedial action to avoid severe losses.

BIBLIOGRAPHY

Baker, W. L. 1972. Eastern forest insects. *USDA For. Ser, Misc. Publ.* no. 1175.

Balch, R. E. 1939. The outbreak of the European spruce sawfly in Canada and some features of its bionomics. *Journ. Econ. Entomol.* 32:412–418.

Becker, G. C., et al. 1966. The taxonomy of *Nediprion rugifrons* and *N. dubiosus* (Hymenoptera: Tenthredinoidea: Diprionidae). *Annals Ent. Soc. Amer.* 59:173–178.

Bennett, W. H. 1958. The Texas leaf-cutting ant. *USDA For. Ser., For. Pest Leaf.* no. 23.

Coppel, H. C., and Benjamin, D. M. 1965. Bionomics of the nearctic pine-feeding Diprionids. *Ann. Rev. of Entomol.* 10:69–96.

Drooz, A. T. 1971. Larch sawfly. *USDA For. Ser., For. Pest Leaf.* no. 8.

Furniss, R. L., and Carolin, V. M. 1977. Western forest insects. *USDA For. Ser. Misc. Publ.* no. 1339.

Graham, S. A. 1956. The larch sawfly in the Lake States. *For. Sci.* 2:132–160.

Hard, J. S., et al. 1976. Hemlock sawfly. *USDA For. Ser., For. Insect and Disease Leaf.* no. 31.

Hewitt, C. G. 1912. The larch sawfly. *Can. Dept. Agr. Expt. Farms, Bull.* no. 10, series 2 (*Ent. Bull.* no. 5).

Johnson, W. T., and Lyon, H. H. 1976. *Insects that feed on trees and shrubs.* Ithaca, N.Y.: Cornell University Press.

Knerer, G., and Atwood, C. E. 1973. Diprionid sawflies: Polymorphism and speciation. *Sci.* 179:1090–1099.

Lyons, L. A. 1977. On the population dynamics of *Neodiprion* sawflies. In Insect ecology—papers presented in the A. C. Hodson ecology lectures. *Univ. Minn. Agric. Expt. Sta., Tech. Bull.* 310:48–55.

Morris, R. F. 1949. Differentiation by small mammal predators between sound and empty cocoons of the European spruce sawfly. *Can. Entomol.* 81.

Neilson, M. M., and Morris, R. F. 1964. The regulation of European spruce sawfly numbers in the maritime provinces of Canada from 1937 to 1963. *Can. Entomol.* 96:773–784.

Rose, A. H., and Lindquist, O. H. 1973. Insects of eastern pines. *Dept. of Environ. Can. For. Serv.,* Publ. no. 1313.

Wilson, L. F. 1966. Introduced pine sawfly. *USDA For. Ser. For. Pest Leaf.* no. 9.

———. 1971a. European pine sawfly. *USDA For. Serv., For. Pest Leaf.* no. 98.

———. 1971b. Jack-pine sawfly. *USDA For. Serv., For. Pest Leaf.* no. 17.

———. 1971c. Walkingstick. *USDA For. Ser., For. Pest Leaf.* no. 82.

———. 1971d. Yellow-headed spruce sawfly. *USDA For. Ser, For. Pest. Leaf.* no. 69.

———. 1977. A guide to insect injury to conifers in the Lake States. *USDA For. Ser., Agric. Handbook* no. 501.

Sucking Insects

All the insects discussed in Chapters 12 and 13 feed on the tissues of trees, and all of them have mandibulate mouthparts in the larval stages. In addition to these chewing insects, there is a large and important group of insects that feeds on plant fluids. Their mouthparts are of the sucking type, in which the mandibles and maxillae have become slender, bristlelike organs enclosed in a sheath, the *labium*. The mouthparts thus form a beak, which is used to pierce the tissues and suck the fluid from them. The sucking insects attacking trees belong to two orders, the Hemiptera and the Homoptera.

The effect of sucking insects on trees is usually inconspicuous; only a few species seem able to kill trees directly. Nevertheless, the trees suffer distinctly injurious effects. Because their work is not conspicuous, the sucking insects of forest trees have received comparatively little consideration. On orchard and ornamental trees, however, they are acknowledged to be exceedingly important enemies. As the intensity of forest management increases, the importance of these insects will receive more consideration. It is quite likely that sucking insects do as much actual damage to forest trees as they do to orchard and ornamental trees.

Sucking insects may injure trees directly by reducing the trees' supply of food and water and, at the same time, producing necrotic spots that block translocation. Some of the sucking insects indirectly disseminate tree diseases by transporting the disease-causing organisms from tree to tree; phloem necrosis of elm is an example. In some cases, the insects are evidently mechanical carriers; but in others, they are essential intermediate hosts. For example, the mosaic diseases are transmitted from plant to plant by sucking insects; in some cases, this is the only way by which these diseases can be carried from one host to another. How important the insect-borne diseases may ultimately prove to be in the forest time alone will tell.

A third way in which sucking insects may injure trees is mechanically, by ovipositing in them (Fig. 14-1). Some species, for example adult cicadas (Cicadidae), do not generally feed on trees at all but seriously injure trees by weakening the branches with egg slits.

The species of sucking insects that attack trees are so numerous that space will not permit a full consideration of this important group. Therefore, the discussion will be confined to a few representative types of the more important families. References to others will be found in the Bibliography.

HEMIPTEROUS INSECTS

The order Hemiptera embraces a large number of sucking insects. The members of this order are the "true" bugs, and it is only to this group that the name *bug* can be correctly applied.

Plant Bugs (Hemiptera: Miridae)

The members of one hemipterous family, the Miridae, are known as the *plant bugs*. Every tree of every species is infested with its share of these insects, but they seldom occur in sufficient numbers to be serious pests of forest trees. Some of them, however, are always sufficiently abundant to be conspicuous wherever

Figure 14-1 Scars on an aspen stem resulting from oviposition by tree hoppers (Cicadelloidea), insects that feed chiefly on the fluids of herbaceous plants. *(From Graham and Knight, Principles of Forest Entomology, 4th ed., Fig. 56, p. 268.)*

their host trees are abundant, and they are sometimes very injurious in nurseries. An example of these potentially injurious plant bugs is the tarnished plant bug, *Lygus lineolaris.* The eggs of this plant bug are deposited in slits cut by the females with their ovipositors in the twigs and small branches of the trees. The young bear a close resemblance to the adults except that they have no wings. They run about actively on the leaves and twigs, obtaining their food by puncturing the tissues and sucking the sap.

The metamorphosis of the tarnished plant bug, like that of other Hemiptera, is gradual. With each succeeding molt, the nymphs become more and more like the adults until the final molt, when they appear as sexually mature, winged adults. This insect is often injurious to ornamentals and to deciduous seedlings in nurseries, especially the oaks, ashes, hickories, and birches. Where spraying the foliage of the trees is practical and safe, most species of plant bugs can be controlled.

Lace Bugs (Hemiptera: Tingidae)

Another important hemipterous family is the Tingidae, or lace bugs. These insects are small in size, usually not more than ⅛ in (3 mm) in length, but they are very striking in appearance. The hemelytra are thin, almost gauzelike in appearance and marked with a network of fine lines. These, combined with the similarly lacelike lateral expansions of the prothorax, give the insects a decidedly lacelike appearance.

The lace bugs lead a very sedentary life. Their entire developmental period may be spent on a single leaf; feeding causes small bleached areas to appear on the upper surface of the leaves and small, dark frass spots to appear on the undersurface of the leaves. Injury done by them is much more likely to be observed than injury caused by the plant bugs, in part because it is usually more localized and in part because the insects themselves are more easily seen.

The lace bugs sometimes occur in extremely large numbers and in some years may, over extensive areas, destroy practically all the foliage of their host trees. However, they are not tree killers. They attack trees of all sizes but appear to prefer saplings or small pole-sized trees. One of the tingids, *Corythuca pallipes,* has been considered the most injurious leaf-feeding insect on yellow birch in certain parts of New York State (Graham and Knight, 1965). These insects are sometimes seriously damaging to hardwood trees growing in nurseries. Birch is one of the common hosts, but basswood, ironwood, maple, oak, sycamore, and willow are also heavily attacked. As a result of heavy lace bug attack, the injured leaves fade in color and later turn brown, producing an effect similar to defoliation.

The control of lace bugs can be accomplished on ornamental trees and in tree nurseries with relative ease because spraying with a contact insecticide is effective. Where water is available, the insects can be held in check by washing the foliage thoroughly with a forcible stream of water from a garden hose. Seldom (if ever) is direct control under forest conditions necessary.

Scentless Plant Bugs (Hemiptera: Rhopalidae)

The boxelder bug, *Leptocoris trivittatus,* is a common scentless plant bug that usually confines its feeding to boxelders. The nymphs feed on developing seeds throughout the summer. In the fall, the adults seek dry, sheltered places in which to overwinter. Wherever boxelders have been planted for shade trees, the adults invade homes to seek overwintering shelter. Inasmuch as boxelder is a dioecious species, prevention and control of this pest lies in favoring the male trees.

HOMOPTEROUS INSECTS

Many members of the insect order Homoptera are important enemies of both forest and ornamental trees; in fact, almost every family of this order has representatives that feed on trees. The most important groups, from the viewpoint of forest entomology, are the aphids, the adelgids, and the scale insects. But members of other groups, for instance, certain species of leafhoppers and spittlebugs, are sometimes injurious. In addition to these, there are the jumping plant lice, which are particularly injurious to orchard trees, sometimes even killing them. Since the 1940s, some of these homopterous insects have become serious pests of forest plantations.

Cicadas (Homoptera: Cicadidae)

The more common species of the Cicadidae are well known to almost everyone; those who do not know these insects by sight are familiar with the strident, rattling song of the male, so often heard on warm summer days. They are sometimes termed *locusts,* which is incorrect. One of the most injurious and best-known members of this family is the periodical cicada, *Magicicada septendecim* (Fig. 14-2).

The periodical cicada has an interesting life history. The full-grown nymphs emerge in spring and early summer from the ground, where they have been passed through the developmental stage. During emergence years, large areas of ground are covered with their exit holes. In places, there may be as many as 100 holes per square foot of surface. The nymphs, heavy-bodied, with broad abdomens and powerful legs, leave their burrows and climb on any convenient object to transform to adults. In years when cicadas are abundant, their cast skins may be observed almost everywhere—on trees, fences, poles, shrubbery, and the sides of buildings.

The female is armed with a strong ovipositor that might be compared to a pair of chisels. She gouges out slits in twigs and small branches in which to deposit her eggs. It is this oviposition injury that constitutes the chief damage to the trees, rather than the sucking of the nymphs. After about 1 week in the egg, the young nymphs emerge, drop to the ground, and promptly burrow into the soil. There, at a depth of about 0.6 m, each one hollows out an individual

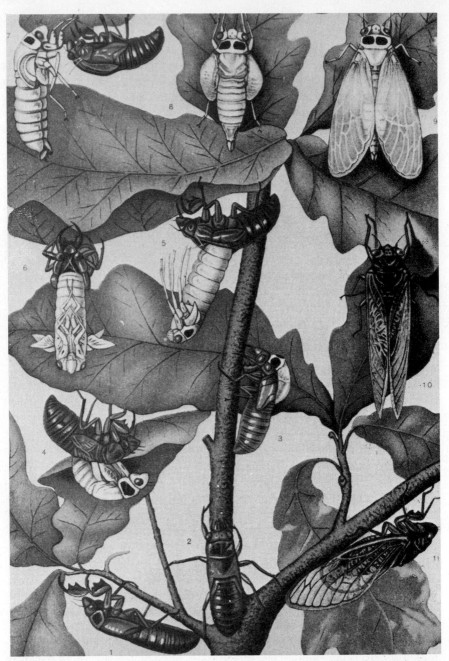

Figure 14-2 The transformation of the periodical cicada, *Magicicada septendecim,* beginning with the nymphs emerging from the ground and climbing upward (1 to 2), the emergence of the newly formed adult (3 to 7), and then the forming of the wings and coloration (8 to 11). *(Riley, USDA.)*

cell next to a small root of some woody plant, inserts its stylet, and begins its long period of sedentary feeding.

The periodical cicada lives for 17 years in the nymphal stage, the longest nymphal stage for any sucking insect. Most other insects with long life cycles have overlapping broods, some of which emerge each year, but this is not true of the periodical cicada within a geographic area. All the individuals in a locality emerge simultaneously. The broods of this insect are well known to entomologists. Each one is numbered, and the time and geographic distribution of each emergence can be accurately forecast. In the South, there is a life-cycle type that requires only 13 years for its development.

The periodical cicada appears in great swarms, particularly in the regions of eastern North America that are heavily wooded with hardwood trees. At such times, it appears that every available twig is filled with egg slits. As a result, many twigs and small branches are killed. This is not particularly injurious to mature forest trees, but to young stands or trees in nurseries or orchards, it is very serious.

The control of this insect by direct means is unnecessary in the forest. In limited areas, as in a nursery, the population of cicadas could be reduced by pruning off and burning all twigs containing egg slits before the eggs hatch. In orchards, the normal spraying program usually prevents damage by the invading cicadas. Natural enemies are important aids in reducing the number of cicadas. Birds, particularly those of the blackbird group, consume the newly emerged adults in large quantities. Immediately prior to emergence, when the nymphs are close to the surface of the ground, they are found and destroyed by scratching birds, skunks, moles, and pigs.

Spittlebugs (Homoptera: Ceropidae)

Spittlebugs are very common insects. The great majority of them feed on herbaceous plants and shrubs, but a few attack trees. Two species attacking pine have received much attention: the pine spittlebug, *Aphrophora parallela,* and the Saratoga spittlebug, *A. saratogensis.* The pine spittlebug attacks all pine species in the eastern United States and Canada but seldom causes recognizable injury to any except Scots pine growing in plantations. It is a pine pest in both the nymphal and the adult stages.

The eggs of the pine spittlebug are also deposited in the twigs of the host. In early spring, they hatch, and the nymphs suck the fluids from the twigs and small branches. The usual feeding place is on the growth of the previous year. The nymphs secrete a foamy fluid, resembling spittle, that covers and hides them and provides protection from heat and dessication (Fig. 14-3). This white spittle on the twigs is the most conspicuous symptom of their attack. When the nymphs are excessively abundant, the dripping of liquid from the trees often resembles the sound of rain.

At each point of attack, a necrotic spots develops in the growing tissues;

Figure 14-3 Frothy mass of spittle covering the nymphs of the pine spittlebug,
Aphrophora parellela. Injury by this insect is often associated with dry weather
or stagnation of dense stands. *(From
Graham and Knight, Principles of Forest
Entomology, 4th ed., Fig. 57, p. 273.)*

this later becomes infiltrated with resin, thus preserving in the wood a permanent record of the injury. When these necrotic areas are numerous, they
obstruct the normal flow of fluids, and the trees fade, lose foliage, and sometimes die.

Mortality resulting from pine spittlebug injury is usually confined to
stagnating stands. Thinning plantations before the trees become overly dense
will prevent mortality.

The Saratoga spittlebug, on the other hand, feeds on conifers only as an
adult; but during that brief period, it causes severe injury to the conifers
throughout the United States, especially to plantations of red and jack pine in
the Lake States.

The eggs of the Saratoga spittlebug are laid on the pine twigs in late
summer. This species overwinters in the egg stage. In the spring, the nymphs
hatch and drop to the ground, where they may feed on any of a number of
shrubs and herbs. Blackberry and sweet fern are the favored hosts, and only
where there is a combination of one or the other of these shrubs with the pine
can enough adults develop to cause injury to the pines. The nymphs feed
beneath the surface of the litter near the root collar of the alternate host.
Therefore, they can easily escape attention. However, when the litter is pushed
aside, the foamy masses of spittle containing the nymphs are clearly observ-
able.

On reaching full growth in July, the nymphs transform to adults and
move to the pines, where they feed and oviposit on the twigs. At each point
where an adult inserts its stylet through the bark, a necrotic spot develops (Fig.
14–4). When these spots become numerous, the twig dies, and the needles turn
red. In the course of several years, the flagging of twigs may involve the entire
tree and cause its death. More often, however, the tree becomes deformed but
still lives.

Because the Saratoga spittlebug requires an alternate host, rapid crown

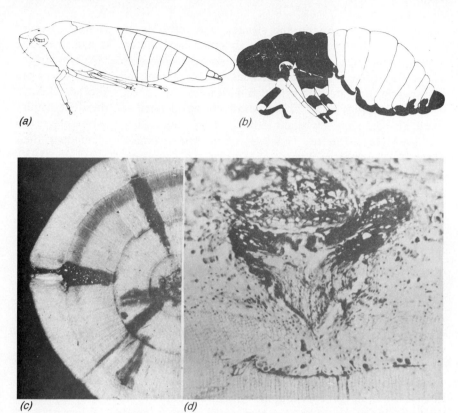

(a)

(b)

(c)

(d)

Figure 14-4 The feeding scars of the adult Saratoga spittlebug, *A. saratogensis (a),* disrupt translocation within the tree. Four-year-old red pine twig with feeding scars *(c)*, and center of feeding scar *(d)*. The nymphs *(b)* feed on shrubs and herbs, especially sweet fern. *(U.S. Forest Service photo.)*

closure will shorten the time that a plantation is susceptible to injury. The risk rating of a red pine site based on the percentage of ground occupied by the alternate host plants has been developed (Wilson, 1971). Before crown closure, the insect can be effectively controlled with insecticides during the adult flight period. However, repeated spraying may be required until the alternate hosts have been destroyed by shading.

Aphids (Homoptera: Aphididae)

Among the homopterous insects, the aphids or plant lice stand out prominently as an injurious family. They are abundant in numbers of individuals and of species, and they are so generally distributed that it is scarcely possible to find a tree of sapling size or larger that is not infested with them to some degree.

The aphids are usually very small, soft-bodied insects with pear-shaped bodies. The legs are long and slender, and most species possess a pair of cornicles on the dorsal side of the abdomen near the caudal end; these are tubelike structures, sometimes erroneously called *honey tubes.* Aphids may be either winged or wingless. When wings are present, all four of them are transparent, delicate, and provided with a few simple or branched veins.

When aphids occur in comparatively small numbers, the direct injury caused is comparatively slight although they may reduce the rate of growth and the vigor of the tree. However, when they become very abundant, their injury to the trees is often great. During the dry years of 1918 and 1930, millions of board feet of Sitka spruce were killed by the spruce aphid, *Elatobium abietinum,* along the tidelands of the Pacific Northwest.

Aphids of the genus *Cinara* have been found to infest both the roots and the main stem of Douglas-fir seedlings. The height growth may be only one-fourth to one-half that of uninfested seedlings (Fig. 14-5).

Trees injured by aphids may succumb to secondary insects or fungous diseases they could have resisted if growing vigorously.

All aphids excrete a sweet material called *honeydew,* which is highly prized as food by other insects. However, the honeydew that falls on cars, benches, or lawn furniture is extremely sticky and once hardened is almost impossible to remove. A sooty mold usually develops in the honeydew, reducing the aesthetic value of shade and ornamental trees.

It is a common sight to see ants busily collecting this honeydew. In some instances, interesting symbiotic relationships have arisen between ants and aphids. The ants care for the aphids and, in return, receive honeydew. The ants are actually beneficial in that they remove the honeydew. When ants are excluded, Johnson (1965) found the aphids became stuck in their own excretions. The ants also protect the aphids from predators. Recently, a predaceous lacewing was discovered that is a mimic of the woolly alder aphid, *Eriosoma lanigerum,* which protects it from predation by the ants (Eisner et al., 1978).

Various aphids have different habits. Some of them live on the bark of the trunk and large branches; others confine their feeding activities to the leaves and green tips; still others feed on the roots of trees. Other species provide shelter for themselves by causing the leaves on which they feed to curl about them. Still others live unprotected on the surface of the trees. Some aphids are provided with glands that secrete a flocculent, waxlike material, which collects over the insects and affords them some protection from their enemies and from the weather.

The life cycles can be complex, having asexual and bisexual reproduction, obligate and alternate hosts, dissimilar forms on alternate hosts, variation in number of generations as a result of weather and available food, and winged and wingless forms. The forester is advised to review the literature regarding the biology of a given aphid species with which one may be concerned.

(a)   (b)

(c)   (d)

Figure 14-5 Aphids of the *Cinara* infest both the main stem *(a)* and the roots *(b)* of Douglas-fir, often causing serious growth losses (*c*, normal; *d*, infested). Note carpenter ants tending aphids *(a)*. *(Weyerhaeuser Company.)*

Gall and Woolly Adelgids (Homoptera: Phylloxeridae)

The species of adelgids belonging in the Phylloxeridae are commonly called *gall, woolly,* or *bark adelgids,* depending on the waxy secretions on the insects or the galls that form on the tree. Formerly, the adelgids were commonly named *aphids;* however, that common name is now restricted to the homopterous family Aphididae. An awareness of this taxonomic revision is essential when reviewing the literature pertaining to these insects. The adelgids are closely related to the aphids, yet they differ in several major respects. For instance, the adelgids have no cornicles, and both parthenogenetic and sexually perfect females lay eggs.

Gall Adelgids   Some of the adelgids produce galls resembling cones on the expanding shoots of spruces. These insects are called *spruce gall adelgids.* In order to complete their sexual cycle, they require an alternate host, a fir, Douglas-fir, a larch, or a pine, depending on the insect species. However, on either the primary or secondary host, the insects may pass through the asexual cycle year after year. With some species such as the eastern spruce gall adelgid, *Adelges abietis,* the secondary host is unknown.

The control of adelgids is practically the same for all species. In the forest, no control is practiced, in spite of the serious injury they may cause. Undoubtedly, the silvicultural practices discussed in Chapter 10 that encourage tree and stand vigor are of value in preventing aphid damage. Only in nurseries, young plantations, ornamental trees, and orchards does adelgid control receive much consideration. There, spraying is the surest and best means of checking outbreaks of these insects. During the growing season, contact or systemic insecticides usually give satisfactory control, and the overwintering eggs can be destroyed by dormant sprays (see Chapter 11). Stock infested in the nursery should not be set out, either in the forest or in ornamental plantations, without

Figure 14-6 Galls on blue spruce caused by the Cooley spruce gall adelgid, *Adelges cooleyi. (U.S. Forest Service photo.)*

its first being treated to destroy the adelgids. Effective treatment can be accomplished by dipping or by fumigating.

The Cooley spruce gall adelgid, *A. cooleyi,* is a common and troublesome pest of spruces and Douglas-fir wherever these trees are grown close together. If the galls are clipped from small spruces, especially in nurseries and in Christmas tree plantations, infestations can be eliminated if reinfestation from Douglas-fir is prevented. Usually, this species is not considered to be a dangerous forest pest.

Woolly Adelgids One of the most serious forest-insect pests of the firs is the balsam woolly adelgid, *Adelges piceae.* This introduced pest was first recorded from Maine in 1908. The infestation spread rapidly southward in the Appalachian Mountains and westward through southern Maine, New England, and the Maritime Provinces. In the Pacific Northwest, the insect was first reported in 1928: a major outbreak was discovered in 1954 by M. M. Grobin, a forester employed by the Harbor Plywood Corporation, Hoquiam, Washington. The western infestation now extends throughout the Coast and Cascade mountain ranges from Oregon into British Columbia.

The insect is parthenogenetic, with a fecundity of 248. The number of generations varies with latitude and elevation, ranging from two to four generations each year. The wingless females are very small, less than 1 mm in length, and covered with a waxy, white wool. After the nymphs hatch, they wander over the bark, searching for a suitable feeding spot. Living tree tissue must be found within the reach of the nymphal stylet, which is not more than 1 mm in length. Once the stylet is inserted, the nymph remains in place, there to feed, mature, lay eggs, and die (Fig. 14-7).

The tree is damaged when the adelgid injects a substance similar to indoleacetic acid. If small amounts are injected, cellular growth is increased, but large amounts inhibit cellular growth. In either case, the tree's physiology is severely disrupted.

The symptoms on the tree that are visible to the forester begin with the curling and dieback of the current year's shoots, a swelling of the buds and shoots, then a gouting of the shoots. The crown of the tree begins to thin because few needles form on the infested shoots while the normal shedding of older needles continues. On smooth-barked trees, bole infestations may become very dense, the bole actually appearing white. If the wood is exposed, it will be reddish and brash because the tracheids are extremely thick-walled. This condition is termed *rotholz* (redwood).

The rates of mortality of the western firs vary with the species, the most susceptible being subalpine fir. Death may occur within 3 to 5 years after attack. Grand fir may endure infestations for as long as 15 years before dying. Pacific silver fir is intermediate between subalpine and Grand fir. Noble fir, Shasta red fir, and white fir may become gouted but appear to be resistant when growing in the forest.

Figure 14-7 The balsam woolly adelgid, *Adelges piceae,* inserts its stylets between the cells *(a)*. The conspicuous accumulation of waxy threads indicates a large population *(b)*. The injury to the twigs creates a condition called *gouting (c,* normal; *d,* gouted twigs of Pacific silver fir). *(Weyerhaeuser Company; U.S. Forest Service; Virginia Polytechnic Institute and State University.)*

The rate of mortality for Pacific silver fir is associated with the quality of the site. On the better sites, this fir has smooth bark with living tissue close to the surface. On poorer sites, however, the bark is thick, corky, and living tissue is beyond the adelgid's stylet. Thus, an increased density of adelgids occurs on the better sites. This fact is the exception which tests the principle that the better the site, the less will be the insect damage. Each species of insect is unique.

The future of firs in the Pacific Northwest is grim; these species are well on the way to being eliminated from commercial production by the balsam woolly adelgid. This loss is serious because the firs in many areas constitute a large percentage of the stand volume. The continued spread of the balsam woolly adelgid threatens the firs in California and British Columbia. In the Southern Appalachians, an aesthetic loss of Fraser fir has occurred in the high-altitude spruce-fir forests. These stands are located in very important recreational areas: the Shenandoah and Great Smoky Mountains national parks and southward along the Blue Ridge Parkway to Mount Mitchell, North Carolina (Fig. 14-8).

The prevention of the spread of the balsam woolly adelgid is difficult; during the crawling stage, the nymphs can be spread by the wind and birds. The shipment of fir seedlings, ornamentals, logs, and Christmas greens and trees could also contribute to the spread of this pest.

Control in the forest has proved impossible to date, despite a major predator-importation program. The use of chemicals on ornamentals has not proven economically feasible. Although genetic variations and crosses of trees have not been attempted, prospects for control through this approach are not promising.

One must constantly inspect the firs from the ground and be prepared to salvage immediately. Initially, survey efforts should be concentrated on the best sites during the fall. The deterioration of dead firs is rapid; within 3 to 5 years, more than 50 percent of the volume of Pacific silver fir is beyond any use for lumber or pulp. Following logging, the area should be burned to destroy the unmerchantable firs and, site permitting, favor other species.

The balsam woolly adelgid is one of our most serious forest insects. This pest has not received the attention and effort the situation demands.

Soft Scales (Homoptera: Coccidae)

The soft scales belong in the family Coccidae. Some of them secrete very little wax and are therefore called *naked scales*. Others secrete enough wax to form an egg sack; still others secrete wax in abundance. The legs of many scales are not reduced to the point of uselessness, at least not until the female becomes filled with eggs. Therefore, the insects can change their location when partly grown. These scales excrete large quantities of honeydew, which coats the leaves and branches of the infested trees. A fungus growing in this honeydew gives the trees an unsightly black appearance and is an aid in the identification of infestation at a distance.

The cottony maple scale, *Pulvinaria innumerabilis,* is a good example of the soft scales (Fig. 14-9). It attacks maples, particularly soft maples, throughout their range and when abundant sometimes kills trees. Prior to cold weather, the female is fertilized and spends the winter in a partly developed

(a)   *(b)*

Figure 14-8 The balsam woolly adelgid, *A. piceae,* has spread throughout the Coast and Cascade ranges of Oregon and Washington and into British Columbia and the Appalachians. (*a*) Infested grand fir near Coquille, Oregon, and (*b*) Frasier fir, Mount Mitchell, North Carolina. *(Virginia Polytechnic Institute and State University.)*

condition on the small branches of the tree. Females complete their development early in the summer. Then they are over 4 mm in length and broadly ovate in form. Each female deposits her eggs beneath a white covering of waxy threads, the so-called *egg sack.* These white masses standing out prominently against the blackened bark of the trees attract some attention.

The specific name, *P. innumerabilis,* well describes its fecundity. A comparatively unproductive female will lay 500 eggs. Some may produce several thousand. The nymphs leave the egg sacks in July or August and crawl to the leaves. There they feed until fall. Most of them locate on the underside of the leaves along the large veins. The males develop more rapidly than the females

Figure 14-9 The white, waxy egg cases of the cottony maple scale, *Pulvinaria innumerabilis*. This conspicuous soft scale (Coccidae) reduces the vigor of the tree, often causing yellowing of the foliage and death of lower branches. *(Samuel A. Graham, University of Michigan.)*

and emerge as winged adults in late summer or autumn. At that time, mating takes place, even though the females are not fully developed. After fertilization and before the leaves drop in the fall, the females move from the leaves and make their way back to the twigs and small branches, where they pass the winter, complete their development, lay their eggs, and die.

When trees are infested with soft scales, the foliage turns yellow, the vigor is reduced, and the lower branches of the tree begin to die. Although infestations seldom last 2 consecutive years, severe outbreaks may kill small or weakened trees within 1 year. Insect predation, especially by the ladybird beetles, is effective. Serious losses in Christmas tree plantations and shade trees can be avoided with insecticides.

Armored Scales (Homoptera: Diaspididae)

The oystershell scale, *Lepidosaphes ulmi,* belongs to the family Diaspididae, the armored scales. This family is characterized by the scalelike covering beneath which the insect lies. This scale is composed in part of molted skins and in part of waxy or resinous excretions of the insect. The oystershell scale is well named, for its general shape is that of an elongated oyster shell. The full-grown scale is about 3 mm long and brown in color. When these insects are abundant, the small branches of the infested trees may be almost completely encrusted with the scale (Fig. 14-10).

The oystershell scale is known to almost everyone who works with woody

Figure 14-10 White ash seedling heavily infested with oystershell scale, *Lepidosaphes ulmi.* Seedlings of white ash growing under shady conditions are almost certain to be infested and killed. *(Samuel A. Graham, University of Michigan.)*

plants. This insect attacks a great variety of fruit trees and deciduous forest trees and shrubs. In the forest, the ashes, poplars, willows, and maples are probably the most susceptible trees. Seedlings of white ash growing under shady conditions are almost certain to be infested and killed. This insect is distributed throughout the temperate regions of the world and either is native to America or was introduced in very early colonial days. It now occurs throughout temperate North America.

The male oystershell scales are said to be very rare. The exact proportion of males to females has never been determined, but it is so low that thousands of scales may be examined without finding a male. The eggs of the female are deposited in late summer or early autumn and pass through the winter without hatching. In the early spring, the nymphs hatch and wander about on the bark of the host tree. At this time, they may creep onto birds, other insects, or anything else that comes in contact with the trees. In this way, they may be carried from tree to tree or even from locality to locality.

After wandering, the nymphs insert their mouthparts and proceed to feed on the fluids of the tree. The female scales never move again. The secretion of the scale then begins. As the insect grows, the scale is increased in size so that the body is always entirely covered. When mature, the female has degenerate legs and eyes; she is little more than a reproductive sack. The male undergoes a metamorphosis; he emerges as a winged insect, with antennae and compound eyes, the mouthparts replaced by a second pair of rudimentary eyes. The adult male takes no food and is short-lived, but during his short life, he presumably fertilizes many females.

By the time the female is full grown, her body is filled with eggs. As she lays them, her body shrinks in size until almost the entire cavity beneath the scale is occupied by eggs. Shortly after oviposition, the female dies; consequently, by late autumn, there are no females alive. The number of eggs deposited by each female varies from 20 to slightly more than 100. In the North, there is only one generation each year; whereas in the South, where there are two generations, the eggs of the first brood are deposited early and hatch during the month of July.

Other important species of armored scales pass through life cycles similar to that of the oystershell scale. The scurfy scale, *Chionaspis furfura,* and the pine needle scale, *C. pinifoliae,* both develop in a manner quite similar to that of the oystershell scale.

Another serious pest is the San Jose scale, *Quadraspidiotus perniciosus,* which was introduced from the Orient. First recorded at San Jose, California, it now occurs throughout the United States. For many years, this insect was one of the most serious pests of orchards, but the release of parasites brought about an effective biological control (Chapter 9).

Eriococcid Scales (Homoptera: Eriococcidae)

The European elm scale, *Gossyparia spuria,* will serve to illustrate the scales of the family Eriococcidae. When first introduced into America, the European elm scale caused much mortality among the elms and was considered most dangerous. It was first reported in the United States in New York State in 1884, but apparently, it had been established for a number of years prior to that time. In 1894, it was reported in California and now is widely distributed throughout the United States and Canada wherever elms are present.

The rapid spread of the European elm scale throughout the United States is surprising. In some instances, it might have been carried by birds; but in most cases, it was carried to new localities on infested nursery stock. The former popularity of elm as a shade tree and the lack of inspection and certification of nursery stock were doubtless factors of importance contributing to its rapid spread.

Now that it has become established, the European elm scale, although still an injurious pest, does not cause so much mortality among the elms as it did when first introduced. This situation can be explained in part by the increased activities of parasites and predators and in part by the increased resistance of the surviving elms. This pest serves as an illustration of the principle that a newly introduced pest is likely to be more injurious when it is first introduced than it will be later.

Other members of the Eriococcidae are serious pests. The beech scale, *Cryptococcus fagisuga,* was introduced into Nova Scotia about 1890 and has now spread southward as far as the Ohio Valley. The species is parthenogenetic, and the crawlers are spread by the wind. Bole infestations rupture the smooth bark, providing infection courts for fungi, especially *Nectria* spp. The mortality of beech has been enormous. Direct chemical control is not feasible in the forest, and to date, biotic control is not sufficient. Lethal temperatures of –35°C kill the beech scale, but survival occurs below the snow line.

Margarodid Scales (Homoptera: Margarodidae)

The scales belonging to the family Margarodidae have great variation in habit and host. Some species feed in exposed positions, others in gall-like pits. Some feed on one tree species; others feed on many hosts.

The red pine scale, *Matsucoccus resinosae,* is an example of this family. The pest was first discovered in Connecticut in 1946 and has slowly spread to New York and New Jersey. There are two generations each year, and the newly emerged nymphs are active crawlers easily disseminated by the wind. At present, there are no effective controls, either biotic or chemical. The spread of the red pine scale has apparently been limited by cold weather; temperatures of –23°C are fatal.

Red pine is the only known native host, but the scale has also been found on Chinese, Japanese red, and Japanese black pines. (*Matsu* is the Japanese word for pine.) Fortuitously, the distribution of the scale has been limited to red pine planted south of its natural range (Fig. 14-11). However, within the infested plantations, tree mortality has been extreme. The economic impact is especially severe because many of the plantations are on watershed properties.

The symptoms of an infestation begin with the woolly patches of scales in bark crevices, then the slight yellowing and stunting of new growth on an occasional branch. With time, the amount of affected foliage increases throughout the crown, and the color of the needles fades to yellow and finally to red (Fig. 14-11). The bark on the bole and branches becomes swollen and cracks. The infested plantations should be salvaged, and tree species other than red pine established.

Figure 14-11 The accidentally imported red pine scale, *Matsucoccus resinosae,* is destroying red pine planted south of its range in Connecticut. The northward spread appears to be limited by low winter temperatures. *(Connecticut Agriculture Experiment Station, New Haven.)*

Other *Matsucoccus* species include the pine twig gall scale, *M. gallicola,* often serious on young pines in the South, and the Prescott scale, *M. vexillorum,* which has caused branch killing of ponderosa pine in the Southwest.

Control of Scale Insects

The control of scale insects begins with prevention. The tree species must be selected for the site, maintained in a vigorous condition by adequate watering and fertilizing, and kept free from soil compaction and atmospheric pollution. Infestations are often associated with environmental conditions, especially atmospheric conditions of heavy dust and smoke, that obviously stress the host plant. Ponderosa pine is heavily attacked by the pine needle scale, *C. pinifoliae,* along dusty roads, as well as when exposed to smoke from mills. Keen (1952) suggested that the stomata are choked by particulate matter, rendering the trees susceptible to the scale insects.

There is also an association of scale infestations with tree species planted outside of their natural range or on adverse sites. This association is especially noticeable regarding ornamentals and trees planted in municipal and recreational areas. Outbreaks are common on trees planted where the topsoil has been removed or extremely compacted, where automobile exhaust fumes accumulate, or where water shortages exist because of paving and drains.

Every possible precaution should be taken to prevent the introduction of aphids and scales into uninfested localities. Planting stock should be carefully inspected, and if scales are found, the stock should be rejected. Once infested with scales, a tree or forest stand will usually remain infested, and only by means of natural factors can these pests be held in check. Fortunately, there are many valuable agencies of natural control. Foresters must remember the importance of biological control, especially regarding the scale insects; spectacular results have been attained against these pests. The importation of the vedalia beetle, *Rodolia cardinalis,* a coccinellid from Australia, in the late 1800s saved the citrus industry of California from the cottonycushion scale, *Icerya purchasi* (Margarodidae).

Predaceous insects, such as the ladybird beetles, feed on scales in both the larval and the adult stages. The larvae of the lacewings are also important enemies. Among the most important of the predatory groups are the mites, although not all mites that are observed under or near scales are necessarily predaceous. Some of them are scavengers.

Like most other insects, competition among themselves appears to be a limiting factor. When the suitable parts of a tree become heavily encrusted with scales, the fecundity of the individual is apparently much reduced. In still more severe cases, the death of leaves or twigs on which the scales are feeding may bring about an acute food shortage and ultimate starvation for many.

Chemical methods of direct control must often be used. Spraying and fumigating are the two most important means of control. These methods are not, of course, applicable in the forest, but they can be used to advantage on nursery stock.

BIBLIOGRAPHY

Hemiptera

Plant Bugs: Miridae

Knight, H. H. 1941. The plant bugs, or Miridae, of Illinois. *Ill. Nat. Hist. Surv. Bull.* no. 22.
———. 1968. Taxonomic review: Miridae of the Nevada test site and the western United States. *Brigham Young Univ. Sci. Bull. Biol.* series 9.
Usinger, R. L. 1945. Biology and control of ash plant bugs in California. *J. Econ. Entomol.* 38:585–591.

Lace Bugs: Tingidae

Drake, C. J., and Ruhoff, F. A. 1965. Lacebugs of the world: A catalog. *U.S. Nat. Museum Bull.* no. 243.
Graham, S. A., and Knight, F. B. 1965. *Principles of forest entomology,* 4th ed. New York: McGraw-Hill Book Company.
Johnson, W. T., and Lyon, H. H. 1976. *Insects that feed on trees and shrubs.* Ithaca: Cornell University Press.
Wade, O. 1917. The sycamore lacebug (*Corythucha ciliata* Say.). *Okla. Agr. Expt. Sta. Bull.* no. 116.

Scentless Plant Bugs: Rhopalidae

Tinker, M. E. 1952. The seasonal behavior and ecology of the boxelder bug, *Leptocoris trivittatus. Minnesota Ecology* 33:407–414.

Wollerman, E. H. 1971. The boxelder bug. *USDA For. Serv., For. Pest Leaf.* no. 95.

Homoptera

Cicadas: Cicadidae

Periodical cicada, *Magicicada septendecim* (L.)

Felt, E. P. 1905. *Insects affecting park and woodland trees.* New York State Museum Memoir no. 8, 231–237.

Marlatt, C. L. 1907. The periodical cicada. *USDA Bur. Entomol. Bull.* no. 71.

Simons, J. N. 1954. The cicadas of California. *Calif. Insect Bull.* 2:153–192.

U.S. Department of Agriculture. 1966. Periodical cicadas. *USDA Leaf.* no. 540.

Spittlebugs: Cercopidae

Saratoga spittlebug, *Aphrophora saratogensis* (Fitch)

Anderson, R. F. 1947. The Saratoga spittlebug. *J. Econ. Entomol.* 40:695–701.

Ewan, H. G. 1961. The Saratoga spittlebug. *USDA Tech. Bull.* no. 1250.

Gruenhagen, R. H., et al. 1947. Burn blight of jack and red pine following spittle insect attack. *Phytopath* 37:757–772.

MacAloney, H. J., and Wilson, L. F. 1971. The Saratoga spittlebug. *USDA For. Serv. For. Pest Leaf.* no. 3.

Secrest, H. C. 1944. Damage to red pine and jack pine in the Lake States by the Saratoga spittlebug. *J. Econ. Entomol.* 37:447–448.

Wilson, L. F. 1971. Risk-rating Saratoga spittlebug damage by abundance of alternate-host plants. *USDA For. Serv. Res. Note* NC 110.

———, and Kennedy, P. C. 1968. Suppression of the Saratoga spittlebug in the nymphal stage by granular Baygon®. *J. Econ. Entomol.* 61:839–840.

Spittlebug, *Aphrophora* spp.

Kelson, W. E. 1964. The biology of *Aphrophora permutata* and some observations on *Aphrophora canadensis* attacking Monterey pine in California. *Pan-Pacific Entomol.* 40:135–146.

Speers, C. H. 1941. The pine spittlebug (*Aphrophora parallela* Say). *N.Y. St. Coll. For. Tech. Publ.* no. 54.

Aphids: Aphididae

Bissell, T. L. 1978. Aphids on Juglandaceae in North America. *Maryland Agric. Exp. Sta.* MP 911.

Dawson, A. F. 1971. Balsam twig aphid. *Can. For. Serv. Pest Leaf.* no. 36.

Eisner, T., et al. 1978. "Wolf-in-sheep's clothing" strategy of a predaceous insect larvae. *Sci.* 199:790–794.

Grobler, J. H. 1962. The life history and ecology of the woolly pine needle aphid, *Schizolachnus pini-radiatae* (Dawson). *Can. Entomol.* 94:35–45.

Holms, J., and Ruth, D. S. 1968. Spruce aphid in British Columbia. *Can. For. Serv. For. Res. Lab., Victoria, B. C. For. Pest Leaf.* no. 16.

Johnson, N. E. 1965. Reduced growth associated with infestations of Douglas-fir seedlings by *Cinara* species. *Can. Entomol.* 97:113–119.

Kennedy, J. S., et al. 1962. *A conspectus of aphids as vectors of plant viruses.* London: Commonwealth Institute of Entomology.

Lindquist, O. H., and Miller, W. J. 1970. Aphids of Ontario forests. *Can. For. Serv. For. Res. Lab. Ontario, Infor. Rept.* 0-X-141.

Richards, W. R. 1972. *The Chaitophorinae of Canada (Homoptera: Aphididae).* Entomol. Soc. Can. Memoir no. 87.

Saunders, J. L. 1969. Occurrence and control of the balsam twig aphid on *Abies grandis* and *A. concolor. J. Econ. Entomol.* 62:1106–1109.

Smith, C. F. 1972. Bibliography of the Aphididae of the world. *N. Car. Agric. Exp. Sta. Tech. Bul.* no. 216.

———, and Parron, C. S. 1978. An annotated list of Aphididae (Homoptera) of North America. *N. Car. Agric. Exp. Sta. Tech. Bul.* no. 255.

Gall Adelgids: Phylloxeridae

Butcher, J. W., and Haynes, D. L. 1960. Chemical control of the eastern spruce gall aphid with observations on host preference and population increase. *J. Econ. Entomol.* 53:979–982.

Cummings, M. E. P. 1959. The biology of *Adelges cooleyi. Can. Entomol.* 91:601–617.

———. 1962. A monomorphic cycle of *Adelges cooleyi* living only on spruce. *Can. Entomol.* 94:1190–1195.

Lindquist, O. H. 1951. The Adelgids on forest trees in Ontario with key to galls on spruce. *Proc. Entomol. Soc. Ontario* 102:23–27.

Plumb, G. H. 1953. The formation and development of the Norway spruce gall caused by *Adelges abietis* L. *Conn. Agr. Exp. Sta. Bull.* no. 566.

Saunders, J. L., and Barstow, D. A. 1970. *Adelges cooleyi* control on Douglas-fir Christmas trees. *J. Econ. Entomol.* 63:150–151.

Tjia, B., and Houston, D. B. 1975. Phenolic constituents of Norway spruce resistant or susceptible to the eastern spruce gall aphid. *For. Sci.* 22:180–184.

Wilford, B. H. 1937. The spruce gall aphid (*Adelges abietis* L.) in southern Michigan. *Univ. Mich. Sch. Nat. Res. Circ.* no. 2.

Woolly Adelgids: Phylloxeridae

Pine leaf adelgids (=chermid =aphid)

Balch, R. E., and Underwood, G. R. 1950. The life history of *Pineus pinifoliae* and its effects on white pine. *Can. Entomol.* 82:117–123.

Cummings, M. E. P. 1962. The biology of *Pineus similis* on spruce. *Can. Entomol.* 94:395–408.

Dimond, J. B., and Bishop, R. H. 1968. Susceptibility and vulnerability of forests to the pine leaf aphid, *Pineus pinifoliae. Maine Agr. Exp. Sta. Bull.* no. 658.

Hoff, R. J., and McDonald, G. I. 1977. Differential susceptibility of 19 white pine species to woolly aphid (*Pineus coloradensis*). *USDA For. Serv. Res. Note* INT 225.

Hoffman, C. H., et al. 1947. A twig droop of white pine caused by *Pineus. J. Econ. Entomol.* 40:229–231.

Lowe, J. H. 1965. Biology and dispersal of *Pineus pinifoliae* (F.). Ph.D. Diss., Yale University.

Balsam woolly adelgid (=aphid): *Adelges piceae* **(Ratz.)**

Amman, G. D. 1962. Seasonal biology of the balsam woolly aphid on Mt. Mitchell, North Carolina. *J. Econ. Entomol.* 55:96–98.

———. 1970. Distribution of redwood caused by the balsam woolly aphid in Fraser fir of North Carolina. *U.S. Dept. Agr. For. Serv. Res. Note* SE 135.

———, and Speers, C. F. 1965. Balsam woolly aphid in the Southeast Appalachians. *J. Forestry* 63:18–20.

———, and Talerico, R. L. 1967. Symptoms of infestation by the balsam woolly aphid displayed by Fraser fir and bracted balsam fir. *USDA For. Serv. Res. Note* SE 85.

Balch, R. E. 1952. Studies of the balsam woolly aphid, *Adelges piceae* (Ratz.) and its effects on balsam fir, *Abies balsamea* (L.) Mill. *Can. Dept. Agric. Publ.* no. 867.

Bryant, D. G. 1976. Distribution, abundance, and survival of the balsam woolly aphid, *Adelges piceae,* on branches of balsam fir. *Can. Entomol.* 108:1097–1111.

Carroll, W. J., and Bryant, D. G. 1960. A review of the balsam woolly aphid in Newfoundland. *Forestry Chron.* 36:278–290.

Clark, R. C., and Brown, N. R. 1961. Studies of predators of the balsam woolly aphid. V. *Can. Entomol.* 93:1162–1168.

Doerkson, A. H., and Mitchell, R. G. 1965. Effects of the balsam woolly aphid upon wood anatomy of some western true firs. *Forest Sci.* 11:181–188.

Foulger, A. N. 1968. Effect of aphid infestation on properties of grand fir. *Forest Prod. J.* 18:43–47.

Greenbank, D. O. 1970. Climate and the ecology of the balsam woolly aphid. *Can. Entomol.* 102:546–578.

Heikkenen, H. J. 1966. Balsam woolly aphid on grand fir. *J. Forestry* 64:546–547.

Johnson, N. E., and Heikkenen, H. J. 1958. A method for field studies of the balsam woolly aphid. *J. Econ. Entomol.* 51:540–542.

———, and Wright, K. H. 1957. The balsam woolly aphid problem in Oregon and Washington. *USDA For. Serv. Res. Pap.* PNW 18.

———, et al. 1963. Mortality and damage to Pacific silver fir by the balsam woolly aphid in southwestern Washington. *J. Forestry* 61:854–860.

MacAloney, H. J. 1935. The balsam woolly aphid in the Northeast. *J. Forestry* 33:481–484.

Mitchell, R. G. 1966. Infestation characteristics of the balsam woolly aphid in the Pacific Northwest. *USDA For. Serv. Res. Paper* PNW 35.

————, and Wright, K. H. 1967. Foreign predator introductions for control of the balsam woolly aphid in the Pacific Northwest. *J. Econ. Entomol.* 60:140–147.

————, et al. 1961. Seasonal history of the balsam woolly aphid in the Pacific Northwest. *Can. Entomol.* 93:794–798.

————, et al. 1970. Balsam woolly aphid. *USDA For. Serv. Pest Leaf.* no. 18.

Page, G. 1975. The impact of balsam woolly aphid damage on balsam fir stands in Newfoundland. *Can. J. For. Res.* 5:195–209.

Rudinsky, J. A. 1957. Notes on the balsam woolly aphid. *Weyerhaeuser Co. Bull.* (unnumbered).

Shea, K. R., et al. 1962. Deterioration of Pacific silver fir killed by the balsam wooly aphid. *J. Forestry* 60:104–108.

Varty, I. W. 1956. Adelges insects of silver firs. *Gr. Brit. For. Comm. Bull.* no. 76.

Scales, General

Miller, D. R., and Kosztarab, M. 1979. Recent advances in the study of scale insects. *Ann. Rev. Entomol.* 24:1–27.

Armored Scales: Diaspididae

Brown, C. E. 1958. Dispersal of the pine needle scale, *Phenacaspis pinifoliae. Can. Entomol.* 90:685–690.

Cummings, M. E. P. 1953. Notes on the life history and seasonal development of the pine needle scale. *Can. Entomol.* 85:347–352.

Edmonds, G. F., Jr. 1973. Ecology of black pineleaf scale. *Environ. Entomol.* 2:765–777.

Ferris, G. F. 1955. *Atlas of the scale insects of North America.* Stanford: Stanford University Press.

Flake, H. W., and Jennings, D. T. 1974. A cultural control method for pinyon needle scale. *USDA For. Serv. Res. Note* RM 270.

Griswold, G. H. 1925. A study of the oyster-shell scale. Cornell University Agric. Exp. Sta. Memoir no. 93.

Herbert, R. B. 1920. Cypress bark scale. *USDA Dep. Bull.* no. 1223.

Keen, F. P. 1952. Insect enemies of western forests. *USDA Misc. Publ.* no. 273.

Luck, R. F., and Dahlsten, D. L. 1974. Bionomics of the pine needle scale, *Chionaspis pinifoliae,* and its natural enemies at South Lake Tahoe, California. *Entomol. Soc. Am. Ann.* 67:309–316.

McKenzie, H. L. 1956. The armored scale insects of California. *Calif. Ins. Surv. Bull.* no. 5.

Nielson, D. G., and Johnson, N. E. 1973. Contribution to the life history and dynamics of pine needle scale, *Phenacaspis pinifoliae,* in central New York. *Entomol. Soc. Am. Ann.* 66:34–43.

Talerico, K. L., et al. 1971. *Fiorinia externa* Ferris, a scale insect of hemlock. *USDA For. Serv., For. Pest Leaf.* no. 107.

Wallner, W. E. 1965. Biology and control of Fiorinia hemlock scale, *Fiorinia externa* (F.). Ph.D. Diss., Cornell University.

Eriococcid Scales: Erioccoidae

Brower, A. E. 1949. The beech scale and beech bark disease in Acadia National Park. *J. Econ. Entomol.* 42:226–228.

Ehrlich, J. 1934. The beech bark disease, a Nectria disease of *Fagus* following *Cryptococcus fagi* (Baer.). *Can. J. Res.* (spec. no.) 10:593–692.

Herbert, F. B. 1924. The European elm scale in the West. *USDA Dep. Bull.* no. 1223.

Shigo, A. L. 1962. Another scale insect on beech. *USDA For. Serv., Northeast For. Exp. Sta. Pap.* no. 168.

———. 1963. Beech bark disease. *USDA For. Serv., For. Pest Leaf.* no. 75.

Soft Scales: Coccidae

Blackman, M. W., and Ellis, W. O. 1916. Cottony maple-scale. *N.Y. State Coll. For. Bull.* 16:26–107.

Burns, D. P., and Donley, D. E. 1970. Biology of the tulip tree scale *Toumeyella liriodendri. Entomol. Soc. Ann.* 63:228–235.

Donley, D. E., and Burns, D. P. 1971. The tulip tree scale. *USDA For. Serv., For. Pest Leaf.* no. 92.

Kattoulas, M. E., and Koehler, C. S. 1965. Studies of the irregular pine scale. *J. Econ. Entomol.* 58:727–730.

Rabkin, F. B., and LeJeune, R. R. 1954. Some aspects of the biology and dispersal o the pine tortoise scale (*Toumeyella numismaticum*). *Can. Entomol.* 86:570–575.

Wilson, L. F. 1971. Pine tortoise scale. *USDA For. Serv., For. Pest Leaf.* no. 57.

Margarodid Scales: Margarodidae

Anderson, J. F., et al. 1976. The red pine scale in North America. *Conn. Agr. Exp. Sta. Bull.* no. 765.

Bean, J. L. 1956. Red pine scale. *USDA For. Serv., For. Pest Leaf.* no. 10.

———, and Godwin, P. A. 1955. Description and bionomics of a new red pine scale, *Matsucoccus resinosae. Forest Sci.* 1:164–176.

McCambridge, W. F., and Pierce, D. A. 1964. Observations on the life history of the pinyon needle scale, *Matsococcus acalyptes. Entomol. Soc. Am. Ann.* 57:197–200.

McClure, M. S. 1976. Colonization and establishment of the red pine scale, *Matsucoccus resinosae,* in a Connecticut plantation. *Environ. Entomol.* 5:943–947.

———. 1977. Population dynamics of the red pine scale, *Matsucoccus resinosae:* The influence of resinosis. *Environ. Entomol.* 6:789–795.

McKenzie, H. L., et al. 1948. The Prescott scale (*Matsuccoccus vexillorum*) and associated organisms that cause flagging injury to ponderosa pine in the Southwest. *J. Agric. Res.* 76:33–51.

Morrison, H. 1928. A classification of the higher groups and genera of the coccid family, Margarodidae. *USDA Tech. Bull.* no. 52.

Parr, T. J. 1939. *Matsucoccus* sp., a scale insect injurious to certain pines in the Northeast. *J. Econ. Entomol.* 32:624–630.

Meristematic Insects (Terminal Insects)

The meristematic tissues of the tree provide food and shelter for many insect species. The term *meristematic tissue* is interpreted here to include not only the cambium layer and the growing tips and roots but also the adjacent soft portion of the xylem and phloem of the stem. In numbers of insect species, the meristem feeders are exceeded only by the leaf eaters.

GROUPS OF MERISTEMATIC INSECTS

In a discussion of insects feeding on the meristematic tissues, insects can be grouped according to their taxonomic position, the condition of their host, or the part of the tree on which they feed.

Taxonomic Position

Meristematic insects represent several orders. In the Diptera, there are a number of cambial miners in living trees. They do not kill trees, but their activities are responsible for certain xylem defects called *pith flecks*. Only occasionally, as in high-value hardwoods, are these insects numerous enough to be of economic importance. Some of the pitch midges (Cecidomyiidae) feed on the cambium and phloem of coniferous trees. The pitch exuding from the tree covers the larvae and serves as a protection for them. Although numerous, this family contains no dangerously injurious species. There are also numerous dipterous species that inhabit the cambium and phloem of dead trees or logs but that are not true meristem feeders. Either they feed on fungi or other microorganisms, or they prey on other meristem feeders. Thus, the Diptera living in meristematic tissues are not a very important group economically.

Of all the meristematic insects, those that feed on the cambium and adjacent soft tissues of xylem and phloem are by far the most injurious. These insects have been called *cambium insects,* but their tunneling is in the phloem; current usage tends toward the name *phloem insects* for this group. The larger proportion of the phloem insects belong to the order Coleoptera. The families of beetles most commonly feeding on phloem are the (1) weevils (Curculionidae), (2) bark beetles (Scolytidae), (3) long-horned borers (Cerambycidae), and (4) flatheaded borers (Buprestidae). Each of these families contains species highly specialized for life in the phloem, where many spend their entire developmental period. Others require the phloem tissues only during the early developmental stages and later are able to complete their development in the xylem.

Certain species of the Coleopterous family Curculionidae feed on the meristematic tissues of the stem and branches. Some of them, for example, the pales weevil, *Hylobius pales,* attack conifers and are known as *reproduction weevils.* When numerous, these insects materially reduce the merchantable value of plantations either at the time of planting or while being grown for Christmas trees, pulpwood, or sawtimber.

Many Lepidoptera are important meristematic insects. The family Tortricidae (=Olethreutidae) contains a number of typical meristematic insects; the tip moths furnish many illustrations.

Host Condition

Meristematic insects have sometimes been divided into two groups: primary and secondary. Unfortunately, these terms have been used rather loosely in previous entomological work; therefore, it is important that the terms be defined. A *primary* insect species is one able to attack a healthy, living tree and complete its normal development therein. A *secondary* insect species, in contrast, is incapable of attacking and completing normal development in a healthy tree. The practical application of these two terms is sometimes difficult because an observer may fail to recognize that a tree has ceased to be healthy. As a result, confusion has arisen, and different observers may arrive at different conclusions about the same insect species.

Feeding Place

Grouping the meristematic insects according to the part of the tree affected seems most convenient for discussion purposes. The divisions used in this chapter and the following three chapters are:

1   Insects in the terminal parts of trees
   a   tips and shoots
   b   roots
2   Insects in cones and seeds
3   Insects in the phloem of the trunk and branches
4   Insects in both phloem and xylem

In the following discussion of meristematic insects, only a few species can be considered. References to some others are included in the Bibliography for the convenience of foresters who will need to expand their study of these insects when future problems arise.

TERMINAL INSECTS

Tip and Shoot Insects

The tip and shoot insect group contains a multitude of insect species, some sufficiently abundant to be serious pests. They seldom kill the trees, but they can reduce the rate of terminal growth or cause deformities that affect the growth and merchantable value of timber, ornamentals, and Christmas trees. Only when these insects are active year after year, killing all or nearly all the new growth each year, can they kill trees.

Tip and Shoot Moths [Lepidoptera: Tortricidae (=Olethreutidae)] The lepidopterous family Tortricidae (=Olethreutidae) has many important forest pests, especially within the genus *Rhyacionia*. This genus is widely distributed throughout the range of pines in the Northern Hemisphere; there are about 35 known species.

Moths in the genus *Rhyacionia* are small, with reddish orange forewings cross-marked with gray bands or gray forewings with white bands (Fig. 15-1). Flight is usually crepuscular (twilight) or nocturnal (night). Most species are univoltine (i.e., having one generation a year); others are bivoltine or multivoltine, especially in the South. Oviposition occurs throughout the periphery of the tree. The first-instar larvae usually mine the needles; the later instars feed inside the buds and growing shoots. Pupation may occur within the larval gallery, but some species drop to the ground and pupate in cocoons. Overwintering usually occurs during the pupal stage.

When the lateral buds are destroyed, there is a reduction in the number of branches. The destruction of the terminal bud may cause an offset, fork, or post horn in the bole of the tree. There is usually no loss of height growth.

(a)

Figure 15-1 The Nantucket pine tip moth, *Rhyacionia frustrana,* is a small tortricid moth (a) that oviposits on southern pines. The trees are damaged by larval feeding in the buds and twigs (b). *(U.S. Forest Service photo.)*

(b)

Substantial losses do occur in the quality of the bole and to ornamentals and Christmas trees. The mining in the newly formed shoots, causing the needles to die and fade, creates an appearance that is alarming to the forester, Christmas tree grower, and/or homeowner.

The tip moths of significance to American forests will now be discussed. A recent review of the systematics and literature of the genus Rhyacionia has been made by Powell and Miller (1978). We shall follow their nomenclature and discuss the more important allopatric (separate-area) pairs of species. These pairs vary slightly in appearance and biology and feed on different species of pine.

Rhyacionia frustrana—Rhyacionia bushnelli These are eastern and midwestern species in the United States. *R. frustrana* was discovered on Nantucket Island, hence the common name, Nantucket pine tip moth. The insect is a pest of numerous pines species, especially loblolly, shortleaf, and Virginia pine in the South. *R. bushnelli* attacks ponderosa pines in the Great Plains. In the past, *R. bushnelli* has been often identified as the Nantucket pine tip moth (Graham and Baumhoffer, 1927; Powell and Miller, 1978).

In the southeastern United States, where four or five generations occur each season, the Nantucket pine tip moth has caused great concern. However, slash and longleaf pines are not seriously affected. Yates (1966) has demonstrated that the oleoresin from loblolly and shortleaf pines will crystallize within a short period after flow commences, owing to larval feeding. In contrast, no such effect on either slash or longleaf pines could be demonstrated. This difference may explain the relative resistance of the latter two species. The larvae can cope with the relative dryness of the crystallized resin, whereas they cannot survive a covering of uncrystallized resin.

In some localities, *R. bushnelli* has done very little damage; in others, it has been very injurious, especially to ponderosa pine seedlings and small saplings.

Rhyacionia rigidana—Rhyacionia subtropica These tip moths attack the southern pines. In the United States, the range of the pitch pine tip moth, *R. rigidana,* extends from Maine to Florida and westward to Missouri and Texas; the subtropical tip moth, *R. subtropica,* is found in the lower coastal plains of South Carolina and Georgia and along the Gulf Coast. *R. rigidana* usually infests pitch, shortleaf, and loblolly pines; *R. subtropica* appears to prefer slash pine as its host. Both species are multivoltine, overwintering as pupae.

Through the eastern United States, *R. rigidana* and *R. frustrana* are often found in the same pine plantations and occasionally in the crowns of mature trees. Recently, Berisford (1974) discovered that these sympatric (same-area) species are isolated not only by incompatible genitalia but also by an inhibition of the males' response to the female sex pheromones of the other species. In addition, the sex pheromones are released at different times during the evening. The generations of these species are asynchronous throughout the summer, in spite of both species emerging from overwintering pupae at the same time in the spring. *R. rigidana* has two generations in the upper Georgia Piedmont; *R. frustrana* has three.

The knowledge that two similar species with varying biologies exist within the same pine plantation is critical for timing chemical control measures.

Western Tip Moths The genus *Rhyacionia* attains its greatest diversity in the western United States. To date, 12 species have been identified; 11 of these species are known to infest ponderosa pine throughout the West (Fig. 15-2). Powell and Miller (1978) suggest that the increased speciation is the result of the wide range of hosts and geographically isolated habitats.

Unfortunately, little is known of the speciation and biologies of the western tip moths. However, with the intensifying of western forestry, especially genetic programs and the establishment of pine plantations, our knowledge of these insects will, of necessity, increase.

Rhyacionia buoliana This is an introduced species with the common name European pine shoot moth. First discovered on Long Island in 1914, the pest has now spread southward to the 40th parallel and westward to Wisconsin and Missouri. This species has also been accidentally introduced into the Pacific Northwest. The continued spread may well be limited; the insect is unable to survive the severe winter temperatures characteristic of most of the northern pine areas. A minimum temperature of $-8°C$ results in the nearly total mortality of the European pine shoot moth.

Nevertheless, in many northern areas, this univoltine species occurs far beyond the expected limit. Its persistence is explained by the protective

(a)   *(b)*

Figure 15-2   Young ponderosa pine can be damaged by the southwestern pine tip moth, *R. neomexicana;* boles are forked (*a*) or crooked (*b*). *(U.S. Forest Service photo.)*

snow covering that, in most winters, shelters the insect in young plantations (Fig. 5-1) or on certain ornamentals, such as Mugo pine, planted along highways. Sooner or later, a severely cold period without snow cover will destroy the outposts beyond the insect's normal range, but they will develop again.

The European pine shoot moth, *R. buoliana,* probably will continue to be a serious pest in pine plantations in southern New England, the central states, and other localities with a similar climate (Fig. 15-3). The damage to red pine, the preferred American host, can be extremely severe in young plantations (Fig. 15-3). Pine species that are less susceptible, such as Austrian, ponderosa, and Scots, are also damaged when planted in close proximity to infested red pine (Fig. 5-5).

Other Tip and Shoot Insects   Numerous other meristematic insects attack the terminal parts of trees. Some of them are decidedly injurious, although most of them are not usually important.

The injury by the lepidopterous shoot borers of the genuses *Eucosma* and *Dioryctria* resembles that caused by the *Rhyacionia,* the difference being that the larvae do not mine the buds but bore downward in the pith of larger shoots.

(a)   *(b)*

Figure 15-3 This red pine plantation on the Eli Whitney Forest, Yale University, (*a*) was severely damaged by the European pine shoot moth, *R. buoliana,* (*b*) during the 1930s drought. This plantation was destroyed by the red pine scale, *Matsucoccus resinosae,* in the 1970s. *(U.S. Forest Service and Virginia Polytechnic Institute and State University.)*

The damage to conifers is the loss of bole quality when the terminals are destroyed.

Numerous small bark beetles belonging to *Pityophthorus, Scolytus, Pityogenes, Orthotomicus, Ips* and buprestid or cerambycid borers such as the genera *Agrilus, Callidium,* and *Neoclytus* are frequently observed working in the terminal parts. Most of these are secondary and need not be considered here. References to some of the more important species will be found in the Bibliography.

Twig Borers

The twig borers are serious pests of deciduous ornamentals and shade trees in both the eastern and the western United States and Canada. Numerous insect species are involved, most belonging to the order Coleoptera. The most obvious symptoms of tree injury are either the dropping of green leaves when the petiole is severed or mined and/or the fading of leaves and dropping of twigs when the larvae mine the pith or cambium. These pests are most often reported by homeowners.

The aesthetic value of ornamental and shade trees of numerous hardwood species has been ruined by the twig pruner, *Elaphidionoides villosus* (Fig. 15-4). This cerambycid oviposits in slits cut in the apex of twigs. The larvae develop by feeding downward in the pith. In late summer, they sever the twig with circumferential galleries just under the bark. The twig breaks and falls; the larvae overwinter in the fallen twig. The ground under heavily infested trees is covered with twigs. Fortunately, the trees are seldom killed.

Other frequently reported pests include the maple petiole borer, *Caulocampus acericaulis,* a sawfly introduced from Europe that causes an early summer dropping of maple leaves; the dogwood twig borer, *Oberea tripunctata,* which mines newly formed twigs; and in the West, the Pacific oak twig girdler, *Agrilus angelicus,* which causes patches of dead twigs, especially in California live oak.

Twig borers are serious pests in southern cottonwood nurseries and plantations. The larvae of the cottonwood twig borer, *Gypsonoma haimbachiana,*

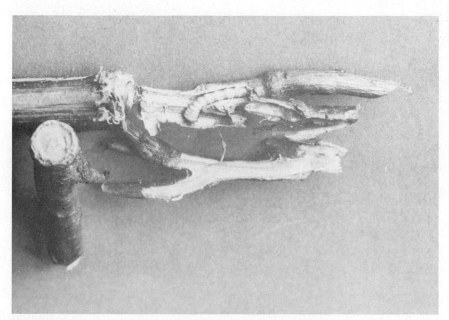

Figure 15-4   The twig pruner, *Elaphidionoides villosus,* is a pest of many hardwood species. The larvae cut circumferential galleries to sever the twig, then pupate on the ground. *(U.S. Forest Service photo.)*

Table 15-1   Host Tree Genera of 11 North American Species of *Pissodes**

Pissodes	Tree genera
similis Hopkins	*Abies*
(= *utahensis* Hopkins)	
strobi (Peck)	*Pinus, Picea*
(= *sitchensis* Hopkins)	
(= *engelmanni* Hopkins)	
approximatus Hopkins	*Pinus, Picea*
(= *canadensis* Hopkins)	
nemorensis Germar	*Pinus*
terminalis Hopping	*Pinus*
schwarzi Hopkins	*Larix, Picea, Pinus*
(= *yosemite* Hopkins)	
radiatae Hopkins	*Pinus*
fasciatus LaConte	*Pseudotsuga*
rotundatus LeConte	*Pinus, Picea, Tsuga*
(= *nigrae* Hopkins)	
(= *alascensis* Hopkins)	
dubius Randall	*Abies*
(= *fraseri* Hopkins)	
(= *piperi* Hopkins)	
affinis Randall	*Pinus*
(= *curriei* Hopkins)	

* After Smith and Sugden (1969).

bore within the buds and pith, destroying terminal growth. This insect is found throughout the East, attacking all *Populus* spp. The curculio weevils, which will be discussed in Chapter 16 as cone and seed insects, feed as adults on the twigs and leaf petioles of walnuts. Shoots may be killed back to the preceding year's growth; infrequently, trees are defoliated.

Terminal Weevils (Coleoptera: Curculionidae)   This group of meristematic insects feeds within the terminals of conifers. These pests do not kill trees, but the form of the bole is distorted. When height growth continues from one or more of the lateral branches, the bole is either crooked or forked. There may be an economic or aesthetic loss whether the tree is used for lumber, pulp, landscape planting, or as a Christmas tree. The terminal weevils belong to the coleopterous genus *Pissodes* of the family Curculionidae, the weevils. The genus formerly contained many species, but cytological testing by Smith and Sugden (1969) of 20 of the more important species reduced the number to 11 acceptable species (Table 15-1).

White Pine Weevil   The most conspicuous meristematic insect of eastern white pine and Sitka spruce is the white pine weevil, *Pissodes strobi.*[1]

[1]The Sitka spruce weevil was formerly classified as *P. sitchensis.* It is now *P. strobi,* but the common name has been retained.

The adults are reddish brown weevils with distinct white markings on the elytra (Fig. 15-5a). The adults pass the winter in the litter beneath the trees. In the spring, they emerge from hibernation to feed on the buds and inner bark of the previous year's leaders of young pines and spruces. The females then deposit their eggs in a chamber chewed out of the phloem. The larvae work downward beneath the bark, destroying the phloem as they proceed. Full growth may be attained before the next whorl of branches is reached, but in some instances, the larvae may pass below the second or even the third whorl before reaching full development. When full grown, the larvae construct pupal cells in the xylem or pith, there transforming to pupae. In late July and August, the young adults emerge and feed for a short time on the inner bark of the new growth before hibernating.

The symptoms of *Pissodes* attack begin in the early spring, when the adults feed on the terminal and upper branches. The resin exuding from feeding punctures will glisten in the sun, then harden to whitish beads. As the larvae develop within the previous year's terminal, the expanding current year's terminal will droop. Within a short time, the needles fade to yellow, then to red. When the adults emerge, their exit holes suggest that the terminal had been hit with shotgun pellets. The dead terminals may remain on the trees for many years (Fig. 15-5b).

Although the white pine weevil is primarily a pest of sapling pines and spruces, its activities are not confined entirely to such trees. Occasionally, large trees may be attacked. The age at which the trees are most susceptible is between 3 and 15 years. After the 20-year mark is passed, the proportion of infested leaders falls rapidly and soon becomes of little consequence in closed stands. However, in a few areas, the weevil continues to attack regardless of the height of the trees.

The terminal growth of the tree is resumed by one or more of the lateral branches. There is an offset in the vertical axis of the bole which is usually overcome within 4 years. Often two laterals will assume height growth, thus forming a forked tree. The result in either case is a deformed bole. Lumber sawn from weeviled trees may be reduced in total value by an average of 25 percent. The wood, both above and below the crook in the bole, is heavily compressed reaction wood. There is also an increase in encased knots, wane, and fungous development.

The weevil also affects the total cubic bole volume produced. The inherent difference between the length of the terminal and the lateral branches results in a loss in height growth when the laterals take over height growth. Height may be reduced by as much as 10 percent. The loss of height and diameter is especially serious when the rotation age of the stand is fixed. Sitka spruce bole volumes in coastal Oregon and Washington are reduced 30 percent when the rotation age is 60 years; also eastern white pine may lose as much as 20 percent of the total cubic bole volume.

(a)   (b)

Figure 15-5   Repeated attack by the white pine weevil, *Pissodes strobi,* (*a*) can eliminate the commercial timber value of eastern white pine (*b*). Note remnant weeviled leader (*arrow*). *(Graham and Knight, Principles of Forest Entomology, 4th ed., Fig. 66, p. 299, and New York State Museum and Science Service.)*

Reproduction Weevils (Coleoptera: Curculionidae)   Another important group of meristematic pests of young pine of all species are the reproduction weevils. In the intensively managed southern forests, they may well be the most important insect pests of young pine plantations. These weevils are normally secondary pests of mature pine, but the adults can be extremely injurious to seedlings.

Pales Weevil   The most common reproduction weevil is the pales weevil, *Hylobius pales* (Fig. 15-6). The adults are brown with light-colored markings on the elytra. This is a nocturnal species that breeds in freshly cut logs, stumps, and injured trees. The eggs are laid in the bark during the early spring months. The larvae, resembling the larvae of the bark beetles, develop in the cambium region. When full grown, they form pupal cells by cutting depressions in the surface layers of the sapwood and roofing them over with shredded wood. This type of pupal case is often called a *chip cocoon*.

The adults emerge in the spring months in the North and from May to November in the South. There are at least two overlapping generations in the South, but 2 years are needed to complete one generation in the North. Adults may live as long as 2 years. In cutover pine areas, the small pine seedlings provide the only fresh phloem available for adult feeding. For this reason, the young seedlings are attacked and destroyed by adult weevils. One weevil can kill a seedling during one night's feeding.

The adults exhibit a positive attraction to the volatiles released by pines, especially the monoterpene, α-pinene. When pines are pruned, sheared, or felled, pales weevils are attracted by the volatilization of α-pinene and may fly into the area in great numbers. The flight capability is known to be at least 7 mi (11.3 km).

Pitch-eating Weevil   The pitch-eating weevil, *Pachylobius picivorus,* is similar in appearance and habit to the pales weevil, with which it is commonly found. However, the pitch-eating weevil is slightly larger and longer-lived. This species is more prevalent in the Deep South, often comprising more than 90 percent of the populations of reproduction weevils in the Gulf Coast states. The adults are active the year around. They do not enter diapause; instead, they only become less active during December and January.

Western Reproduction Weevils   In California, the pine reproduction weevil, *Cylindrocopturus eatoni,* is a serious pest of conifers, especially ponderosa pine seedlings in plantations with large amounts of weeds and brush. The adults feed on the needles and shoots, leaving puncture wounds. The larvae mine within the bark. The first symptom of damage is the fading of needles in the fall.

Throughout the Pacific Northwest, the Douglas-fir twig weevil, *C. furnissi,* attacks the shoots of young Douglas-fir. The adults feed on the shoots, then oviposit within the bark. The larvae mine the phloem tissue, killing the shoots. This weevil has become a serious pest in Christmas tree plantations, especially on dry sites and during drought years.

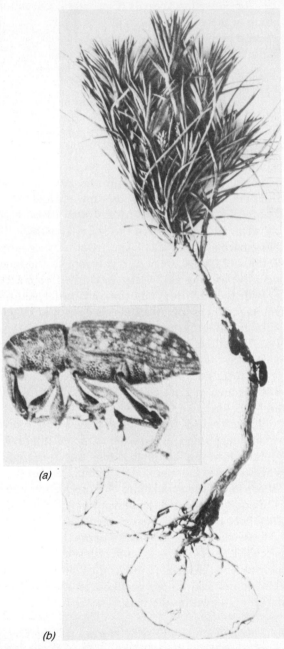

(a)

(b)

Figure 15-6 The pales weevil, *Hylobius pales* (*a*) is a serious pest in the establishment of southern pine plantations. The adults feed on phloem tissue (*b*) and may destroy a seedling during one night's feeding. *(U.S. Forest Service photo.)*

ROOT INSECTS

The terminal insects considered thus far are those that injure above-ground parts. There are other terminal insects that feed on the small rootlets beneath the ground. These root-feeding species are usually, but by no means always, enemies of seedlings. Trees growing in nurseries and saplings in the forest are damaged as a result of the activities of root-feeding insects that eat the fine roots and rootlets. Large roots are also fed upon. For example, some species of weevils burrow into the soil as adults and deposit their eggs in the roots; others deposit their eggs above ground and their larvae mine into the below-ground parts.

White Grubs (Coleoptera: Scarabaeidae)

The most important insects feeding on roots belong to the white grub group represented by species of the genus *Phyllophaga*. Various species belonging to other genera within this family also feed on the roots of trees but are not nearly so important.

The white grubs are the larvae of the beetles commonly called *May* or *June* beetles. The adults are about 8 mm in length, heavy-bodied, and brown or black in color. They emerge from the ground in early spring to feed nocturnally on the foliage of trees. At times, the beetles are so abundant that they may completely strip stands of oak. These nocturnal beetles are strongly attracted to light.

The females deposit their eggs in the ground close to their food plants. The degree of larval infestation depends more on proximity to food for the adults than on the kind of vegetative ground cover. The life cycle may show some variations from species to species, but in general, the young hatch in summer and feed for the remainder of the season on organic material and roots near the surface of the soil. When winter comes, most species burrow into the ground and hibernate. The next spring, they work their way upward toward the surface and continue to feed on the roots of trees or herbaceous vegetation. With the arrival of fall, they again burrow into the ground for hibernation; and the next spring, they continue to feed on roots. The larval growth during this final season of development is completed by midsummer. Then the larvae make pupal cells in the ground, where they transform to pupae. After an interval of a few weeks, transformation to the adult takes place; the beetles remain in the ground until the following spring. Thus, 3 full years are usually required to complete the life cycle.

In the South, the cycle may be reduced to 2 years; but in the extreme northern part of the United States and in a large part of Canada, completion of the cycle requires 4 or possibly 5 years. In the West, the May beetles are abundant in Arizona, and their life cycle varies from 2 to 4 years (Fig. 15-7).

White grubs are sometimes exceedingly injurious in forest nurseries. Seed-

(a)   *(b)*

Figure 15-7 The May or June beetles seriously damage pine seedlings throughout North America. In the West, the tenlined June beetle, *Polyphylla decemlineata,* adults feed on ponderosa pine needles (*a*), and the grubs (larvae) feed on the roots (*b*). *(U.S. Forest Service photo.)*

ling damage almost always occurs in locations within a few hundred yards of the food plants of the adults. Oak is one of the favorite adult foods for several important species.

In forest plantations, the damage from white grubs is sometimes tremendous. In Michigan during the severe droughts of the 1930s, the grubs completely destroyed many young plantations. It is not uncommon for grubs to destroy from 10 to 25 percent of the planted seedlings in a single season.

Agrilus Borers (Coleoptera: Buprestidae)

The aspen root borer, *Agrilus horni,* attacks and kills small aspens, usually less than 20 mm in diameter, in the Lake States (Fig. 1-3). The eggs are deposited during an extended period in summer, on the bole a short distance above the ground. The young larvae tunnel in the phloem downward into a root for 15 cm or more from the root crown, then work back to the above-ground stem, cutting a spiral gallery around the root and the lower part of the stem. The pupal cell is formed in the center of the stem. This *Agrilus* is one of the common pests of aspen suckers if they are growing either on a relatively poor site or in a stand of low density. The injury to the roots is usually associated with, and supplemented by, root-rotting fungi.

Chemical control is impractical in the forest. If injury by the aspen root *Agrilus* is to be avoided, stand density of between 4000 and 12,000 suckers per acre should be maintained until the trees are in their fifth year.

Root Weevils (Coleoptera: Curculionidae)

Several root weevils belonging to the genus *Otiorhynchus (=Brachyrhinus)* are sometimes very injurious to coniferous trees in the nursery. For example, the strawberry root weevil, *Otiorhynchus ovatus,* is a general feeder that attacks many different kinds of plants. The larvae live in the soil and eat bark from the small roots. This stripping of the roots can result in the death of a plant. Almost all species of conifers are attacked, but the beetles in a given locality may have a special preference for certain species. Hemlock, white cedar, and various spruces are attractive to them.

Foresters are often asked to solve insect problems of ornamental shrubs. Perhaps the most common problem in the South is the black vine weevil *O. sulcatus* feeding on azaleas and rhododendrons. This *Otiorhynchus* species strips the roots during the larval stage. The adult feeds nocturnally on the leaves, leaving notches or holes in the leaf margin.

Other Root Insects

In addition to the white grubs and weevils, there are a number of insects of lesser importance that feed on the roots of trees. Some species of the order Diptera attack the root meristem, but they are of minor economic importance. For example, some of the cecidomyid gall insects attack the small roots and produce galls.

Soil-inhabiting insects such as wireworms (Coleoptera: Elateridae) will feed on the small roots and, if numerous, will cause considerable injury to seedlings, either in the forest or in the nursery. Like the white grubs, they feed during the larval stage on vegetable material, including plant roots. The adults of this family are commonly called *click beetles.* They are heavily chitinized, flattened beetles, resembling some of the buprestids in form. The larvae are slender, elongate, and heavily chitinized, hence the name *wireworm.* The legs are feebly developed.

These insects are found most commonly in heavy, moist soils that contain much plant material. Small trees growing in such locations are most likely to suffer injury from these pests. Soil insecticides have proved to be a satisfactory way to control wireworms, but the precautions discussed in Chapter 11 should be observed. During moist seasons, moderate injury to the tree roots is repaired promptly by vigorous root growth. Then young trees are able to become established in spite of considerable injury. However, in dry years, even light damage may cause tree death. Because root-eating insects are hidden in the soil, much of their injury escapes observation. Not until the damaged trees show above-ground evidence of their activities are these insects noticed.

PREVENTION

The prevention of damage by the terminal insects begins with selecting a tree species suitable for the site and keeping within the range of the chosen species. Then the trees must be kept in a vigorous condition, free from competition. Insect-caused damage resulting in economic and aesthetic losses would be sharply curtailed if foresters did not plant excessively droughty soils with pines of the subsections Sylvestres and Australes. The subsection Contortae, for example lodgepole, jack, and Virginia pine, is much better adapted for dry sites.

Tip and Shoot Moths

The tip and shoot moths are most often serious pests of young pines growing in the open. They are seldom found on young pines growing in the understory, such as the natural regeneration in the shelterwood system.

These insects are usually pests on pines planted beyond their range or on dry sites. For example, when slash pine is planted within the range of south Florida slash pine, the seedlings and saplings are seriously damaged by the subtropical pine tip moth, *R. subtropica* (Bethune, 1963). Similarly, loblolly pine planted on northwest Florida sand hills is seriously damaged by *R. subtropica* yet local sand pine is not affected (Burns, 1975). However, when sand pine is planted in the Georgia-Carolina sand hills, the young trees are attacked by tip moths. The planting of pines in the Nebraska sand hills and the subsequent heavy infestation of tip moths provides another example.

Outbreaks of the European pine shoot moth, *R. buoliana,* on red pine in Michigan have been associated with soil moisture stress during the needle-mining larval stage (Heikkenen, 1963) . Damage to the tree increases on trees developing shallow root systems and when consecutive years with hot, dry summers occur. However, when trees are deeply rooted, they do not become infested, regardless of the weather. The majority of red pine plantations damaged by *R. buoliana* have been planted beyond the tree's range.

Twig Borers

The prevention of twig borer injury to ornamentals can best be accomplished by maintaining the vigor of the tree, pruning infested twigs when possible, and destroying the leaves and twigs that fall to the ground. In the forest, prevention is out of the question.

Terminal Weevils

Studies in New England, New York, and Minnesota have shown conclusively that naturally seeded eastern white pine, growing in dense stands, is less subject to weevil damage than trees growing in open stands. However, many white pine plantations have been unduly weeviled because of open spacing or off-site planting. Unless this damage can be repaired, the trees will produce little (if any) valuable wood. Good results can be obtained by pruning crop trees so that they will recover from weevil injury. This method of rehabilitating damaged stands has proved very profitable. The trees should be pruned as soon as possible following weevil attack. The forester should, at this time, select the crop trees; there may well be 100 lightly damaged trees distributed throughout an acre.

Eastern white pine growing under aspen or paper birch and Sitka spruce growing under red alder are seldom attacked by terminal weevils. The hardwood shade reduces the temperatures on and around the terminal below the optimum for weevil feeding and oviposition.

The association of weevil damage with dry sites is well known. In New York, Connola and Wixson (1963) concluded that white pine can be grown with a minimum of weevil damage when soil mottling and/or hardpan are not present to a depth of 3 ft. These symptoms of poor internal drainage of the soils occurred primarily on heavy soils in southern New York, rather than on light soils in the Adirondack region.

The increased incidence of weevil damage near the southern range of a tree species also occurs in the coastal forests of the Pacific Northwest; weevil damage to Sitka spruce is serious in Oregon and Washington. In the South, loblolly and slash pine, planted on dry, sandy sites, have been infested with the deodar weevil, *Pissodes nemorensis.*

Reproduction Weevils

In the southern United States, contiguous cutting operations in pine forests have resulted in a great increase in the number of reproduction weevils. The weevils are initially attracted by the volatiles released by the new cuttings, where they feed on fresh phloem. The next generation of adults also feeds on and may kill most of the nearby natural or planted seedlings. Then the weevils are attracted into an adjacent cutting block.

The reproduction weevils are also serious pests in Christmas tree plantations that are selectively harvested or sheared. The insects are attracted into the plantations when trees are sheared or harvested. Then feeding occurs on the established trees and newly planted seedlings.

Prevention of damage is difficult. At present, we do not have a predictive technique; the weevils may or may not appear. Although they are not common in pine plantations established after harvesting stands of pure hardwoods, the weevils may appear following logging of mixed pine-hardwood and pure pine stands.

Foresters should walk through pine plantations frequently during the first year and a half of establishment to estimate the percentage of seedling survival.

Weevil damage is difficult to see; the small, dead seedlings fade in color and fall into the dry weeds.

If planting is delayed for a season after cutting, the weevils will have emerged and left the locality. The reason is that there is no host attractant in the area; the freshly cut pine surfaces will have glazed over. The delay in planting need be done only once in managed forests; the time lag in the establishment of plantations is then in annual synchrony with the acreages being harvested. This is an important factor in the management of pine plantations. The forester must compare the costs of delayed planting with the costs of losing 1 year's growth (also the original and repeated costs of site preparation and planting) as prorated against the expected returns over a 20 or 30 year rotation.[2]

Root Insects

The prevention of root-feeding insects is crucial when forest nurseries are established. The tree species that are the preferred hosts of the adult beetles inhabiting the area should be removed from the adjacent forest. The plant cover, especially the grasses, must be destroyed and allowed to remain fallow for several years or killed with chemicals. Prior to planting, the soil should be fumigated.

CONTROL

The direct use of insecticides to control meristematic insects is not warranted in the forest. Regarding the establishment of plantations, planting stock should be treated routinely with insecticides, either as a dip or root slurry (see Chapter 11). The planting crews and foresters must know they are working with poisons and be carefully supervised. Chemical treatment after the plantations are established is seldom warranted.

Within residential and recreational areas, these pests seldom (if ever) kill trees. Direct control measures should consist of the removal and burning of infested tips and twigs. Highly valued ornamental trees and shrubs can be protected with soil applications of systemic insecticides or with foliage sprays of the contact insecticides.

The intensive management of Christmas tree plantations permits the shearing and destruction of infested tips and the application of contact and systemic insecticides with mist blowers.

The control of meristematic insects in seed production areas and seed orchards is discussed in Chapter 16.

[2]Personal communication, R. L. Hedden, Weyerhaeuser Company.

BIBLIOGRAPHY

Tip and Shoot Insects

Tip Moths, *Rhyacionia* spp.

MacKay, M. R. 1959. Larvae of the North American Olethreutidae. *Can. Entomol. Supp.* no. 10.

Powell, J. A., and Miller, W. E. 1978. Nearctic pine tip moths of the Genus *Rhyacionia:* Biosystematic review. *USDA For. Serv. Agric. Handbook* no. 514.

Rhyacionia adana Heinrich

Martin, J. L. 1960. Life history of the pine tip moth, *Rhyacionia adana* Heinrich, in Ontario. *Can. Entomol.* 92:725–728.

Rhyacionia buoliana (Schiffermuler), European pine shoot moth

Carolin, V. M., and Daterman, G. E. 1974. Hazard of European pine shoot moth to western pine forests. *J. Forestry* 72:136–140.

Heikkenen, H. J. 1963. Influence of site and other factors on damage by the European pine shoot moth. Ph.D. Diss., University of Michigan (Univ. Micro. No. 64-6689).

Miller, W. E. 1967. The European pine shoot moth: Ecology and control in the Lake States. *For. Sci. Mono.* no. 14.

Rhyacionia frustrana (Comstock), Nantucket pine tip moth

Berisford, C. W., and Kulman, H. M. 1967. Infestation rate and damage by the Nantucket pine tip moth in six loblolly pine stand categories. *For. Sci.* 13:428–438.

Graham, S. A., and Baumhoffer, L. G. 1927. The pine tip moth in the Nebraska National Forest. *J. Agric. Res.* 35:323–333.

Harms, W. R. 1969. Sand pine in the Georgia-Carolina sandhills: Third-year performance. *USDA For. Serv. Res. Note* SE 123.

Yates, H. O., III. 1966. Susceptibility of loblolly and slash pines to *Rhyacionia* spp. oviposition, injury, and damage. *J. Econ. Entomol.* 59:1461–1464.

Rhyacionia neomexicana (Dyar), southwestern pine tip moth

Jennings, D. T. 1975. Life history and habits of the southwestern pine tip moth, *Rhyacionia neomexicana* (Dyar). *Ann. Entomol. Soc. Am.* 68:597–606.

Lessard, G., and Jennings, D. T. 1976. Southwestern pine tip moth damage to Ponderosa pine. *USDA For. Serv. Res. Pap.* RM-168.

Rhyacionia rigidana (Fernald), pitch pine tip moth

Berisford, E. W. 1974. Species isolation mechanisms in *Rhyacionia frustrana* and *R. rigidana. Ann. Entomol. Soc. Am.* 67:292–294.

Miller, W. E., and Neiswander, R. B. 1959. The pitch pine tip moth and its occurrence in Ohio. *Ohio Agric. Exp. Sta. Res. Bull.* no. 840.

————, and Wilson, L. F. 1964. Composition and diagnosis of pine tip moth infestations in the Southeast. *J. Econ. Entomol.* 57:722–726.

Rhyacionia subtropica Miller, subtropical pine tip moth

Bethune, J. E. 1963. Pine tip moth damage to planted pines in south Florida. *USDA For. Serv. Res. Note* SE-7.

Burns, R. M. 1975. Tip moth control pays off. *For. Farm.* 34:13.

McGraw, J. R. 1975. The biology of *Rhyacionia subtropica* Miller. Ph.D. Diss., University of Florida.

Merkel, E. P. 1963. A new southern pine tip moth. *J. Forestry* 61:226–227.

Rhyacionia zozana (Kearfott), ponderosa pine tip moth

Stevens, R. E. 1966. The ponderosa pine tip moth, *Rhyacionia zozana,* in California. *Ann. Entomol. Soc. Am.* 59:186–192.

Shoot Borers, *Eucosma* spp.

Eucosma gloriola Heinrich, eastern pine shoot borer

DeBoo, R. F., et al. 1971. The eastern pine shoot borer, *Eucosma gloriola* in North America. *Can. Ent.* 103:1473–1486.

Eucosma sonomana Kearfott, western pine shoot borer

Stevens, R. E., and Jennings, D. T. 1977. Western pine-shoot borer: A threat to intensive management of ponderosa pine in the Rocky Mountain area and Southwest. *USDA For. Serv. Gen. Tech. Rept.* RM-45.

Stoszek, K. J. 1973. Damage to ponderosa pine plantations by the western pine-shoot borer. *J. Forestry* 71:701–705.

Reproduction Weevils

Hylobius pales (Herbst), pales weevil

Bullard, A. T., and Fox, R. C. 1969. Field observations on the pales weevil, *Hylobius pales,* in the South Carolina Piedmont. *Ann. Entomol. Soc. Am.* 62:722–723.

Carter, E. E. 1916. *Hylobius pales* as a factor in the reproduction of conifers in New England. *Proc. Soc. Am. For.* 11:297–307.

Finnegan, R. J. 1959. The pales weevil, *Hylobius pales* (Herbst) in southern Ontario. *Can. Entomol.* 91:664–670.

Manwan, I. 1964. The biology of the pales weevil, *Hylobius pales* (Herbst) in Arkansas. M.S. Thesis, University of Arkansas.

Pierson, H. B. 1921. The life history and control of the pales weevil (*Hylobius pales*). *Harvard For. Bul.* no. 3.

Speers, C. F. 1958. Pales weevil rapidly becoming a serious pest of pine reproduction in the south. *J. Forestry* 56:723–726.

———, and Ebel, B. H. 1971. Pales and pitch-eating weevils: Ratio and period of attack in the south. *USDA For. Serv. Res. Note* SE-156.

———, and Rauschenberger, J. L.1971. Pales weevil. *USDA For. Serv., For. Pest Leaf.* no. 104.

Thomas, H. A., and Hertel, G. D. 1969. Responses of the pales weevil to natural and synthetic host attractants. *J. Econ. Entomol.* 62:383–386.

Root Insects

Hylocius aliradicus Warner

Ebel, B. H., and Merkel, E. P. 1967. *Hylobius* weevil larvae attack roots of young slash pines. *For. Sci.* 13:97–99.

Hylobius radicis Buchanan, pine root-collar weevil

Finnegan, R. J. 1962. The pine root collar weevil, *Hylobius radicis* Buch. in southern Ontario. *Can. Ent.* 94:11–17.

———. 1964. Control of the pine root-collar weevil, *Hylobius radicis. J. Econ. Entomol.* 55:483–486.

Schaffner, J. F., Jr., and McIntyre, H. L. 1944. The pine root-collar weevil. *J. Forestry* 42:269–275.

Wilson, L. F. 1967. Effects of pruning and ground treatments on population of the pine root-collar weevils. *J. Econ. Entomol.* 60:823–827.

Cylindrocopturus eatoni Buchanan, pine reproduction weevil

Callahan, R. Z. 1960. Resistance of pines to the pine reproduction weevil, *Cylindrocopturus eatoni. J. Econ. Ent.* 53:1044–1048.

Stevens, R. E. 1971. Pine reproduction weevil. *USDA For. Serv., For. Pest Leaf.* no. 15.

Pachylobius picivorus (Germar), pitch-eating weevil

Franklin, R. T. 1963. The reproduction weevil problem. *For. Farmer* 23, no. 1:12, 18.

———, and Taylor, J. W. 1970. Biology of *Pachylobius picivorus* in the Georgia Piedmont. *Can. Ent.* 102:962–968.

Thatcher, R. C. 1960. Influence of the pitch-eating weevil on pine regeneration in east Texas. *For. Sci.* 6:354–361.

Weevils, *Pissodes* spp.

Pissodes approximatus Hopkins, northern pine weevil

Finnegan, R. J. 1958. The pine weevil, *Pissodes approximatus* Hopk. in southern Ontario. *Can. Entomol.* 90:348–354.

————, and Godwin, P. A. 1967. Northern pine weevil, *Pissodes approximatus* Hopk. *Can. Dept. For. Pub.* no. 1180:145–147.

Wilson, L. F., and Schmiege, D. C. 1965. Pine root-collar weevil. *USDA For. Serv., For. Pest Leaf.* no. 39.

Pissodes nemorensis Germar, deodar weevil

Ollieu, M. M. 1971. Damage to southern pines in Texas by *Pissodes nemorensis. J. Econ. Entomol.* 64:1456–1459.

Pissodes strobi (Peck), white pine weevil

Belyea, R. M., and Sullivan, C. R. 1956. The white pine weevil: A review of current knowledge. *Forestry Chron.* 32:58–67.

Brace, L. G. 1971. Effects of white pine weevil damage on tree height, volume, lumber recovery and lumber value in eastern white pine. *Canad. For. Serv. Pub.* no. 1303.

Connola, D. P., and Wixson, E. C. 1963. White pine weevil attack in relation to soils and other environmental factors in New York. *New York St. Museum Sci. Serv. Bull.* no. 389.

Graham, S. A. 1926. The biology and control of the white pine weevil, *Pissodes strobi* Peck. *Cornell Univ. Agr. Expt. Sta. Bull.* no. 449.

Harman, D. M. 1971. White pine weevil attack in large white pines in Maryland. *Ann. Entomol. Soc. Am.* 64:1460–1462.

Hastings, A. R., and Godwin, P. A. 1970. White pine weevil. *USDA For. Serv., For. Pest Leaf.* no. 21.

Hopkins, A. D. 1907. The white pine weevil. *USDA Bur. Entomol. Circ.* no. 90.

Kulman, H. M., and Harman, D. M. 1967. Unsuccessful attacks by the white pine weevil in Virginia. *J. Econ. Entomol.* 60:1216–1229.

MacAloney, H. J. 1932. The white pine weevil. *USDA Circ.* no. 221.

Smith, S. G., and Sugden, B. A. 1969. Host trees and breeding sites of native North American *Pissodes* bark weevils with a note on synonymy. *Ann. Entomol. Soc. Am.* 62:146–148.

Sullivan, C. R. 1961. The effect of weather and the physical attributes of white pine leaders on the behavior and survival of the white pine weevil, *Pissodes strobi* Peck, in mixed stands. *Can. Entomol.* 93:721–741.

Taylor, R. L. 1930. The biology of the white pine weevil, *Pissodes strobi* Peck, and a study of its insect parasites from an economic viewpoint. *Entomol. Am.* (n.s.) 10:1–86.

Pissodes strobi (=*sitchensis*) (Peck), Sitka spruce weevil

Carlson, R. L. 1966. The effect of the Sitka spruce weevil on Sitka spruce. M.F. Thesis, University of Washington, College of Forest Research.

Gara, R. I., et al. 1971. Influence of some physical and host factors on the behavior of the Sitka spruce weevil, *Pissodes sitchensis,* in southwestern Washington. *Ann. Entomol. Soc. Am.* 64:467–471.

Silver, G. T. 1968. Studies on the Sitka spruce weevil, *Pissodes sitchensis,* in British Columbia. *Can. Entomol.* 100:93–110.

Pissodes terminalis Hopping, lodgepole terminal weevil

Drouin, J. A., et al. 1963. Occurrence of *Pissodes terminalis* Hopping in Canada: Life history, behavior and cytogenetic identification. *Can. Entomol.* 95:70–76.

Stark, R. W., and Wood, D. L. 1964. The biology of *Pissodes terminalis* Hopping in California. *Can. Entomol.* 96:1208–1218.

Pissodes schwarzi (=yosemite) Hopkins, Yosemite bark weevil

Stevens, R. E. 1966. Observations on the Yosemite bark weevil in California. *Pan-Pac. Ent.* 3:184–189.

Twig Borers and Girdlers

Proteoteras willingana (Kearfott), Boxelder twig borer

Peterson, L. O. T. 1958. The boxelder twig borer, *Proteoteras* willingana (Kearfott). *Can. Ent.* 90:639–646.

Agrilus angelicus Horn, Pacific oak twig girdler

Brown, L. R., and Eads, C. O. 1965. A technical study of insects affecting the oak tree in southern California. *Calif. Agr. Exp. Sta. Bul.* no. 810.

Gypsonoma haimbachiana (Kearfott), cottonwood twig borer

Morris, R. C. 1960. Biology of *Gypsonoma haimbachiana,* a twig borer in eastern cottonwood. *Ann. Ent. Soc. Am.* 60:423–427.

Chapter 16

Meristematic Insects (Cone and Seed Insects)

In the forest, the meristematic insects that feed on cones and seeds are relatively unimportant economically; the abundance of seed produced far exceeds the number required for adequate reproduction. Most tree species produce heavy seed crops periodically, the heavy crops being interspersed by years when seed production is light.

Much attention has been directed recently toward the improvement of tree varieties by breeding genetically superior individuals. Superior genetic material has been selected and vegetatively propagated in seed orchards in which the trees are widely spaced, cultivated, fertilized, and protected from domestic stock and wild browsing animals. The first American seed orchard was established in the South in the early 1950s (Dorman, 1976); by 1974, there were more than 243 seed orchards with 9263 acres in the United States, over 80 percent of the acreage being in the South (U.S. Forest Service, 1974) (Table 16-1). These orchards and the forest nurseries are now meeting much of the demand for planting stock (U.S. Forest Service, 1978).

The acreage in seed production areas has also increased in recent years. These are areas of high-quality trees of a certain species selected for the purpose of increasing seed production. The stands are improved by cutting the

Table 16-1   Acreage of Seed Orchards in the United States, 1974*

Region	State	Federal	Industrial	Total
North	563	176	16	755
West	20	640	250	910
South	2731	1150	3717	7598
Total	3314	1966	3973	9263

* Derived from U.S. Forest Service (1974).

less desirable individual trees and removing the competing vegetative under-growth to release the selected trees. A seed production area is treated in much the same way as a seed orchard and will require special attention in order to control insect pests (Fig. 16-1). This conclusion was reached by a study of preharvest losses in slash pine seed production areas in the Southeast (DeBarr and Barber, 1975) and red pine in the North (Mattson, 1978). Losses between flowering and maturing of the cones could, without protection, reach the 60 to 70 percent level.

The most important orders and genera of cone and seed insects are:

Coleoptera
 Conophthorus, cone beetles
 Curculio, nut weevils
Diptera
 Contarinia, cone midges
 Rhagoletis, husk flies
Hemiptera
 Leptroglossus, seed bugs
 Tetrya, seed bugs
Hymenoptera
 Megastigmus, seed chalcids
Lepidoptera
 Cydia, seedworms
 Dioryctria, coneworms
 Eucosma, cone borers
Thysanoptera
 Gnophothrips, thrips

These are the genera most frequently encountered, but they are by no means the only destroyers of tree seed. For example, infestations of aphids, scales, tip moths, and many defoliators also materially reduce the production of tree seed.

INSECTS FEEDING ON FLOWERS

The forester must be aware that the insects which feed on the shoots and twigs will reduce the number of flower-bearing branches. These meristematic insects were discussed in Chapter 15; of special importance are the tip and shoot

Figure 16-1 The study of insects damaging seed production is greatly facilitated by the use of hydraulic hoists. *(U.S. Forest Service photo.)*

moths and the terminal weevils. The larvae of some of the sawflies, loopers, and budworms also feed on flowers. Prevention and control of these pests is important in the management of seed orchards and seed production areas.

Thrips (Thysanoptera: Phlaeothripidae)

The thrips are members of the primitive order Thysanoptera. These small, slender insects are usually wingless and have rasping mouthparts. The thrips are well known for the damage they cause to garden flowers, but until recently, they were not considered important on forest trees. However, the intensive research now being conducted in southern seed orchards has shown that the thrips are very destructive.

The life history of the slash pine flower thrips, *Gnophothrips fuscus,* is not completely known at present. We do know that this cryptic species hides in the bracts and scales of slash pine cones, feeding on the female buds and flowers from January to mid-February. The infested flowers exude resin, shrivel, harden, and die. If an infested flower happens to survive, the mature cone is crooked, with concave areas having deformed scales. In slash pine seed orchards, flower mortality has reached 45 percent, and seed yields from infested cones are reduced by more than 50 percent.

INSECTS FEEDING ON CONES

Many insects that feed on the shoots, buds, and flowers also feed on the cones, especially those of the lepidopterous genus *Dioryctria*. The different species of *Dioryctria* vary greatly in their habits; they may feed on green tips, flowers, either young or old cones; mine woody shoots; attack the phloem on the boles of sapling-sized trees; or inhabit fusiform rust cankers. This discussion will be confined to the southern pine coneworm, *D. ametella,* and the webbing cone-worm, *D. disclusa.*

Coneworms (Lepidoptera: Pyralidae)

The pine coneworms are holarctic. There are approximately 20 species known at present; the recognized numbers are slowly increasing, and the probability is high that the classification will be revised as the research effort expands.

In the southern United States, there are now seven *Dioryctria* species known, the most important being the southern pine coneworm, *D. ametella* (Fig. 16-2). This species has several overlapping generations per year. The adults are nocturnal, mating in the early morning hours. During the early season, the larvae tend to feed on the shoots and flowers and also on the cones and galls infected with fusiform rust. The larvae of later generations feed within the cones. This species usually overwinters in the first instar within the infested cone.

Cones damaged by this species are readily identified by the mixture of resin and frass surrounding the larval entrance tunnel. With time, the infested portion of the cone will turn brown.

In the northern United States and Canada, pine cones are attacked by the webbing coneworm, *Dioryctria disclusa.* This species has only one generation per year. The adults are nocturnal and are active during midsummer; oviposi-tion occurs beneath bark scales on needleless twigs. The newly emerged larvae do not feed; instead they spin hibernacula, in which they overwinter. With the warm weather of late spring, the larvae feed primarily in staminate flowers; after pollen shedding, the later-stage larvae feed on second-year cones.

The infested cones have a single larval entrance hole at the base of the cone. The larva feeds on the entire seed-bearing portion of the cone. The ejected frass collects in a silk web, forming a large, reddish brown mass. One larva requires two cones to complete its development and then pupates, usually within the cone. Most infested cones are destroyed.

Cone Borers (Lepidoptera: Olethreutidae)

The cone borers belong to the genus *Eucosma.* They mine the shoots and cones of conifers throughout North America. In the southern United States, the most destructive species are the shortleaf pine cone borer, *E. cocana,* and the white pine cone borer, *E. tocullionana.* In the North, red pine is the preferred host of the red pine cone borer, *E. minotorana.* The cones of the western pines are infested with *E. rescisscrianna.*

The *Eucosma* spp. apparently have one generation per year. The adults emerge in the spring and oviposit on the cone stalks. The early-instar larvae feed gregariously in immature cones, but later instars are solitary cone feeders. Pupation occurs at midsummer; the larvae leave the cones and spin cocoons in the soil, where they overwinter.

Figure 16-2 The southern pine coneworm, *Dioryctria amatella* (*a*), is a serious pest of loblolly seed production. The larvae (*b*) feed in shoots and first-year cones (*c*) and also second-year cones (*d*). *(U.S. Forest Service photo.)*

The larvae destroy the cones by boring straight toward the cone axis, where they feed on the immature seeds. Feeding continues until all seeds are destroyed; the larvae then migrate to fresh cones. The larvae expel their frass within the cone.

The symptoms of damage of *Eucosma* feeding are numerous entrance and emergence holes on the cone surface, little (or no) evidence of frass, and the shrinking and browning of infested cones.

Cone Beetles (Coleoptera: Scolytidae)

The cone beetles, members of the scolytid genus *Conophthorus,* seriously damage the cones of pines throughout most of North America except the South. The *Conophthorus* beetles have one generation per year, with the adults emerging in the spring, usually from second-year cones. The adults fly to the crowns of nearby pines, where they feed on the shoots and cones. Shortly afterward, the females bore into the cone base or stem, exhibiting a preference for second-year cones. The female mines and oviposits within or along the cone axis, depending on the insect species. The larvae feed on the seeds and cone scales, then pupate within the cone. The infested cones die and usually fall to the ground.

The overwintering habits of *Conophthorus* species vary. The sugar pine cone beetle, *C. lambertianae,* leaves the cone after oviposition and attacks fresh cones or shoots, overwintering within shoots in the crown or in shoots or cones that have fallen. The white pine cone beetle, *C. coniperda,* overwinters in the pupal stage within fallen cones. The red pine cone beetle, *C. resinosae,* overwinters as an adult, usually within mined shoots and buds that have fallen to the ground (Fig. 16-3).

The symptoms of cone beetle attack become apparent in the summer and fall. The attacked cones fall to the ground, reduced in size, discolored, with the entrance and emergence holes characteristic of the species. The interior of the cone is filled with frass. In the tree crown, the infested current year's shoots and needles fade and turn red. The infested shoots and buds usually drop to the ground during the fall.

INSECTS FEEDING ON NUTS AND FRUITS

Little is known about the insects feeding on the nuts and seeds of hardwoods in the forest. The consensus is that insects can cause serious losses in pecan, hickory, and walnut orchards (Miller, 1973). In the forest, oak mast production, a valuable wildlife food source, is often destroyed. The major weevil pests of pecan and walnut belong to the genus *Curculio,* and the weevils damaging the acorns of oaks belong to the genus *Conotrachelus.*

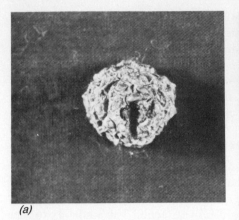

(a)

Figure 16-3 The red pine cone beetle, *Conophthorus resinosae,* chews an entrance groove in the base of the cone (*a*). Infested cones are stunted; the lower cone is normal (*b*). *(U.S. Forest Service photo.)*

(b)

Nut Weevils (Coleoptera: Curculionidae)

The nut weevils occur throughout North America. The genus *Curculio* contains 27 species, about evenly divided between the eastern and western United States. The life histories and habits are similar. The adults are active throughout the growing season. When nuts and acorns begin to ripen, the female bores a hole in the shell and oviposits several eggs. After the infested nut or acorn falls to the ground, the fully grown larvae usually burrow into the soil. Each larva spins a pupal cell, yet may remain in diapause for 1 to 3 years before pupating.

The more common nut weevils in the East are the pecan weevil, *Curculio caryae,* and the black walnut curculio, *Conotrachelus retentus.* The filbert weevil, *Curculio occidentalis,* a native acorn insect, has become a pest of commercial filberts in the Pacific Northwest.

The acorns of most Eastern oak species provide food and shelter for the *Conotrachelus* weevils, which are similar in appearance and habit to *Curculio* weevils except the adults emerge in the fall and overwinter in the litter. The *Conotrachelus* also inhabit the nuts as well as fresh wounds on the boles of numerous hardwood species.

Fruit and Husk Flies (Diptera: Tephritidae)

The fruit and husk flies belong to the dipterous genus *Rhagoletis.* The larvae mine within the fruits of such trees as apple and cherry; many are serious economic pests of commercial fruit orchards. The walnut husk fly, *Rhagoletis completa,* breeds within the husks of walnuts. The larvae reduce the husks to a foul-smelling black slime that discolors the shell and nutmeat. Although the quantity of nut production is not affected, the quality is destroyed. The presence of this pest can be detected with flight traps, and populations can be decreased by planting late-maturing walnut varieties. The flies cannot oviposit after the husks have hardened.

The closely related genus *Ceratitis* contains the Mediterranean fruit fly, *C. capitata,* the devastating pest of citrus fruits. This species was eradicated from Florida at enormous expense. However, the possible reintroduction of fruit flies remains a constant threat to the citrus industry. Foresters should be well aware of the need for quarantines on the importing and transporting of fruits, nuts, and seeds.

INSECTS FEEDING ON SEEDS

Seedworms (Lepidoptera: Tortricidae (=Olethreutidae)

The seedworms of the lepidoterous family Tortricidae (=Olethreutidae), genus *Cydia* (=Laspeyresia), are distributed throughout North America, with some species attacking conifers and other hardwoods. This genus differs from other Tortricidae in that the larvae feed predominately on seeds.

There is usually one generation per year, but some larvae may remain in diapause for one year or more. The females oviposit directly on the cones or nuts in the spring. The larvae bore into and feed within the seeds. Approximately seven slash pine seeds are needed for larval development of the slash pine seedworm, *C. anaranjada.* The mature larvae bore into the cone axis, where they overwinter; pupation occurs in the spring within the cone (Fig. 16–4).

There are no external symptoms of seedworm attack on the cones. The infestation, however, is readily apparent when the cone is cut in half longitudinally. The destroyed seeds are filled with frass, and mature larvae or pupae are within the cone axis. Annual seed loss has been estimated to range from 2 to 20 percent.

The commercial harvest of hickory nuts and pecans is often ruined by the hickory shuckworm, *C. caryana.* This pest occurs throughout eastern North America, having one generation in the north and as many as four in the south. The larvae bore into and feed within nuts and pupate within the shucks. Overwintering occurs within the nuts, which fall to the ground.

Seed Chalcids (Hymenoptera: Torymidae)

The seed chalcids are hymenopterous insects belonging to the genus *Megastigmus.* The small chalcid wasps infest the seeds of numerous shrubs and trees. Two important species are the balsam fir seed chalcid, *M. specularis* (Fig. 16–5), and the Douglas-fir seed chalcid, *M. spermotrophus.*

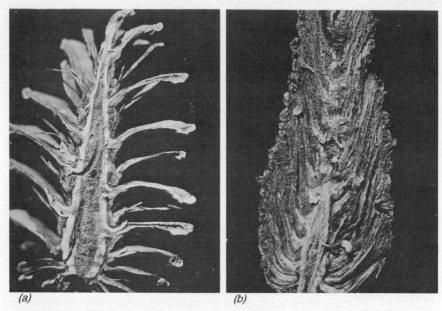

(a)   (b)

Figure 16-4 The larvae of the seedworms, *Cydia* (= *Laspeyresia*) spp., feed only on seeds (a) but pupate in the cone axis (b). (U.S. Forest Service photo.)

The balsam fir seed chalcid emerges in the spring. The female oviposts directly within the seed of the young cones. Although a number of eggs may occur within a seed, only one larva matures. With the disintegration of the cones, the infested seeds fall to the ground. Larvae overwinter within the seeds and pupate in the spring.

The Douglas-fir seed chalcid is similar in appearance and habit to the balsam fir seed chalcid. This species has been introduced into Europe.

There are no symptoms of seed chalcid attack prior to adult emergence (Fig. 16–5). The presence of the seed chalcids can be determined only by dissecting or radiographing the seed. These pests have been found to destroy as much as 60 percent of an annual seed crop. The unwanted importation of seed insects warrants adequate quarantine and treatment and certification of seeds.

Seedbugs (Hemiptera: Coreidae, Pentatomidae)

The important effect of insects on production of forest tree seed is reflected in the discovery of seed damaged by the true bugs. The most common of these species are in the family Coreidae, the leaffooted pine seedbug, *Leptoglossus corculus* (Fig. 16–6), and the western conifer seedbug, *L. occidentalis*. The family Pentatomidae contains the shield-backed pine seedbug, *Tetrya bipunctata*.

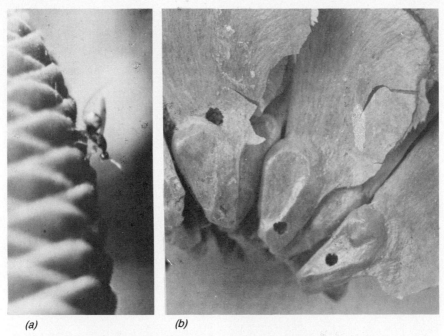

(a)   (b)

Figure 16-5 The balsam fir seed chalcid, *Megastigmus specularis,* oviposits within the seed (*a*). The only symptom of attack is the adult emergence hole in the seed(*b*). *(Canadian Forestry Service.)*

The seedbugs may well be responsible for many of the aborted conelets, empty seeds, and molds that occur in seed-germination tests. Although the adults are large bugs, they are very active and thus are not often seen. However, their importance can be verified by the fact that seed yields are increased after cones are caged with screen wire.

The seedbugs are multivoltine, are active throughout the growing season, and move freely between conelets and cones. The nymphs and adults feed by inserting their stylets into the seed and siphoning the fluids from the developing embryos within both the first- and the second-year cones.

The only symptom of attack by seedbugs is a microscopic hole (<0.06-mm-diam) in the cone scale and seed. Their damage is best detected in the mature seed by radiography (Fig. 16–6).

Cone Midges (Diptera: Cecidomylidae)

The gall-forming midges of the dipterous genus, *Contarinia,* are represented by one of the more serious pests of Douglas-fir seeds, the Douglas-fir cone midge, *C. oregonensis* (Fig. 16–7). This small fly (<3.6 mm long) emerges in the spring and oviposts in the female flowers of Douglas-fir when they are open

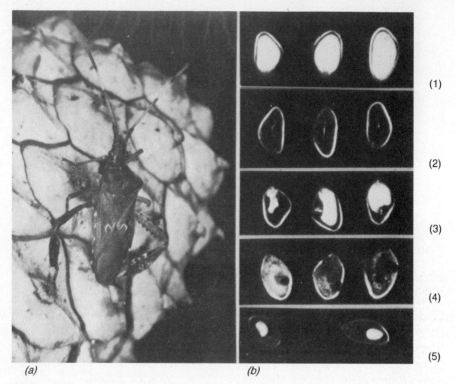

(a)   *(b)*

Figure 16–6 The leaffooted pine seedbug, *Leptoglossus corculus* (*a*), has recently been discovered to be an extremely serious pest of seeds. Seed damage is best detected by radiography (*b*): (1) sound, (2) empty and damaged by (3) seedbug, (4) seedworm, and (5) chalcid. *(U.S. Forest Service photo.)*

for pollination. The larvae tunnel into the young cone scales, and galls form near the ovules. The U-shaped larva develops within its gall during the summer; in the fall, the mature larva jumps out of the rain-soaked cone into the ground litter, where it spins a cocoon. In the spring, the larva pupates within its cocoon, and within a few weeks the adults emerge.

Seed losses caused by cone midges result from the destruction of developing seeds and the mechanical inability to extract the seeds: the galls fuse the seeds to the cone scales. There are practically no external symptoms of cone midge infestation. However, when the cones are sliced along the axis, the galls' U-shaped larvae, and empty seeds, are readily apparent.

Cone midges have also been found infesting cones of southern pine in the United States and red pine in Canada.

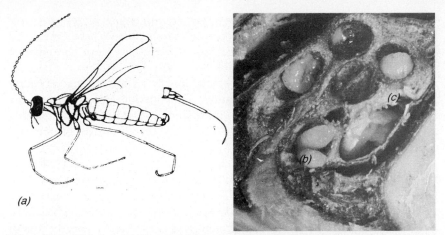

Figure 16–7 The Douglas-fir cone midge, *Contarinia oregonensis* (*a*). Larval feeding (*b*) causes the formation of galls that fuse the seed to the cone scale (*c*). *(Weyerhaeuser Company.)*

IMPACT

The most important known cause of cone and seed mortality is insect attack. The major insect genera that have been discussed are distributed widely in North America. Nut weevils and husk flies are of concern in the eastern hardwood forests. Although cone borers and coneworms are serious problems everywhere, cone beetles are of major importance in the northern and western coniferous forests. Southern pine are especially damaged by thrips and seed-bugs, whereas in the Pacific Northwest seed midges are a serious problem.

Insects have a serious impact on the production of tree seed: annual losses are seldom less than 10 percent, and occasionally entire crops are destroyed. The annual average loss in red pine is estimated by Mattson (1978) to be 25 percent. The impact is compounded by the periodicity of cone crops, good crops occurring every 3 to 10 years. When a good crop does occur, the following year's crop is poor. At present, we cannot predict when a given species will produce an abundant cone crop, especially when our most important pine species require between 2 and 3 years for the cones to mature. Efforts to sustain high annual seed crops may also sustain high insect population levels.

The forester should be well aware of the great variation in the population dynamics of cone and seed insects in place, in time, and within a given tree. The insect populations and percentage of damage are regulated by the periodicity and size of the food supply, the annual cone crop. There is usually a general increase in insect populations when cone crops increase annually; conversely, the percentage of cone and seed loss decreases. When a poor cone year follows a bumper cone crop, however, the emerging increased insect populations will decimate what few cones and seeds exist. Where cone crops have the least annual variation, insect populations may have the greatest impact.

The developing strobili are subject to numerous interacting mortality factors. The life table is the best method of identifying and evaluating the mortality factors in a given area. This information is essential for establishing the economic loss thresholds and the feasibility of preventive and control measures.

A life table for red pine cone production has recently been published by Mattson (1978) (Table 16–2). Mattson discovered that the greatest mortality occurred during the flowering stage. Twenty-six percent of the flowers were destroyed by jack pine budworms, *Choristoneura pinus,* dispersing outward from adjacent jack pine stands. The coneworms caused an 18 percent loss of flowers by tunneling within the flower and conelet-bearing shoots.

The survival of overwintering cones was good; only 8 percent were lost when cone-bearing branches were clipped off by pernicious red squirrels. However, serious losses occurred to the 1-year-old cones: 78 percent were

Table 16–2   Factors Affecting Red Pine Cone Yield

x	Lx	d_xF	dx	$100\,qx$	Sx
Age interval	Cone count	Mortality factor	Cone loss per tree	% loss	Survival rate*
Flowers	16.0	Defoliators	4.2	26.0	
		Coneworms	2.9	18.1	
		Cone beetles	0.2	1.3	
		Unknown	0.6	3.9	
		TOTAL	7.9	49.3	0.507
Conelets	8.1	Squirrels			
		Unknown			
		TOTAL	0.7	8.3	0.917
1-year cones	7.5	Cone beetles	2.1	27.7	
		Coneworms	3.2	43.6	
		Cone borers	0.5	6.5	
		TOTAL	5.8	77.8	0.222
2-year cones	1.7	Size	0.4	23.9	
		Aborted seed	0.2	9.3	
		Other insects	0.1	8.3	
		Seedworms	$T^†$	0.9	
		TOTAL	0.7	42.4	0.577
Normal cones	0.9		15.1	94.4	0.056

Source: After Mattson (1978).

*Totals are computed from the % loss column.

†*T* means *trace*, or less than 0.1.

destroyed, mostly by coneworms and cone beetles. During the second year, the mortality of the surviving cones reached 42 percent, the result of feeding by cone beetles and coneworms. The total cone yield was reduced by 99 percent.

The economic importance of insects can readily be computed from a life table for the seed yields of slash pine developed by DeBarr and Barber (1975) (Table 16–3).

DETECTION

The detection of insects injurious to seed production is based on the tree species concerned and on a general knowledge of the life history and habits of associated insect genera within a given area. In intensively managed seed orchards, there is no substitute for direct sampling within the tree crowns; this is best achieved with hydraulic hoists (Fig. 16–1). Sampling may also be attempted from scaffolding and ladders.

The detection of insect populations in the future may well include light and pheromone traps and baits such as monoterpenes.

Within seed production areas, rough estimates of flower and cone production and damage can be made by using binoculars (7 X 50) and by examining the shoots and cones that have fallen to the ground.

Foresters assigned to purchase cones are well advised to construct a

Table 16-3 Factors Affecting Slash Pine Seed Yield

x	Lx	d_xF	dx	100 qx	Sx
Age interval	Seed count	Mortality factor	Seed loss per tree	% loss	Survival rate*
Conelets	31,830	Coneworms	3,500	11.0	
1967		Abortion	955	3.0	
		Thrips	95	0.3	
		Cone rust	65	0.2	
		Missing	3,630	11.4	
		TOTAL	8,245	25.9	0.741
Cones	23,585	Coneworms	2,390	10.1	
1968–1969		Abortion	1,590	6.7	
		Missing	1,530	6.5	
		TOTAL	5,510	23.3	0.767
Seeds	18,080	Seedworms	335	1.8	
1969		Seedbugs	1,660	9.2	
		Empty	1,680	9.3	
		TOTAL	3,675	20.3	0.797
Harvest	14,400		17,430	54.7	0.453

Source: After DeBarr and Barber (1975). Based on observations on 12 trees; assuming 100 seeds per cone and 12,000 seeds per pound.

*Totals are computed from the % loss column.

cone-slicing knife (Winjim and Johnson, 1960) or a hydraulic cone cutter (Fatzinger and Proveaux, 1971). Accurate counts of filled and damaged seeds can be made quickly from the cut surface after the cone is halved longitudinally.

The detection of insect-damaged seeds and nuts is best accomplished with radiography (Belcher, 1973). Although adult emergence holes are obvious, the presence of cryptic larvae will influence germination tests. The forester can determine the percentage of sound seed within 5 minutes by placing the seeds on radiographic paper and submitting them to soft x-ray exposure, the roentgen dosage varying with the species. The exposed paper is then placed in an instant processor. The percentage of sound seed is determined for every seed lot tested at the U.S. Forest Service's Eastern Tree Seed Laboratory, Macon, Georgia. The test is available to commercial interests at cost.

PREVENTION AND CONTROL

The prevention of seed loss due to insects begins with the location and selection of trees. The problem of adjacent species is serious. Mattson (1972) has shown that the diversity of insect species increases with the acreage of pine (either red or jack) located within one-third mile of red pine seed production areas. Similar tree genera and alternate hosts of serious diseases should not be grown within one-third mile of each other; this is especially true of red oaks in the South, which are the alternate hosts for fusiform rusts. Foresters involved with seed orchards or production areas may also have problems with the frequency of violent windstorms, killing frosts, and ice storms. In general the areas should be reasonably dispersed and remote from public use; have good, single-access, locked road gates; and be well posted.

The vigor of the trees is also important. Trees within seed production areas are usually thinned to provide full crown exposure, and seed orchards are planted at 30-ft spacings. These wide spacings permit the removal of competing vegetation, fertilizing, irrigation, and application of insecticides. Foresters should avoid the undue forcing of annual seed crops. Stress problems, such as stem cankers, may occur on trees that have been overly fertilized, watered, root-pruned, and treated with hormones.

The ground should be kept free of all tree material that falls. The serious problem of cone beetles in the North can be effectively prevented by the use of prescribed burning (Miller, 1978). Squirrel damage may be reduced by placement of aluminium sleeves around the bole.

Some cone and seed insects can be controlled with insecticides. The multitude of insect species has previously caused problems regarding the timing and number of applications. However, development of the residual-contact insecticides and the systemics has reduced many problems. Contact insecticides are effective, especially for thrips and seed chalcids when applied with mist blowers when the pines are flowering. The remaining insect genera are effectively controlled with the systemic insecticides when disked into the soil in granular form once each year in the spring.

Careful use of insecticides by qualified operators is justified when the high economic value of the seed is considered relative to (1) the proven destructiveness of the insects, (2) the low cost of insecticide application, and (3) the reduction in environmental hazards afforded by the use of systemic insecticides (Chapter 11).

Insects that infest seeds can be controlled after harvest by fumigation and heat (Hedlin, 1974). Seeds that have not been treated should neither be imported nor exported.

Foresters should continue their strong support of the genetics and breeding of superior trees, the use of certified seed, and the planting of sound nursery stock. Research to prevent and control the mortality factors affecting seed production should receive increased consideration.

BIBLIOGRAPHY

Literature Reviews

Barcia, D. R., and Merkel, E. P. 1972. Bibliography on insects destructive to flowers, cones, and seeds of North American conifers. *USDA For. Serv. Res. Paper* SE 92.

Ebel, B. H., et al. 1975. Seed and cone insects of southern pines. *USDA For. Serv. Gen. Tech. Rept.* SE 8.

Hedlin, A. F. 1974. *Cone and seed insects of British Columbia.* Victoria, B.C.: Can. For. Serv. Pac. For. Res. Cen.

———, et al. 1979. *Cone and seed insects of North American conifers.* N. Am. For. Comm. FAO (In press).

Keen, F. P. 1958. Cone and seed insects of western forest trees. *USDA Tech. Bul.* no. 1169.

Lyons, L. A. 1957. Insects affecting seed production in red pine IV. *Can. Entomol.* 89:264–271.

Miller, W. E. 1973. Insects as related to wood and nut production. *USDA For. Serv. Gen'l. Tech. Rept. N.C.* no. 4.

Osburn, M. R. 1966. Controlling insects and diseases of the pecan. *USDA, Agric. Handbook* no. 240.

Impact

DeBarr, G. L., and Barber, L. R. 1975. Mortality factors reducing the 1967–69 slash pine seed crop in Baker County, Florida: A life table approach. *USDA For. Serv. Res. Paper* no. SE 131.

Mattson, W. J. 1978. The role of insects in the dynamics of cone production of red pine. *Oecologia* 33:327–349.

Yates, H. O., III, and Ebel, B. H. 1978. Impact of insect damage on loblolly pine seed production. *J. Econ. Entomol.* 71:345–349.

Seed Orchards and Production Areas

Dorman, K. W. 1976. The genetics and breeding of southern pines. *USDA For. Serv. Agric. Handbook* no. 471.

U.S. Forest Service. 1974. Forest tree seed orchards. *USDA For. Serv. Mimeo.*

U.S. Forest Service. 1978. 1977 report: Forest planting, seeding, and silvical treatments in the United States. *USDA For. Serv. Mimeo.*

Prevention and Detection

Belcher, E. W., Jr. 1973. Radiography in tree seed analysis has new twist. *Tree Planter Notes* 24:1–5.

Bramlett, D. L. 1977. Cone analysis of southern pines: A guide book. *USDA For. Serv. Gen'l. Tech. Rept.* no. SE 13.

Fatzinger, C. W., and Proveaux, M. T. 1971. A hydraulically operated pine cone cutter. *USDA For. Serv. Res. Note* no. SE 165.

Miller, W. E. 1978. Use of prescribed burning in seed production areas to control red pine cone beetle. *Envir. Entomol.* 698–702.

Winjum, J. K., and Johnson, N. E. 1960. A modified-knife cone cutter for Douglas-fir seed studies. *J. Forestry* 58:487–488.

Meristematic Insects (Bark Beetles and Other Phloem Borers)

This chapter is subtitled bark beetles and other phloem borers because of the great importance and destructiveness of the bark beetles in the family Scolytidae. However, the coverage will include additional insects that infest the phloem region and the outer layers of xylem of trees; like bark beetles, these eat the meristem tissues of the trunks and larger branches. Representatives of several insect orders live in these tissues; the more destructive and common are from the orders Coleoptera and Lepidoptera.

We will limit our discussion to four arbitrary sections: (1) bark beetles, engravers; (2) bark beetles, *Dendroctonus;* (3) phloem borers, Coleoptera; and (4) phloem borers, Lepidoptera. Some of the third and fourth groups may score the wood more deeply than bark beetles and often pupate in the wood rather than in the bark; whereas bark beetles spend their entire developmental cycle in the phloem and bark.

BARK BEETLES, ENGRAVERS

Most of the genera and many of the species of the family Scolytidae are found in North America (Bright, 1976). Each genus of this family has a characteristic manner of working, and there is much variation among the individual species in both this characteristic and their life histories. Almost all members of this family are tree-inhabiting insects.

Certain traits and habits are common to most bark beetles. For instance, all true bark beetles excavate egg galleries in fresh phloem. The larvae work away from the egg gallery, mining in the succulent tissues of the inner bark until they are full grown. At the end of its larval gallery, each larva excavates a pupal cell, usually between the bark and the wood but sometimes in the sapwood or in the outer bark. The pattern formed by the tunnels is so specific in character that an expert can usually identify the species of beetle from the pattern and the tree species (Fig. 17-1).

Most bark beetles are secondary in their attack characteristics in that they infest trees in poor health, suppressed, dying, or dead. At one extreme are those species that attack only trees that are already dead or weakened beyond any possibility of recovery; at the opposite extreme are those species that may attack trees that are completely healthy. This apparently primary type of infestation is most likely under the pressure of outbreaks when large numbers of trees are killed. Some species are in the middle ground and are usually secondary but occasionally may become primary. Only under favorable conditions are these intermediate forms able to attack and kill living trees. The balsam-fir bark beetle, *Pityokteines sparsus,* is an example of a bark beetle that appears never to be primary. A possible example at the other extreme is the mountain pine beetle, *Dendroctonus ponderosae,* which may not be a primary insect at all times but during outbreaks may kill many relatively healthy trees.

In recent years, a considerable effort has been made to identify the various pheromones associated with bark beetles. Much of this effort has concentrated on three genera, *Dendroctonus, Ips,* and *Scolytus,* but it is generally believed that most (if not all) bark beetles may be influenced in their behavior by such attractants (Borden, 1974). Many of these result in aggregation, which is necessary in the colonization process. A number of articles describing the identification of the various chemicals have appeared in the *Journal of Insect Physiology* (Pitman et al., 1969; Renwick and Vité, 1972; Wood et al., 1966; Young et al., 1973); there have also been a number of studies on use of pheromones in management of populations (Pitman, 1971, 1973; Knopf and Pitman, 1972; Furniss et al., 1976). The intensive work on bark beetle pheromones by scientists such as Silverstein, Wood, and their associates cannot be overlooked, but discussion must be limited in a text. The student interested in studying pheromones is advised to review the original literature; those referred to here are only a few of the many papers published in the late 1960s and early 1970s. Additional comments on pheromones may be found in other chapters of this text. The techniques are relatively new and undeveloped; their use in surveys and control may be one of the major developments to be expected in the remainder of this century.

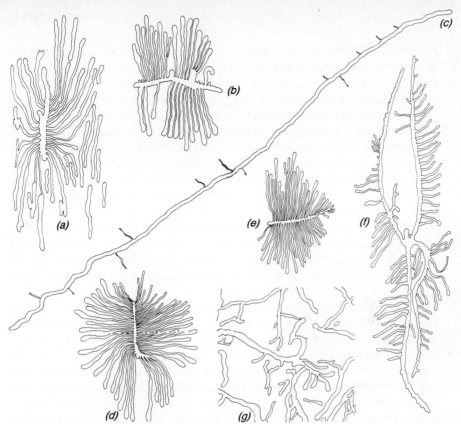

Figure 17-1 Some common forms of bark beetle galleries: (a)*Pseudohylesinus nebulosus*, (b)*Leperisinus oregonus*, (c) *Dendroctonus ponderosae*, (d) *Scolytus multistriatus*, (e) *Hylurgopinus rafipes*, (f) *Ips grandicollis*, (g) *D. frontalis*. (Drawing by Toshiaki Ide.)

Western Engravers

A number of serious local outbreaks of the fivespined and fourspined engravers in the genus *Ips* have occurred in California and Oregon on ponderosa pine. These outbreaks have often been in vigorous stands growing under favorable site conditions and almost invariably have been associated with sporadic logging operations. According to Struble and Hall (1955), the California fivespined ips, *Ips paraconfusus,* passes through four or occasionally five generations during a season, the overwintering beetles emerging in the early spring, the final brood overwintering under the bark of the attacked trees or slash.

In the spring, the overwintering beetles are unable to breed successfully in living trees and therefore must have a supply of fresh slash if they are to multiply successfully. They prefer slash ranging in size from 6 to 24 in (15 to 60 cm) in diameter and cut after January 1. After lying on the ground over winter, fall-produced slash is unsuitable for breeding. When an abundance of fresh slash is available, the beetles increase in number and may kill nearby trees.

The males cut through the bark into the phloem, where they excavate a nuptial chamber. There they are joined by females, usually three in number, each of which excavates a separate egg gallery in the phloem from the nuptial chamber. As they extend their galleries, the females push the frass into the nuptial chamber, from which the male relays it to the outside. Thus, the galleries of the engraver beetles are kept clear of frass. Accumulations of brown frass about the entrance holes are conspicuous evidence of the presence of the beetles beneath the bark. The eggs are deposited in separate niches along the sides of the egg galleries, and the larvae work through the phloem. On completion of their development, the larvae form pupal cells between the bark and the wood. By July, numerous young adults emerge from the slash and seek fresh breeding places. If logging has been continued and fresh slash is therefore available, the emerging broods continue to work in that material. But if the supply of slash is limited, they attack living trees. This is possible because by July the trees, having completed their vigorous spring growth, are unable to repel a mass attack by the beetles.

Treatment of slash in the spring and early summer is therefore the key to successful control of these engravers. Without a supply of freshly cut green material, large populations of the beetle cannot develop. Because slash cut in fall and early winter has become unsuitable for the beetles by spring, thinning operations, clearing for road construction, and insofar as is practicable, logging operations should be conducted between August and January.

Struble and Hall (1955) in summary recommend the following measures for controlling the engraver beetles:

1 When practicable, cut young-growth pine after July 15.
2 Lop and scatter all slash so as to expose the main stem to the sun.

3   In logging or salvage operations, utilize all logs to an 8-in (20-cm) top diameter.

4   Keep accumulations of slash or green logs away from living trees, and fell trees away from dense thickets of young growth.

5   When necessary, kill broods by fire or by application of registered chemicals.

The intent of this discussion has been to use the California fivespined ips, *I. paraconfusus,* as an example but not to claim that it is more or less important than other western species of the genus *Ips.* The various species are described in a series of articles by Hopping (1963 to 1965) and by Lanier (1970); some are discussed in pest leaflets (Massey, 1971; Struble, 1970; Sartwell et al., 1971) and in "Western Forest Insects" (Furniss and Carolin, 1977).

The fir engraver, *Scolytus ventralis,* is an important pest of true firs, *Abies* sp., in the western forests (Stevens, 1971). Trees from pole size to mature saw timber are killed. Trees may be killed in one season or may be weakened through successive attacks. This insect has caused an estimated average annual loss in California alone of 450 million bd ft. This is more than half the estimated annual net growth of firs. Like many other bark beetles, the fir engraver readily attacks wind-thrown firs and green cull logs resulting from logging operations. The wide variation in injury to the host makes the possibility of effective chemical control remote. The removal of decadent and weakened firs may be the best control possibility because vigorous, healthy stands are not usually severely affected. Berryman (1973) has suggested a hypothesis for the population behavior of this species, which he states is determined by the quantity of susceptible trees.

Two additional species of native bark beetles are also destructive to western true firs. These are the silver fir beetles, *Pseudohylesinus granulatus* and *P. sericeus* (Thomas and Wright, 1961). The beetles characteristically attack wind-thrown, felled, or injured trees, but poles and saplings stagnating in dense stands and understory trees are also commonly killed. These insects have generally been considered secondary, but an outbreak in northwestern Washington in mature Pacific silver fir resulted in the loss of 528 million bd ft of timber between 1947 and 1955. Preventive control may be possible through forest management practices that utilize true fir stands while they are young and vigorous.

Eastern Engravers

Of all the eastern engravers, the pine engraver, *I. pini,* is the best-known species though Southern foresters might suggest *Ips avulsus.* It is common throughout the eastern half of the United States and Canada and may occasionally become primary. Its habits and life history are very similar to those of other species of the genus. The adults are small, black beetles about ⅛ in

(3 mm) in length. As in the case with all the members of the genus, the caudal portion of the elytra, called the *declevity,* is concave and bordered on each side with spines. The declevity gives the beetle the appearance of having the rear end of the body cut off sharply. The beetles attack freshly cut or unhealthy pines and, to a lesser extent, spruce. After completing their development, the young adults feed for a short time in the inner bark and then emerge to seek fresh material in which to start a new generation. In the North, there is usually only one generation of this engraver during a season; however, Thomas (1961) reports that the beetle has two generations a year as far north as Ontario. Farther south, because of the longer and warmer season, there may be three or possibly four generations. Some species of engraver beetles pass the winter under the bark of their host trees, but the pine engraver hibernates in the litter on the ground.

Control of the pine engraver, *I. pini,* is usually unnecessary, but occasionally, the insects may become sufficiently numerous to require protective measures, especially during drought periods. As in the case of the western engravers, if logging is carried on periodically, rather than continuously, beetles of the genus *Ips* in the East may temporarily become sufficiently abundant to kill some living timber. The recommendations for the control of the western engravers are equally appropriate for the eastern species. However, in the northern states, slash cut later than October is usually suitable for infestation in the spring.

In southern pine forests, there are three species of *Ips* that are responsible for much annual mortality: *I. calligraphus, I. grandicollis,* and *I. avulsus* (Speers, 1971). The three are similar in habits and life histories but can be readily distinguished by their size and number of elytral teeth. *I. avulsus* is small, 2.1 to 2.6 mm long, and has four teeth on each elytron; *I. grandicollis* is 3.0 to 3.9 mm long and has five teeth; and *I. calligraphus* is 4.0 to 6.0 mm long and has six teeth.

Infestations in green timber are usually sporadic and of short duration. Tree killing often occurs near slash areas, but spots may be found in other locations. The broods develop very rapidly during the summer, making control difficult. Broods may be emerging as the trees begin to fade. Some attacked trees may contain all three species, or there may be only one or sometimes two species present (Fig. 17-2). Three to five females and one male usually occupy each nuptial chamber, with its series of radiating egg galleries. Many generations may occur each year in the South. *I. avulsus* may have 10 or more; whereas the larger species may have only 6. *Ips* beetles seem to be attracted to any stand in which some disturbance has left fresh pine slash, stumps, and weakened or injured trees. Lightning strikes, very frequent in parts of the Southeast, are often the centers for spot killings by *Ips* beetles.

The three species are not alike in their attack characteristics, as was shown by Mason (1970). His studies revealed that *I. avulsus* was much more

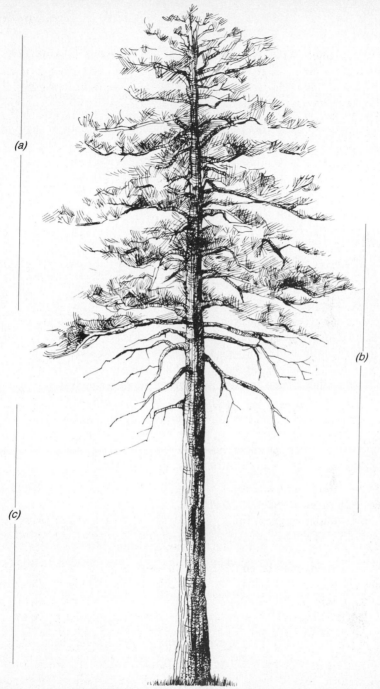

Figure 17-2 When the southern species of *Ips* are in one tree, they are commonly distributed in specific zones: (*a*) *Ips avulsus,* (*b*) *I. grandicollis,* (*c*) *I. calligraphus. (Drawing by Toshiaki Ide.)*

aggressive than *I. grandicollis* and that this was related to a greater tendency toward flight aggregation in the former species.

Bark beetles may serve as vectors for serious tree diseases. Such is the case with the Dutch elm disease, *Ceratocystis ulmi.* The two vectors are the smaller European elm bark beetle, *S. multistriatus,* and the native elm bark beetle, *Hylurgopinus rufipes.* These are both secondary beetles, but in their capacity as vectors, they assume primary importance. The adults of the smaller European elm bark beetle feed on the inner bark on the twigs of healthy elms. Because of this habit, it is the chief vector of the disease. The only known methods of control are directed at destruction of the vectors. Control is impossible in forest areas, so the American elm is rapidly disappearing as an important component of eastern forests. Elms in urban areas may be saved by thorough sanitation procedures along with chemical application to destroy the vectors.

In this brief discussion of engraver beetles, we have concentrated mainly on the genera *Ips* and *Scolytus.* There are many other genera that the forester will encounter while accomplishing fieldwork. For example, Bright (1963) has described the species in the genus *Dryocoetes.* These beetles are common in dead and dying spruce and fir trees. The forester is not expected to be an expert on insect identification, but the identification of engravers on fir is a good example of the need for careful study. The mention here of Bright's work on *Dryocoetes* might lead a forester at a later time to a conclusion that engraver work in fir trees is caused by *Dryocoetes.* But it could be that the balsam-fir bark beetle, *P. sparsus* (Hosking and Knight, 1976), was the insect. Or could it have been the fir engraver, *S. ventralis?* Further study might result in an identification of none of these three. Clearly, proper identification is the most important first step in any form of pest management.

BARK BEETLES, *DENDROCTONUS*

The bark beetles of the genus *Dendroctonus* are by far the most important beetle enemies of living trees. The tremendous annual losses occasioned by the attack of these beetles has already been partially discussed. All species of this genus are not equally injurious, but some of the losses in major outbreaks (Table 17-1) illustrate the seriousness of the problems caused by some of the more destructive species.

Wood (1963) provided a revision of the genus that has been changed very little since completion. One species name, *D. obesus,* has been corrected to *D. rufipennis;* and one new species has been identified, *D. rhizophagus* (Thomas and Bright, 1970).

Some of these beetles, for example the red turpentine beetle, *D. valens,* are almost always secondary. This may change with time because of changes in management or possibly changes in the habits of the insect. The black turpen-

Table 17-1 Timber Mortality in Some Major Outbreaks of Bark Beetles

Insect	Years	States	Mortality (billion fbm)
Western pine beetle,	1921–1937	Oregon	12.6
Dendroctonus brevicomis	1931–1937	California	6.0
Mountain pine beetle,	1911–1935	Idaho and Montana	15.0
D. ponderosae			
Douglas-fir beetle,	1962–1964	Oregon and Washington	3.0
D. pseudotsugae			
Spruce beetle,	1940–1952	Colorado	5.0
D. rufipennis			
Southern pine beetle,*	1960–1976	12 southern states	2.5
D. frontalis			

*The value of these losses in the southern states was estimated at over $200 million (Price and Doggett, 1978).

tine beetle, *D. terebrans,* was once considered a completely secondary insect. Today, it is causing extensive losses throughout the pine belt of the southern states (Smith and Lee, 1972). Most *Dendroctonus* species are pests of mature timber under ordinary circumstances, but when they become epidemic, some will attack indiscriminately any tree above 6 in (15 cm) in diameter, regardless of its condition.

The most important species of the genus are the southern pine beetle, *D. frontalis,* the spruce beetle, *D. rufipennis,* the mountain pine beetle, *D. ponderosae,* the western pine beetle, *D. brevicomis,* and the Douglas-fir beetle, *D. pseudotsugae.* Other species, such as the roundheaded pine beetle, *D. adjunctus,* and the black turpentine beetle, *D. terebrans,* may occasionally become abundant and cause severe injury to the forest.

Character of Attack

Because each species of *Dendroctonus* has its own special habits, it is not easy to generalize concerning either their life history and habits or their economic status. Even the same species may behave differently in various parts of its range, thus requiring variations in control methods. On the other hand, they all have certain features in common.

One characteristic common to all important economic species is their gregarious habit of attack. Selected individual trees or groups of trees are attacked en masse. This procedure, if successful, results in the prompt death of the tree and a minimum flow of resin to hamper the beetles. Although the beetles are able to cope with a moderate amount of resin for a short time, they are overwhelmed or driven from the trees by a continuous or copious flow. For this reason, not even the most aggressive species are able to attack successfully as individuals. Even when a moderately large number of beetles attack a particularly vigorous tree, the attack may not succeed. Such a failure is indicated by the presence of numerous large pitch tubes on the bark.

Typical Life History

The life histories of *Dendroctonus* beetles vary in detail but are similar in many respects. The adults work in pairs and cut egg galleries in the phloem. Unlike the engravers, the beetles of this genus do not keep the egg galleries clear of frass. Some frass is cast out of the entrance tunnel when the gallery is being started, but by the time oviposition has been completed, most of the egg gallery is packed with frass. This characteristic will usually distinguish the galleries of the *Dendroctonus* beetles from those of the engravers. The larvae work away from the egg tunnels, feeding on the phloem. When full grown, they form cells in either the phloem or the outer bark, where they pupate and transform to adults (Fig. 17-3). The winter is passed in all stages except the pupa. The overwintering brood completes development in the spring, so that the peak of adult emergence occurs in July, although some beetles are flying throughout

Figure 17-3 Mature larvae of the spruce beetle, *D. rufipennis*. Note the pupal cells in which the larvae are located. *(U.S. Forest Service photo.)*

the season. In the North and at higher elevations, there is usually one generation or one and a partial second; but in warmer localities, there are several generations during the season. Table 17-2 summarizes the characteristics of the more important *Dendroctonus* beetles.

Necessity for Control

In many localities, the fate of the existing virgin forests depends on the ability of the forester to cope with the *Dendroctonus* problem. At times, the ravages of the beetles have been so damaging that many timberland operators have felt compelled to liquidate their holdings in the shortest possible time in order to save them from the beetles. Unless these reserve stands are removed over as long a time as possible, we cannot obtain, in the next tree generation, the normal distribution of age-classes so essential to good forest economy. Thus, the beetle hazard has been a threat to sustained yield management.

Forest landowners in the West may be persuaded to practice forestry on a sustained-yield basis only if the entomologist can convince them that it is possible to control the *Dendroctonus* beetles. The control measures can be aimed at the prevention of outbreaks or at their direct suppression.

Preventive Practices

In the prevention of bark beetle outbreaks, as discussed in Chapter 10, it was pointed out that tree classifications and risk rating could be used as a guide for the removal of high-risk trees in the mature forest. This sanitation-salvage cutting can be used successfully to reduce the hazards from the less aggressive *Dendroctonus* beetles, such as the western pine beetle *D. brevicomis* (Fig. 17-4). It is applicable to overmature stands and to those on relatively poor sites (Keen and Miller, 1960; Johnson, 1972).

Protection of the younger forests from the attack of the more aggressive beetles, such as the mountain pine beetle, *D. ponderosae,* presents problems of control that have not been completely solved. However, much has been accomplished, so that with proper management, the damage done by these destructive insects can be greatly reduced (Safranyik et al., 1974; Sartwell and Stevens, 1975; Cole and Cahill, 1976; Amman et al., 1977; Schmid, 1977; Anon., 1979). The prompt utilization of recently killed or dying timber will often prevent outbreaks. Also, the application of practices designed to maintain thrifty growth (presented in Chapter 10) should materially reduce the chance of outbreaks in any stand. It is a desirable practice to leave freshly cut logs on the ground during the beetle flight and, after they are infested, to utilize them as promptly as possible in order to destroy the beetles in them.

When to Use Suppressive Practices

When *Dendroctonus* beetle numbers increase significantly, then direct methods of control must be considered if disastrous outbreaks are to be prevented.

Table 17-2 Characteristics of *Dendroctonus* Beetles

Common name	Specific name	Size and color of adults	Egg galleries	Larval mines	Pupal cells	No. of generations	Common host trees
Western pine beetle	*brevicomis*	3.0 to 5.0 mm Dark brown to black	Winding between bark and wood	In phloem, little exposed on inner surface	In outer bark	1 and partial second to 2 and partial third	Ponderosa pine, Coulter pine, lodgepole pine (rarely)
Southern pine beetle	*frontalis*	2.2 to 4.0 mm Brown to black	Winding between bark and wood	Practically all in phloem	Outer bark except thin-barked trees	3 to 5	Shortleaf pine, loblolly pine, and other pines and spruce in Southeast
Mountain pine beetle	*ponderosae*	3.7 to 6.4 mm Brown to black Head with frontal groove	Elongate, usually straight but sometimes sinuous	Grouped alternately on opposite sides of egg gallery, exposed on inner surface	In phloem, usually exposed on removal of bark	1 and a partial second in northern part of range	Lodgepole pine, sugar pine, western white pine, ponderosa pine, whitebarked pine, limber pine, Engelmann spruce, Mexican white pine, piñon

Common name	Species	Size and color	Gallery	Egg/larval arrangement	Location	Generations per year	Hosts
Douglas-fir beetle	*pseudotsugae*	4.0 to 7.0 mm Red to dark brown Hairy	Vertical, sinuous, or straight, usually 12-14 in (30 to 35 cm) long, occasionally 30 in (75 cm) or very short	Fan-shaped groups, mostly in phloem but may engrave wood, on alternate sides of egg gallery	In phloem, may or may not be exposed when bark is removed	1 and a partial second	Douglas fir, western larch, bigcone pine (usually secondary)
Spruce beetle	*rufipennis*	4.5 to 7.0 mm Reddish brown to black	Vertical, straight, 6 to 8 in (15 to 20 cm) long	Grouped on opposite sides of egg gallery, first in phloem, later between bark and wood	Between bark and wood, engraving wood	1 or 1 in two seasons	Mature black spruce, red spruce, white spruce, Engelmann spruce, blue spruce
Red turpentine beetle	*valens*	5.7 to 9.5 mm	Irregular, winding, or straight gallery at base of tree	Grouped along egg gallery and work in mass formation	Between bark and wood	2 or more	All pines and occasionally spruces, larches, and firs, (primary on Monterey pine)
Black turpentine beetle	*terebrans*	5.0 to 8.0 mm	Elongate gallery parallel to grain	Eggs in groups along gallery, larvae feed in groups	Between bark and wood	2 or more	All southern pines, most serious on loblolly, slash, and longleaf

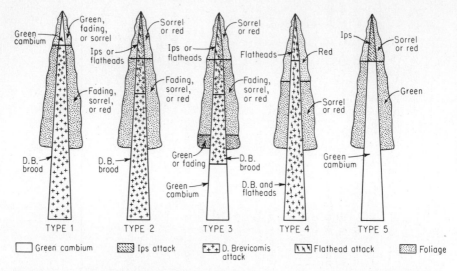

Figure 17-4 Insect attack near the base of a tree does not necessarily indicate either the cause or the progression of mortality by bark beetles. Note that in type 5, only *Ips* is present in the top of the tree. In type 2 *D. brevicomis* may have entered after *Ips* was well established. The color phases suggest that mortality progresses from the top downward and that initial attack was by *Ips*. *(U.S. Forest Service diagram.)*

The changing complex of bark beetle populations emphasizes the necessity for the annual detection and evaluation activities.

There have been many arguments regarding the use of direct control measures against bark beetles. Most managers and entomologists are aware of the secondary characteristics of many of these pests and that trees in a healthy condition are not so likely to be attacked. Nevertheless, losses from bark beetle outbreaks continue to be very large, and preventive methods through silvicultural applications are not readily adopted and applied because of various constraints, the foremost being economics. Some specific recommendations on direct control may be found in literature on specific problems (McCambridge and Trostle, 1972; Smith and Lee, 1972; Schmid and Beckwith, 1975; Massey et al., 1977; Furniss and Orr, 1978).

The southern pine beetle *D. Frontalis* (Fig.17-5), has been one of the more serious pests in this group and is particularly worrisome to foresters who have been increasing their emphasis on high yields from southern pine forests. An intensive program of research has been applied to this pest problem in the Expanded Southern Pine Beetle Research and Applications Program (Leuschner et al., 1977). Coster (1977) has pointed out that chemical methods have generally failed to stop outbreaks of the pest and that the development of indirect methods is a necessity. Preventive methods through silvicultural applications have been presented for control of the southern pine beetle and other bark beetles. Unfortunately, these methods are expensive and are not

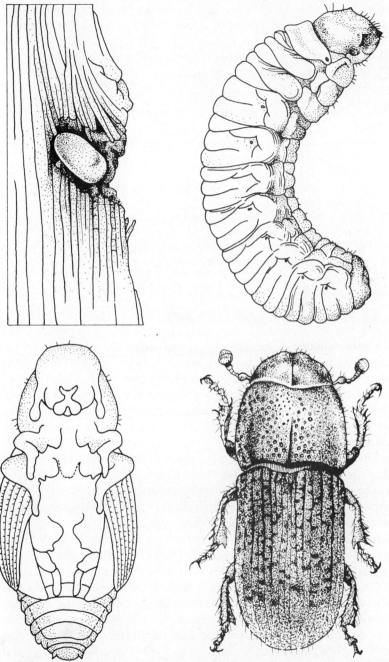

Figure 17-5 The stages of the southern pine beetle, *D. frontalis.* Actual length of beetle is approximately 3 mm. *(Diagram redrawn by Toshiaki Ide.)*

fully proven. Application on a large scale may require the development of less costly methods and the development of markets for low-quality wood products.

A few rules on direct control are appropriate, though the use of preventive methods is always to be preferred.

1 Maintenance control (i.e., removal of trees as they become infested) is justified only
   a on high-value recreational or protection forests or
   b where individual trees can be utilized profitably
2 Suppression control is justified
   a when strong broods in single trees or groups of trees indicate an upward population trend
   b when areas are isolated and can be treated completely in a single season
   c when infested trees can be salvaged by logging and
   d when cost of control will not exceed the value of benefits derived therefrom

Direct control on a maintenance basis may seem a logical procedure to the average forester. Superficially, the yearly costs do not seem extreme, the work time can be allotted in the plans for the year, and no severe losses may occur. In the long run, however, such procedures may be far more costly than other methods. Maintenance work is completely needless in years when population trends are downward. Only in years when population trends are upward is direct control of any value. Therefore, control should be based on biological and economic analysis, rather than on the convenience of planning and application.

Direct control varies greatly with the species. No rule-of-thumb recommendations can be made. In the West, direct control is much more likely to be required in areas infested by the mountain pine beetle, *D. ponderosae*, than in those infested by the western pine beetle *D. brevicomis*. This is because of the difference in aggressiveness between these species. Rules for removal of high-risk trees, which are so effective in protecting overmature stands from the western pine beetle, cannot be applied to the younger stands that characterize the Black Hills and Colorado forests. In general, direct control must be based on (1) careful biological evaluations that reveal a serious risk of high continual losses, followed by (2) an economic evaluation showing that the benefits of control will exceed the costs. Details on silvicultural procedures for prevention of mountain pine beetle outbreaks were summarized in Chapter 10 and are described fully by Amman et al. (1977) and Berryman et al. (1978).

Detection of Infestations

Recognizing *Dendroctonus* infestations is not always an easy matter, and survey or scouting crews must adapt themselves to using different methods to suit the particular cricumstances. Early-season attacks on pine produce easily recognizable color changes in the foliage. The color phases through which infested trees pass are (1) slight paling of the needles, (2) fading to greenish yellow, and (3) turning red (Fig. 17-6). Red tops are obviously the most conspicuous, but trees showing this characteristic no longer indicate a bark beetle hazard; the beetles will have emerged from the trees by that stage. The pale and yellowish tops are signs that indicate the presence of beetles. The red tops are, of course, not marked for treatment, but they are an aid in indicating where to look for newly infested trees.

Late-summer attacks in pine often cause no distinguishable color changes. Therefore, the presence of a late brood can be spotted only by the resinous frass or pitch tubes around the entrance tunnels. Infestations of the spruce beetle *D. rufipennis* are especially difficult to detect, though evidence of wind-felled trees in an area may be an indicator (Fig. 17-7). Often, there is very little color change apparent. In some cases, the needles are shed while the brood is still in the trees. For this reason, the ground survey crews must examine each tree with care to determine whether or not it is infested.

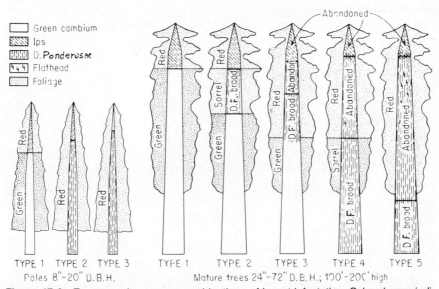

Figure 17-6 Trees may have many combinations of insect infestation. Color phases indicate the order of attack by various insects. From these diagrams, it is apparent that ponderosa pine trees apparently killed by *Dendroctonus* may have previously been attacked by other beetles. *(U.S. Forest Service diagram.)*

Figure 17-7 Severe blowdown in a stand of Engelmann spruce. The spruce beetle, *Dendroctonus rufipennis*, may increase to outbreak numbers in such situations. *(U.S. Forest Service photo.)*

Direct Control Procedure

The objective of direct control of *Dendroctonus* beetles is to reduce the population of the beetles in an area to the point of innocuousness. After an infested area has been detected and evaluated by the survey crews, the infested trees are marked for attention and then treated by one of several methods to destroy the infesting brood. The attacked trees are salvaged or destroyed to protect their neighbors. No method has as yet been devised to save an attacked tree.

Only those trees that actually contain broods are marked for treatment. An attempt is made to mark every infested tree in a control area. In order to do this, the markers must examine every tree for signs of fading, for active entrance holes indicated by frass or pitch tubes, or for both. In order that no trees will be missed, the area is examined in contiguous strips, each spotter examining and marking for treatment the trees on his or her own strip. The width of the strips will vary, but a common width is 2 chains (40 m). The usual procedure is to lay out a grid pattern with string line within which the spotter works.

The spotters are followed by the treating crews. Any season when the beetles are not actively in flight is satisfactory for control. The kind of treatment will depend on the species of tree and the species of beetle.

Cost of Control

From our discussion, it is evident that treatment for destroying a beetle brood must be adjusted to the beetle species and to the species and size of the trees involved. The cost of applying the various kinds of treatment varies greatly; even the same treatment will vary in cost, depending on the size and accessibility of the trees. Therefore, it is impossible to make any generalizations concerning the cost of beetle control.

The forest entomologist, through experience, must estimate the cost of control, and the forest owner must then decide whether the protection afforded by the operation and the values involved will justify the expense. However, the decision cannot always be made in the light of a single ownership. Flights of beetles sometimes result in infestation of new areas several miles away from the places where the beetles developed. Therefore, in the public interest, bark beetle infestations must often be controlled regardless of ownership. Here is one of the situations in which public funds, appropriated under the Cooperative Forestry Assistance Act of 1978 or under the various state forest-pest laws, can play an important role in forest-insect control. Participation of all parties affected, as provided for under pest control laws, provides the only satisfactory solution of the beetle control problem.

Research on bark beetles has produced many contributions during recent years. These serious pests are being thoroughly investigated. Their life histories are generally well known, and knowledge is gradually being accumulated on their overall relationship in the ecosystems of which they are a part. These studies should lead to the more effective regulation of bark beetle populations.

PHLOEM BORERS, COLEOPTERA

There are many species of these insects that spend part or all of their larval stage in the phloem area of trees. Some species, especially among the Ccrambycidae, are not readily separated from those to be discussed in Chapter 18. However, those covered here are perhaps better known for their damage to trees by their activities in the phloem rather than in the wood itself.

Sugar Maple Borer

Among the cerambycid phloem borers, one of the most striking and best-known examples of a primary species is the sugar maple borer, *Glycobius speciosus* (Baker, 1972; Johnson and Lyon, 1976). This insect confines almost all its larval activities to the phloem and adjacent soft tissues. Only when it reaches full growth does it penetrate into the sapwood, there to form its pupal cell.

The adults are beautiful black beetles, almost 1 in (2.5 cm) in length, marked with yellow stripes. They emerge in midsummer and deposit their eggs in slits cut with their mandibles in the bark of sugar maple trees, usually on the trunk. The larvae tunnel through the bark and into phloem. There they pass the winter and in the year following accomplish the greatest injury. Because the larvae tend to tunnel around the tree rather than lengthwise of the trunk, a very few of them working close together may girdle a tree and kill it. Usually, however, the larvae are not gregarious; and consequently, the infested tree does not die immediately. However, the killing of patches of bark opens the way to the attack of other organisms. The larvae attain full growth by the end of the summer following the year of oviposition. At that time, they tunnel into the wood, where the pupal cell is formed; there the second winter is passed in the prepupal stage. The pupal stage occurs the following summer just prior to the emergence of the adults. Thus, a period of 2 years is required for the complete life cycle of this insect.

To control this insect in ornamental trees, it is advisable to destroy badly infested trees. In moderately infested ornamentals, the larvae may be cut out and destroyed. This work should be done as early as possible in the life of the insects so that the resultant wounds will be small and, as a result, quick to heal. The presence of young larvae is indicated by the exudation of frass and sap from the egg slits. In woodlands and groves, the presence of a luxuriant undergrowth is said to reduce the danger of infestation by this beetle. The beetle is a light-loving species; hence, trees in the open are more subject to attack than are trees in closed stands. This fact should be borne in mind when improvement cuttings and thinnings in hardwood forests are contemplated.

Agrilus Beetles

Another large group of coleopterous phloem insects are members of the family Buprestidae, the metallic wood borers or flatheaded borers. Some of these species spend their entire larval period in the phloem, for example, the two-lined chestnut borer, *Agrilus bilineatus,* the bronze birch borer, *A. anxius,* and the bronze poplar borer, *A. liragus.* The *Agrilus* beetles are widely distributed in the United States. The three mentioned here are similar in appearance: brown or black in color, slender, and about ½ in (12 mm) in length. Adults emerge in late spring or early summer and deposit their eggs on the bark of their host trees, cementing the eggs firmly in place. These insects are light-loving. The adults like to bask in the sun; thus, they are likely to deposit eggs on trees in the open and on branches and trunks of forest trees that are exposed to the sun's rays. The larvae emerge from the eggs by boring through the part of the shell that is cemented to the bark, through the outer bark and phloem layers to the cambial region. There they typically cut winding, frass-filled galleries that score the wood. The larvae are elongated, flattened grubs with the head invaginated into the somewhat expanded prothorax. By autumn, having completed their growth, they tunnel into the bark and there form their pupal cells, in which they pass the winter as prepupae. In the spring, they transform to the pupa; and in June, they emerge as adults. Prior to ovipositing, the beetles feed for a time on foliage of the host trees (Fig. 17-8).

The twolined chestnut borer, *A. bilineatus,* infests chestnuts, oaks, and possibly beeches. It is often injurious to ornamental trees. In some localities,

Figure 17-8 Feeding of *Agrilus* adults on trembling aspen leaves. Feeding of this sort is characteristic of many adult buprestid beetles. *(From Graham and Knight, Principles of Forest Entomology, 4th ed., Fig. 78, p. 323.)*

the twolined chestnut borer frequently works with the root rot, *Armillaria mellia,* a parasitic fungus. *Armillaria,* unlike the borer, appears to attack trees growing in woodlands and kills groups of trees. Around the edges of these killings, the twolined chestnut borer finds a desirable place and usually joins with the fungus in continuing the destruction of the trees. A tree that is heavily infested by this insect is almost certain to die; however, the tunnels resulting from a light attack are usually overgrown in healthy, vigorous trees.

Trees weakened by drought, defoliation, or other injuries are most susceptible. This insect is often cited in relation to outbreaks of the gypsy moth, *Lymantria dispar,* and was heavily involved following oak leafroller, *Archips semiferanus,* outbreaks in Pennsylvania in the earlier 1970s. Baker (1972) suggests that watering or fertilizing of valuable shade trees may be helpful in preventing losses.

The bronze birch borer, *A. anxius,* is an especially well-known forest and ornamental-tree pest. In parts of the United States, it has made impossible the growing of birches as ornamentals. Even in the forest, overmature stands, slow-growing birches on poor sites, and trees left on cutover areas are attacked and killed by this insect. Like the twolined chestnut borer, *A. bilineatus,* the bronze birch borer stands on the borderline between the secondary and the primary insects. On cutover lands, many trees that have apparently been killed by this borer have actually died from adverse physical effects. In other instances, subnormal trees that might have survived for years have been attacked and killed.

Table 17-3 Some Beetles of the Genus *Agrilus* that May Cause Problems on Forest and Shade Trees of the United States and Canada

Species	Hosts	Common name
A. bilineatus	Chestnut, oaks, beech	Twolined chestnut borer
A. anxius	Birch (several species)	Bronze birch borer
A. liragus	Aspens and poplar	Bronze poplar borer
A. horni	Aspen	—
A. arcuatus var. torquatus	Hickory and pecan	Hickory spiral borer
A. acutipennis	Oak	—
A. juglandis	Butternut	—
A. difficilis	Honey locust	—
A. otiosus	Numerous hardwood species	—
A. angelicus	Oak	Pacific oak twig girdler

Therefore, the protection of birches from this insect depends, to a considerable degree, on maintaining the trees in a thrifty condition. The effect of exposure to the sun's rays, especially on the south and west sides, deserves special comment. Birches so exposed in the course of logging operations are likely to be attacked and die within a few years. Therefore, in partial cuttings, special care should be exercised to avoid leaving the trunks of birches exposed to the sun. Either birches should be cut, or they should be left shaded by trees less subject to exposure, such as sugar maple.

The bronze birch borer, *A. anxius,* was discussed as a pest of birches and poplars for many years. Barter and Brown (1949) gave a new species name, *A. liragus,* to the form infesting poplars. Thus, *A. anxius* is host-specific to *Betula* spp. The two species can be separated by characters of the male genitalia and by chromosome counts. Barter (1957) provides a more complete description of the insect and its development in hosts of a weakened condition. In 1965, Barter further described the habits of the bronze poplar borer. There are numerous species in the genus *Agrilus* with similar habits (Table 17-3).

Melanophila Beetles

Several buprestid species of the genus *Melanophila* have attracted considerable attention. One of these is the hemlock borer, *Melanophila fulvoguttata* (Fig. 17-9). This insect is a flattened, metallic black beetle with three small white spots on each elytron. The larvae are legless grubs with the widened prothorax characteristic of most buprestids.

The insect has occurred in large numbers on several occasions. The ecological situation described by Samuel A. Graham[1] when discussing one of these periods of increased beetle activity was interesting in that severe damage to hemlock was apparently in progress. However, within a few years, most of the healthy trees recovered. A thorough study revealed that drought was the

[1]Personal communication.

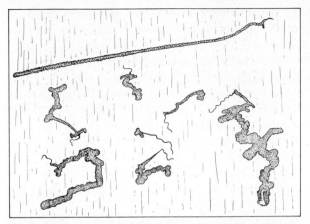

Figure 17-9 Tunnels of the hemlock borer, *Melanophila fulvoguttata*. The long transverse tunnel represents an unsuccessful attack made while the tree was alive. The others are successful attacks made immediately after the tree died. (*From Graham and Knight, Principles of Forest Entomology, 4th ed., Fig. 79, p. 324.*)

real cause of death in most trees that died. The borers attacked after the trees were dying and thus were acting only in a secondary capacity. The forester should always be watching for such situations because many of these phloem borers are the conspicuous evidence of mortality but may often be only the secondary cause.

Tree killing on cutover lands has sometimes been attributed to the borer when the actual cause of death was exposure. Hemlock is a very sensitive tree; on cutover lands, it is affected (as is birch) by exposure to the sun's rays and often dies as a result of exposure or a combination of exposure and borers. Hemlock is also easily injured by trampling about its roots. Such mistreatment is likely to be followed by decadence, borer attack, and death. Therefore, this tree should not be used for shade on camp or picnic grounds; nor should hemlock groves be used for such purposes.

The eggs of the hemlock borer, *Melanophila fulvoguttata,* are deposited in groups deep in the bark crevices, where they are cemented firmly. The young larvae penetrate directly through the bark to the inner phloem. If conditions are unfavorable for them, they do not reach the cambium but cut transverse tunnels in the phloem until they die without seriously injuring the tree. If conditions are favorable, they penetrate to the cambium and construct tortuous, frass-filled galleries just outside the wood until their growth is completed. Then they cut their pupal cells in the outer bark, where they pass the winter as prepupae. The following spring, they pupate and emerge during June and July to mate and oviposit, thus completing the life cycle.

Another species, the California flatheaded borer, *M. californica,* is an important insect in overmature ponderosa pine forests of California and Oregon, especially on the drier sites (Lyon, 1970). It attacks apparently healthy trees in the top. Often, the attack may be unsuccessful, and the attacking larvae either die before reaching the phloem or cut very short galleries before they are killed. These galleries are promptly overgrown by wood. Only when the tree becomes decadent are the larvae able to develop to maturity (Fig. 17-10). Then they kill the tops of trees that might have continued to live.

The borer usually works in association with another pest such as the western pine beetle, *D. brevicomis.* The California flatheaded borer, *M. californica,* often attacks the tree first, predisposing it to further attacks by bark beetles. It is most common in trees of declining vigor or in areas damaged by fire.

Sanitation-salvage logging (a selective cutting of poor-vigor trees) is an effective method of control in merchantable stands. In young stands or special-use areas such as recreational areas, logging may not be feasible. Chemical control has proved successful with penetrating oil sprays.

The California flatheaded borer, *M. californica,* primarily attacks ponderosa and Jeffrey pines. It shows a decided preference for Jeffrey pine when

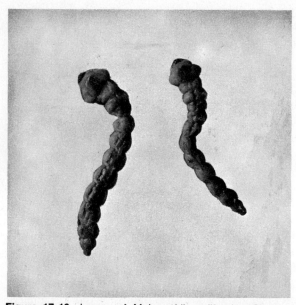

Figure 17-10 Larvae of *Melanophila californica.* Observe the widened prothorax into which the head is invaginated; this is characteristic of all buprestid larvae. *(U.S. Forest Service photo.)*

the stands contain a mixture of the two species. The insect attacks trees of all size classes but is most common in those of declining vigor. Other species of *Melanophila* include *M. drummondi,* the flatheaded fir borer, on fir, larch, and spruce, and *M. notata* on pines.

Chrysobothris Beetles

The flatheaded appletree borer, *Chrysobothris femorata,* is the most common of these insects in North America; it occurs widely throughout Canada and the United States. It attacks a wide variety of deciduous trees in addition to apple. It is especially destructive to newly planted trees and those weakened by drought, defoliation, or other factors (Baker, 1972). Johnson and Lyon (1976) say that regular fertilization and irrigation of most ornamental plants will reduce successful attacks by the pest.

The Pacific flatheaded borer, *Chrysobothris mali,* is a closely related species causing similar injury. It is considered one of the worst problems on newly planted trees and shrubs in the western coastal regions of the United States and Canada. More than 70 plant species are attacked (Johnson and Lyon, 1976). As in the genera *Agrilus* and *Melanophila,* there are numerous species of *Chrysobothris* in the forests and on shade trees in North America.

Weevils

Many weevils of the family Curculionidae are phloem-eating insects. Some of these have been mentioned among the tip and root insects previously discussed, but others feed on the phloem of the bole. Probably the most damaging is the pine root collar weevil, *Hylobius radicis;* its biology and habits are described by Schaffner and McIntyre (1944) and by Wilson and Schmiege (1970). This weevil attacks many species of pines but is particularly destructive in plantations of introduced species, such as Scots, Austrian, or Corsican pines, and in some red pine plantations growing on poor sites.

The young larvae feed in the inner bark at and below the root collar. Feeding continues in this general area, causing much resin flow from the injured bark. Gradually, the soil becomes resin-soaked, so that a layer of pitch-infiltrated soil may be 2 or 3 in (50 or 75 mm) thick near the feeding area. Trees become greatly weakened in the root collar area and are susceptible to windthrow. Small trees may be completely girdled. Generally, mortality is greatest in trees 4 in (10 cm) diameter at breast height (dbh) or less. The life cycle is variable in length, with much overlapping of generations.

Wilson and Schmiege recommend the following practices:

1   Avoid planting the highly susceptible pine species on light, sandy soils when the weevil is known to be in the area.
2   Avoid planting mixtures of pines.

3 Remove open-grown, older, susceptible brood pines before planting young trees.
4 Maintain fully stocked stands of trees.
5 Prune the lower 2 to 4 ft (60 to 120 cm) of branches, and clear away the litter beneath infested trees.

The northern pine weevil, *Pissodes approximatus,* also attacks pines and spruces (Wilson, 1977). The larvae of this species make galleries in the inner bark of the main stem and larger branches. This close relative of the white pine weevil, *P. strobi,* is found from the Atlantic Coast to Manitoba in Canada and south to Minnesota and North Carolina in the United States. The adult is somewhat larger than the white pine weevil. Damage may be severe where there are a quantity of breeding sites such as fresh stumps in Christmas tree plantations. Breeding materials are scarce in natural stands, and damage is rare. The deodar weevil, *P. nemorensis,* is found in the southern states and breeds in cedar and pines, where it damages the laterals and terminals. Small pines may be girdled and killed. The damage is similar to that caused by the white pine weevil, *P. strobi.*

Nitidulidae

Recent studies have revealed a considerable amount of defect in lumber resulting from the work of the sap-feeding beetles of the family Nitidulidae. These small beetles cause a type of defect called *bark pocket* in oaks, black gums, hickories, and yellow poplar. The larvae and adults feed in sap oozing from injured bark, and the larval feeding kills patches of cambium. These areas heal over, leaving a pocket of ingrown bark and stained wood. The losses in southern bottomland oaks are second only to carpenterworm, *Prionoxystus robiniae,* damage and in some cases may be even more injurious (Fig. 17-11).

PHLOEM BORERS, LEPIDOPTERA

Although most of the phloem borers are members of the Coleoptera, there are some members of the Lepidoptera in this group (Solomon and Dix, 1979). Most common of these, perhaps, are the pitch moths. Although there are numerous species in this group, the life cycles and types of injury are very similar for all of them. The well-known pitch mass borer, *Synanthedon pini,* adequately illustrates these lepidopterous phloem insects.

Pitch Mass Borer

The adult pitch mass borer, *S. pini,* is one of the clearwing moths of the family Sesiidae. In shape and coloring, these moths suggest wasps; consequently, members of this family are frequently cited to illustrate protective coloration

Figure 17-11 The Nitidulidae, called *sap beetles,* are responsible for much defect in southern oak. *Top:* Overgrown scars on the bark. *Middle:* View of inner surface of the bark. *Bottom:* Cross section shows permanent defect. *(U.S. Forest Service photo.)*

and mimicry. This insect does not kill the trees but causes defects that lower the value of the lumber made from them.

The adult moths appear in midsummer and deposit their eggs on the bark of the host tree, usually at the edge of a wound. The larvae, which are typical caterpillars, spend 2 or 3 years in the developmental stages. During this period, they feed in the phloem and adjacent tissues, in which each larva excavates a broad chamber near the point where the egg was deposited. After the larvae have become full grown, they transform to the pupal stage in the pitch mass accumulated over the burrow. Just before the adult moth is ready to emerge, the pupa works the anterior portion of its body out of the sticky mass of resin so that the moth can emerge without becoming entangled.

The control of pitch moths is very difficult in the forest. The systematic examination of the trees in order to secure the destruction of larvae by mechanical means has been recommended for certain species, but such an operation would be economically justifiable only on ornamentals or where extremely severe infestations occur in very valuable timber. Because moths seek wounds and scars for oviposition, the prevention of mechanical injury to trees or any treatment that tends to stimulate the rapid healing of wounds will reduce the amount of injury by the pitch moth. Thrifty trees growing on a suitable site are less susceptible than trees growing on poorer sites.

The Sequoia pitch moth, *S. sequoiae,* is a common pest in the West on Monterey pine, ponderosa pine, and sugar pine (Johnson and Lyon, 1976). Habits and damage are similar to those of the pitch mass borer, *S. pini.* Large accumulations of pitch and frass collect at these locations of damage. Trees in a state of decline are especially attractive to this insect. A second western species, the Douglas-fir pitch moth, *S. novaroensis,* causes pitch masses that are similar on spruce, pine, and Douglas-fir in the West.

Zimmerman Pine Moth

The Zimmerman pine moth, *Dioryctria zimmermani,* is similar in appearance to the species of the some genus discussed under seed and cone insects. Like those species that attack cones, the Zimmerman pine moth exhibits great variation in its feeding habits, suggesting that more than a single species may be referred to by the same name. It is responsible for damage to many pine species in the United States and Canada. Although it feeds on the branch tips, the most serious injury to the trees is along the trunk at branch whorls or between whorls. Damage may kill trees, retard the growth, or cause poor form. Mortality of the attacked trees often results from wind breakage at the points of severe infestation. The presence of larvae in the trunk is indicated by exudation of resin mixed with larval frass (Rennels, 1960).

The following recommendations will help to prevent severe damage in new plantations.

1 Plant pines at 6-by-6-ft (1.8-by-1.8-m) spacing or closer.
2 Avoid planting highly susceptible species such as Scots pine, Corsican pine, and Japanese red pine.
3 Promptly replant spots where trees die.
4 If possible, avoid planting within ½ mi (0.8 km) of known infestations.
5 Destroy infested material and heavily infested trees.

These recommendations are mainly based on the fact that dense, closed stands are not generally infested. This pest occurs throughout most (if not all) of northern United States and southern Canada and less commonly in the southern states (Carlson and Wilson, 1976).

Pitch Twig Moths

The northern pitch twig moth, *Petrova albicapitana,* may cause damage to branches of pines. These insects create a hollow, thin-walled pitch blister in which the larva develops (Wilson, 1977; Rose and Lindquist, 1973). Other species causing similar damage include the pitch twig moths *P. comstockiana, P. virginiana, P. houseri, P. burkeana,* and *P. metallica.*

As a group, the pitch twig moths are minor pests of young trees. Young pines from 1 to 5 ft (30 to 150 cm) in height are most heavily infested. Attack at the base of a growing terminal may kill the terminal; if it is not girdled, it may survive as a crooked trunk.

BIBLIOGRAPHY

Amman, G. D., et al. 1977. Guidelines for reducing losses of lodgepole pine to the mountain pine beetle in unmanaged stands in the Rocky Mountains. *USDA For. Ser. Gen'l. Tech. Rept.* INT-36.

Baker, W. L. 1972. Eastern forest insects. *USDA For. Serv. Misc. Publ.* no. 1175.

Barter, G. W. 1957. Studies of the bronze birch borer, *Agrilus anxius* Gory, in New Brunswick. *Canad. Entomol.* 89:12–36.

———. 1965. Survival and development of the bronze poplar borer *Agrilus liragus,* Barter and Brown (Coleoptera: Buprestidae). *Canad. Entomol.* 97:1063–1068.

———, and Brown, W. J. 1949. On the identity of *Agrilus anxius* Gory and some allied species (Coleoptera: Buprestidae). *Canad. Entomol.* 81:245–249.

Berryman, A. A. 1973. Population dynamics of the fir engraver, *Scolytus ventralis* (Coleoptera: Scolytidae). *Canad. Entomol.* 105:1465–1488.

Berryman, A. A., et al. (eds.). 1978. Theory and practice of mountain pine beetle management in lodgepole pine forests. Pullman, Wash.: Washington State University.

Borden, J. H. 1974. Aggregation pheromones in the scolytidae. In *Pheromones,* ed. M. C. Birch, p. 135–160. New York: American Elsevier Publishing Co.

Bright, D. E., Jr. 1963. Bark beetles of the genus *Dryocoetes* (Coleoptera: Scolytidae) in North America. *Ann. Entomol. Soc. Amer.* 56:103–115.

———. 1976. The insects and arachnids of Canada, Part 2. The bark beetles of Canada and Alaska (Coleoptera: Scolytidae). *Canada Dept. of Agric. Publ.* no. 1576.

Carlson, R. B., and Wilson, L. F. 1976. Zimmerman pine moth. *USDA For. Serv., For. Pest Leaf.* no. 106.

Cole, W. E., and Cahill, D. B. 1976. Cutting strategies can reduce probabilities of mountain pine beetle epidemics in lodgepole pine. *J. Forestry* 74:294–297.

Coster, J. E. 1977. Towards integrated protection from the southern pine beetle, *J. Forestry* 75:481–484.

Furniss, M. M., et al. 1976. Aggregation of spruce beetles (Coleoptera) to sendenol and repression of attraction by methylcyclohexenone in Alaska. *Canad. Entomol.* 108:1297–1302.

Furniss, M. M., and Orr, P. W. 1978. Douglas-fir beetle. *USDA For. Serv., For. Insect and Disease Leaf.* no. 5.

Furniss, R. L., and Carolin, V. M. 1977. Western forest insects, *USDA For. Serv. Misc. Publ.* no. 1339.

Hosking, G. P., and Knight, F. B. 1976. Investigations on the life history and habits of *Pityokteines sparsus* (Coleoptera: Scolytidae). *Life Sci. & Agric. Expt. Sta., Univ. of Maine, Orono, Tech. Bull.* no. 81.

Hopping, G. R. 1963a. The natural groups of species in the genus *Ips* DeGeer (Coleoptera: Scolytidae) in North America. *Canad. Entomol.* 95:508–516.

———. 1963b. The North American species in group I of *Ips* DeGeer (Coleoptera: Scolytidae). *Canad. Entomol.* 95:1091–1096.

———. 1963c. The North American species in group II and III of *Ips* DeGeer (Coleoptera: Scolytidae). *Canad. Entomol.* 95:1202–1210.

———. 1964. The North American species in groups IV and V of *Ips* DeGeer (Coleoptera: Scolytidae). *Canad. Entomol.* 96:970–978.

———. 1965a. The North American species in group VI of *Ips* DeGeer (Coleoptera: Scolytidae). *Canad. Entomol.* 97:533–541.

———. 1965b. The North American species in group VII of *Ips* DeGeer (Coleoptera: Scolytidae). *Canad. Entomol.* 97:193–198.

———. 1965c. The North American species in group VIII of *Ips* DeGeer (Coleoptera: Scolytidae). *Canad. Entomol.* 97:159–172.

———. 1965d. The North American species in group IX of *Ips* DeGeer (Coleoptera: Scolytidae). *Canad. Entomol.* 97:422–434.

———. 1965e. The North American species in group X of *Ips* DeGeer (Coleoptera: Scolytidae). *Canad. Entomol.* 97:803–809.

Johnson, P. C. 1972. Bark beetle risk in mature ponderosa pine forests in western Montana. *USDA For. Serv. Res. Paper* Int-119.

Johnson, W. T., and Lyon, H. H. 1976. *Insects that feed on trees and shrubs.* Ithaca: Cornell University Press.

Keen, F. P., and Miller, J. 1960. Biology and control of the western pine beetle. *USDA For. Serv. Misc. Publ.* no. 800.

Knopf, J. A. E. and Pitman, G. B. 1972. Aggregation pheromone for manipulation of the Douglas fir beetle. *Journ. Econ. Entomol.* 65:723–726.

Lanier, G. N. 1970. Biosystematics of North American *Ips* (Coleoptera: Scolytidae), Hopping's group IX. *Canad. Entomol.* 102:1139–1163.

Leuschner, W. A., et al. 1977. An integrated research and applications program. *J. Forestry* 75:478–480.

Lyon, R. L. 1970. California flatheaded borer. *USDA For. Serv. For. Pest Leaf.* no. 24.

Mason, R. R. 1970. Comparison of flight aggregation in two species of southern *Ips* (Coleoptera: Scolytidae). *Canad. Entomol.* 102:1036–1041.

Massey, C. L. 1971. Arizona five-spined *Ips. USDA For. Serv. For. Pest Leaf.* no. 116.

———, et al. 1977. Roundheaded pine beetle. *USDA For. Serv. For. Insect & Disease Leaf.* no. 155.

McCambridge, W. F., and Trostle, G. C. 1972. The mountain pine beetle, *USDA For. Serv. For. Pest Leaf.* no. 2.

Pitman, G. B. 1971. Trans-verbenol and alpha-pinene: Their utility in manipulation of the mountain pine beetle. *J. Econ. Entomol.* 64:426–430.

———. 1973. Further observations on Douglure in a *Dendroctonus pseudotsugae* management system. *Environ. Entomol.* 2:109–112.

———, et al. 1969. Specificity of population: Aggregating pheromones in *Dendroctonus. J. Insect Physiology* 15:363–366.

Price, T. S., and Doggett, C., eds. 1978. *A history of southern pine beetle outbreaks in the southeastern United States.* Southern forest insect working group, Macon, Ga.: Georgia Forestry Commission.

Rennels, R. G. 1960. The Zimmerman pine moth. *Univ. Illinois, Agr. Expt. Sta. Bull.* no. 660.

Renwick, J. A. A., and Vité, J. P. 1972. Pheromones and host volatiles that govern aggregation of the six-spined engraver beetle, *Ips calligraphus. J. Insect Physiology* 18:1215–1219.

Rose, A. H., and Lindquist, O. H. 1973. Insects of eastern pines. *Dept. of Environ., Canad. For. Serv. Publ.* no. 1313.

Safranyik, L., et al. 1974. Management of lodgepole pine to reduce losses from the mountain pine beetle. *Dept. of Environ., Canad. For. Serv., Forestry Tech. Rept.* no. 1.

Sartwell, C., et al. Pine engraver, *Ips pini* in the western states. *USDA For. Serv. For. Pest Leaf.* no. 122.

———, and Stevens, R. E. 1975. Mountain pine beetle in ponderosa pine. *J. Forestry* 73:136–140.

Schaffner, J. V., Jr., and McIntyre, H. L. 1944. The pine root collar weevil. *J. Forestry* 42:269–275.

Schmid, J. M., and Beckwith, R. C. 1975. The spruce beetle. *USDA For. Serv. For. Pest Leaf.* no. 127.

———. 1977. Guidelines for minimizing spruce beetle populations in logging residuals. *USDA For. Serv. Res. Paper* RM-185.

Smith, R. H., and Lee, R. E. III. 1972. Black turpentine beetle. *USDA For. Serv. For. Pest Leaf.* no. 12.

Solomon, J. D., and Dix, M. E. 1979. Selected bibliography of the clearwing borers (Sesiidae) of the United States and Canada. *USDA For. Serv. Gen'l. Tech. Rep't.* 50–22.

Speers, C. F. 1971. *Ips* bark beetles in the south. *USDA For. Serv. For. Pest Leaf.* no. 129.

Stevens, R. E. 1971. Fir engraver. *USDA For. Serv. For. Pest Leaf.* no. 13.

Struble, G. R. 1970. Monterey pine *Ips. USDA For. Serv. For. Pest Leaf.* no. 56.

————, and Hall, R. C. 1955. The California five-spined *Ips. USDA For. Serv., Circ.* no. 964.

Thomas, G. M., and Wright, K. H. 1961. Silver fir beetles. *USDA For. Serv. For. Pest Leaf.* no. 601.

Thomas, J. B. 1961. The life history of *Ips pini* (Say) (Coleoptera: Scolytidae). *Canad. Entomol.* 93:384–390.

————, and Bright, D. E., Jr., 1970. A new species of *Dendroctonus* (Coleoptera: Scolytidae) from Mexico. *Canad. Entomol.* 102:479–483.

Wilson, L. F. 1977. A guide to insect injury of conifers in the Lake States. *USDA For. Serv. Agric. Handbook* no. 501.

Wilson, L. F., and Schmiege, D. C. 1970. Pine root collar weevil. *USDA For. Serv. For. Pest Leaf.* no. 39.

Wood, D. L., et al. 1966. Sex pheromones of bark beetles: I. Mass production, bioassay, source, and isolation of the sex pheromone of *Ips confusus. J. Insect Physiol.* 12:523–536.

Wood, S. L. 1963. A revision of the bark beetle genus *Dendroctonus* Erickson (Coleoptera: Scolytidae). *Great Basin Nat.* 23.

Young, J. C., et al. 1973. Aggregation pheromone of the beetle *Ips confusus:* Isolation and identification. *J. Insect Physiol.* 19:2273–2277.

Wood Destroyers (Standing Trees and Logs)

The large variety of insects that will be discussed in this chapter has made the selection of a title difficult. Some of these insects are meristem insects for at least a part of their life cycle, and some are found only in the wood of trees and logs that have been dead for a considerable time. Thus, there is overlap with Chapter 17, which discussed bark beetles and other phloem infesting pests, and also with Chapter 19, which considers insects on wood that has been manufactured and is in use for some purpose. All these insects do bore in wood itself, and they enter at some stage through the bark on the tree or log.

We have separated these insects into three arbitrary groupings according to the type of damage and ecological requirements:

1 *Pests of living trees.* These borers may attack living trees, and most spend a part of their cycle in the meristematic tissues. Many may also attack recently dead trees or logs.

2 *Pests of dead or dying trees and logs.* These insects may occasionally be found in living materials, but most are found in dead or dying trees or logs. Most of these do not require a period of time in meristem tissues.

363

3 *Ambrosia beetles.* This special group of wood borers may attack either living or dead trees and logs. They live entirely in the wood.

PESTS OF LIVING TREES

Some of the outstanding examples of insects that attack living trees are the locust borer, *Megacyllene robiniae,* the poplar borer, *Saperda calcarata,* the carpenterworm, *Prionoxystus robiniae,* the linden borer, *S. vestita,* the western cedar borer, *Trachykele blondeli,* and the red oak borer, *Enaphalodes rufulus.* All of these species work in the phloem for a time after hatching and later penetrate into the wood (Fig. 18-1). Their work in the meristematic tissues is of relatively short duration and is not sufficiently extensive to result in the death of the trees. Therefore, the damage caused by these insects is in the reduction in quality of logs resulting from their boring and the weakening of standing trees by their work.

Locust Borer

The locust borer, *M. robiniae,* is one of the best-known examples of the phloem wood insects that attack living trees (Wollerman, 1970; Baker, 1972). It is a member of the family Cerambycidae. At one time, in the eastern United States, the black locust was considered to be one of the best trees for posts, poles, and railroad ties because of its rapid growth and the durability of its wood when it is in contact with the ground. Many plantations of this species were set out in Pennsylvania and in the Ohio and Mississippi valleys, often on very poor sites, where growth was slow and trees were generally in poor condition.

A large proportion of these plantations were abandoned as valueless because of the ravages of the locust borer, *M. robiniae.* Black locust is especially desirable for producing a quick cover on eroded lands and is widely used in plantations for soil protection. As a result of the increasing number of plantings for protection purposes, there was a corresponding increase in the economic importance of the locust borer.

The adults of the locust borer, *M. robiniae,* are beautiful black beetles, ½ to ¾ in (12 to 16 mm) in length, with narrow yellow crossbands on the elytra. The beetles emerge in late summer and autumn. They deposit their white eggs singly in cracks and crevices in the rough bark of black locust trees. The numerous moist spots on the bark at points where the newly hatched larvae are working are conspicuous on heavily infested trees in the spring. Young trees under 2 in (5 cm) in diameter and older trees over 6 in (15 cm) in diameter appear to be comparatively immune to attack; but once a tree is infested, it may remain subject to infestation, regardless of size. Thickness and surface character of the bark ordinarily appear to be the factors that determine the susceptibility of the tree to attack. The bark should be sufficiently old to be rough but not so thick that the young larvae cannot bore through it. This

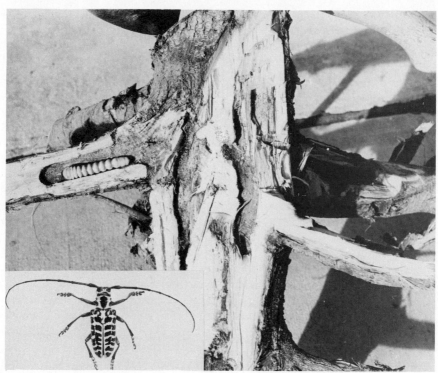

Figure 18-1 The cottonwood borer, *Plectrodera scalator,* one of the Cerambycidae that infest living trees. The damage to the stump area of a living tree was the result of larval feeding. The adult (*insert*) is a colorful black and white insect approximately 30 mm in length. *(U.S. Forest Service photos.)*

limitation of attack to certain specific age-classes aids greatly in formulating plans for the control of this insect.

Eggs are deposited in bark crevices during late summer and early fall. These hatch very soon, and the young larvae cut their way through the bark to the inner phloem. There they feed until cold weather forces them into hibernation. The winter is passed in the larval tunnels. In the spring, the larvae continue their work in the phloem for a short time before entering the wood. Examination of infested trees will readily disclose whether the larvae have entered the wood or are still in the phloem. While they are working in the phloem, they cast a brown-colored frass out of openings through the bark; whereas after they enter the wood, they push a yellowish-colored frass from their burrows. Throughout the life of the larva, an opening from its burrows to the outside is maintained through which the frass is ejected. For a time, the entrance hole serves this purpose; but later, other passages are cut to the outside as convenience demands. The larvae, after completing their growth in late July and August, excavate cells in the wood in which they transform to the pupal stage and, later in the same season, to the adult. Thus, the life cycle is completed in 1 year.

Wollerman (1970) presents several suggestions for control of the locust borer, *M. robiniae.* Silvicultural methods should be used for prevention of damage in forested areas. Slow-growing young stands are very susceptible to borer attack. These stands can be cut back before the borers become serious. The vigorous sprouts that follow will generally be less subject to attack. Severely infested stands can also be improved by clear-cutting during the dormant season. The sprouts that follow should be thinned. This procedure has produced good second crops of trees with little injury. Thinning is also beneficial to lightly injured stands on better sites. Removal of infested individuals will reduce the borer population and help to protect the more desirable residual trees.

Oak Borers

Donley and Terry (1977) state that wood-boring and cambium-mining insects together cause a greater loss of high-grade hardwood than all other damaging agents combined. Lumber degrade studies from five southern states showed that losses caused by oak borers averaged $25/1000 bd ft of oak lumber produced. Losses in two oak species from a variety of oak insects are cited in Tables 18-1 and 18-2.

Two oak borers are especially damaging to southern hardwoods. The red oak borer, *E. rufulus* (Coleoptera: Cerambycidae), seriously damages various oaks by creating large, extensive galleries in the wood. Damage can be reduced greatly by proper stand improvement. Most borer larvae die if the host trees to be cut in thinnings are previously poisoned while the larvae are still in the phloem (Hay, 1962).

The red oak borer, *E. rufulus,* has a 2-year life cycle, with emergence of

Table 18-1   Losses in Overcup Oak from Three Bottomland Stands where Light, Moderate, and Heavy Insect Damage Occurred, Expressed in Terms of Percentage of Sawn Lumber Grade

	Infestation severity					
	Light		Moderate		Heavy	
Lumber grade*	Potential	Actual	Potential	Actual	Potential	Actual
FAS[†]	3.2	0.6	10.1	3.1	45.0	0.9
No. 1 common	36.5	22.6	44.8	22.2	31.5	9.3
No. 2 common	28.6	40.0	31.0	31.0	8.3	20.3
No. 3A common	18.6	20.9	9.3	18.7	9.2	23.8
No. 3B common	13.1	15.9	4.7	24.9	6.0	45.7

Source: Adapted from Morris (1964).

* Lumber grades are standardized. The highest-quality FAS grade for hardwood lumber is much more valuable than lower grades in the list.

[†] FAS is a lumber grade.

Table 18-2   Losses in Black Oak from Two Upland Oak Stands where Medium and Heavy Insect Damage Occurred, Expressed in Terms of Percentage of Sawn Lumber Grade

| | Infestation severity | | | |
| | Medium | | Heavy | |
Lumber grade*	Potential	Actual	Potential	Actual
FAS[†]	2.8	2.3	4.0	.1.9
No. 1 common	33.6	16.9	33.1	17.6
No. 2 common	54.3	51.6	24.9	20.0
No. 3A common	8.0	18.4	27.6	45.3
No. 3B common	1.3	10.8	10.4	15.2

Source: Adapted from Morris (1964).

* Lumber grades are standardized. The highest-quality FAS grade for hardwood lumber is much more valuable than lower grades in the list.

[†] FAS is a lumber grade.

adults occurring during odd-numbered years in the central states. Eggs are laid during the summer on the bark of trees over 2 in (5 cm) in diameter. The young larva spends its first year in the inner-bark region. In July and August of the second year, it burrows through the sapwood and for 6 to 10 in (15 to 25 cm) into the heartwood. Once the larva has started its burrow into the sapwood, it is too late to kill the insect by killing the tree. Thus, trees to be cut should be poisoned between peak emergence in July of odd-numbered years and the first of June in even-numbered years to assure larvae mortality.

The white oak borer, *Goes tigrinus* (Coleoptera: Cerambycidae), tunnels in the trunks and larger branches of young trees (Solomon and Morris, 1965). This pest has a 3- to 4-year life cycle; the tunnels produced are long and of large diameter. Mature larvae are 25 to 35 mm long and 8 mm in diameter. Stands growing on poor sites with drainage problems are frequently infested. Woodpeckers may be the major natural control of these and other oak pests (Solomon and Morris, 1971).

Western Larch Borer

The western larch borer, *Tetropium velutinum* (Coleoptera: Cerambycidae), has been recognized as a serious pest problem in the Rocky Mountains and the Pacific coastal regions of North America (Ross, 1967). The life cycle of this insect is generally 1 year, though a partial second generation may occur. This insect may kill weakened western larch and hemlock, but it also causes significant damage to western larch logs by its penetration into the sapwood.

Western Cedar Borer

The western cedar borer, *Trachykele blondeli* (Coleoptera: Buprestidae), is a beautiful metallic emerald green buprestid that is very injurious, especially to western red cedar in certain localities (Burke, 1928). This is one of the species that gains entrance into the wood through the unbroken bark. The adults lay their eggs in crevices on the branches of living trees, and the larvae bore their long, flattened tunnels in both the sapwood and the heartwood until they are fully grown. This requires several years, during which time they leave the branches and work in the tree trunk, sometimes extensively. When fully grown, they return to the branch to pupate. There the pupal cell is constructed within ½ in (12 mm) of the surface. Transformation to the pupa occurs in midsummer and to the adult stage about 3 weeks later. These adults remain in the pupal cells until the following spring, at which time they emerge, feed on the foliage, mate, and oviposit.

The kind of injury caused by this insect is objected to by consumers of posts and poles, but actually, it causes little (if any) reduction in the usefulness of these materials. However, it does ruin the wood for shingles, siding, boat planking, and other uses in which tightness is essential. No control for the western cedar borer, *T. blondeli,* in the forest has been developed. The habits of oviposition are such that the adults or eggs are difficult to reach with an insecticide. The fact that certain stands are more severely injured than others indicates that silvicultural control may ultimately be possible. It must be admitted, however, that such practice might not appeal to companies in the pole business. Poles are only one product from these lands, and the pole-producing companies generally purchase directly from the land managers. Thus, they are not typically involved directly in land management practices.

Other Buprestid Borers

Among the other buprestid borers that attack the wood of living trees is the golden buprestid, *Buprestis aurulenta.* This western species attacks the pitchy wood in fire scars, lightning scars, and injuries caused by logging machinery. The galleries are filled with pitchy frass. Repeated attacks may completely destroy the wood, weakening the trees so that they may be easily wind-thrown. Furniss and Carolin (1977) report that this buprestid occasionally attacks timbers and boards in buildings.

A similar species, the turpentine borer, *B. apricans,* was formerly a serious pest in stands of southern pines used for the production of naval stores (Baker, 1972). Deep chipping of the faces, burning, and subsequent checking of the wood created conditions favorable for oviposition by the beetles. Repeated attacks so weakened the trees that they were subject to breakage by windstorms.

The turpentine borer, *B. apricans,* may still be found in fire scars and other injuries that kill the bark and expose the bare wood at the base of standing trees; but in turpentine orchards where modern methods are used, it is practically nonexistent. Instead of the narrow, deeply chipped faces, usually four on a tree, and frequently scorched by fire, the modern face is somewhat wider, no more than two to a tree; and the chipping is shallow, each strip being little deeper than the bark. The gum flowing down the face is drained into a container through metal troughs attached to the tree. These are fastened with double-headed nails that can easily be pulled out of the wood when the trough is removed. In this way, the face is continuously covered by a coating of gum that prevents checking; and no nails are left in the face after the gum harvest is completed, which is usually after 5 years.

Previously, the nails left in the faces made the butt section of a turpentined tree unmerchantable. Therefore, that section, usually infested with borers, was left in the woods. Today, the butt section is utilized along with the rest of the tree. Modern methods thus largely eliminate places suited for oviposition by the beetles and also remove from the woods parts of the trees that previously were a hazard.

Poplar Borer

Another important phloem wood beetle is the poplar borer, *S. calcarata* (Coleoptera: Cerambycidae). It is distributed throughout the United States and Canada wherever its host trees are found. This insect is injurious to forest trees both directly and indirectly. Directly, it reduces the value of the trees it attacks by cutting large tunnels in the wood. When they are numerous, the tunnels may so weaken the tree that windbreak occurs. Occasionally, when they are extremely numerous, these beetles can kill a few trees; but generally, the poplar borer is not a tree killer.

The poplar borer, *S. calcarata,* does even more important economic damage by its indirect effects than by its direct injury. Other injurious insects are attracted to trees that have been attacked by this beetle. For instance, several Buprestidae, such as *Dicera prolongata* and *Poecilonota cyanipes,* deposit their eggs in the old egg scars of the poplar borer. The carpenterworm, *Prionoxystus robiniae* (Lepidoptera: Cossidae), also finds that the scars made by the poplar borer offer desirable locations for oviposition. Thus, this beetle may be the indirect cause of other insect injury. Certain other habits of the poplar borer are conducive to another type of secondary injury. This insect keeps its tunnels comparatively free of frass by pushing this waste material out of openings through the bark. These open tunnels offer a ready means of access deep into the wood for the inocula of wood-rotting fungi. *Fomes igniarius,* an important heartrot that causes a large proportion of the defects found in aspen and poplar, frequently gains entrance through these openings. It has been shown that a large proportion of heartrot infections in some regions have arisen from insect tunnels. Also, in certain years, the egg niches provide favorable infection courts for the *Hypoxylon* canker fungus (Graham and Harrison, 1954). Thus, the poplar borer and the carpenterworm are indirectly responsible for a great deal of damage.

The adult beetles are rather large, often over 1 in (2.5 cm) in length, and gray with yellow markings. Emerging from their pupal cells in late July and

August, the female beetles lay their eggs in slits that they gnaw in the bark. The period of incubation is long, 20 to 25 days. The larvae, as soon as they hatch, tunnel into the phloem, where they feed until overtaken by cold weather. The following spring, they enter the wood, where they spend the remainder of their developmental period tunneling through the wood. In the autumn, when full grown, they hollow out a pupal cell in the sapwood. This cell is blocked off by a plug of shredded wood from the tunnel in which the insect previously lived. The winter is passed in the prepupal stage within the pupal cell. Early the following summer, the larvae pupate; after 25 to 30 days in that stage, they transform to the adult, cut their way to the surface, and emerge. Thus, the normal developmental period is 2 full years. In Colorado, however, at elevations ranging from 6000 to 9000 feet (1800 to 2700 m), this species requires 3 full years to complete its life cycle. In warmer climates, on the other hand, it is probable that a developmental period shorter than the normal is possible.

Control measures for this insect are usually necessary only on the poorer quality sites. A thrifty, closed stand is seldom damaged severely. A severely infested stand can be recognized by the black oval spots on the bark that overgrow the egg niches. These marks last indefinitely.

Carpenterworm

The carpenterworm, *P. robiniae,* is one of the common phloem wood insects that may well be cited as a lepidopterous representative of this group (Fig. 18-2). It is distributed generally throughout the United States and the southern

(a)   (b)

Figure 18-2 The carpenterworm, *Prionoxystus robiniae.* (*a*) The life stages, adult female wingspread 75 mm. (*b*) Attacks on Nuttall oak during a drought period. *(U.S. Forest Service photo.)*

part of Canada. It was first described as a pest of black locust, hence its specific name, but it also feeds on a variety of other hardwood species. Some of the trees that are frequently attacked are oaks, elms, and poplars. Although the carpenterworm seldom kills trees, it is nevertheless a primary insect, and most of its damage is done to living trees.

The adults are large gray moths with a wingspread of about 3 in (75 mm). They emerge in early summer and deposit their eggs on the bark of the host tree. Wounds or scars on the trunk or larger branches provide the most attractive oviposition places. The larvae bore into the phloem, where they excavate shallow galleries. After a brief period of feeding in the phloem, they penetrate into the sapwood and later into the heartwood. The remainder of the larval life is spent in the wood. The larvae are typical caterpillars in form and are pinkish white in color. Their galleries are long and winding and are kept free of frass so that the larvae can pass freely back and forth. In order to maintain free passages, the borings are pushed out through holes in the bark. These holes provide an easy avenue of entrance for the spores of wood-rotting fungi; consequently, these organisms are almost always associated with the work of the carpenterworm.

The length of the life cycle is variable (Hay and Morris, 1970), requiring 2 to 4 years, depending on location. In the Deep South, 2 or 3 years are required; in the central and western states, 3 years; and in the northern states and southeastern Canada, 4 years.

At the end of the larval period, a pupal cell is formed in the wood near the surface. Before pupation, a hole is cut almost through the bark, and the tunnel behind the cell is closed with a plug of shredded wood. The larvae then transform to the pupal stage, and later the moths emerge.

Solomon and Hay (1974) have prepared an annotated bibliography on the insect that highlights the research on severity of damage and possibilities of control. The carpenterworm, *P. robiniae,* has been recognized in recent years as a major factor in the management of southern hardwood forests.

Clearwing Moths

The clearwing moths (Lepidoptera: Sesiidae) are generally regarded in forested areas as rather interesting members of the community; a few are recognized as occasionally damaging. These become much more significant to people in the urban environment, where several species are very serious pests (Johnson and Lyon, 1976). The larvae of these moths feed on a variety of ornamental plants and often cause serious losses. Schread (1971) states that maintenance of healthy growing conditions is the most effective method of preventing damage. Because larvae enter commonly through wounds, mechanical damage is an invitation for possible infestation. Some examples of the pests in this group are:

1   Dogwood borer, *Synanthedon scitula*
2   Lesser peachtree borer, *S. pictipes*
3   Rhododendron borer, *S. rhododendri*
4   Ash borer or lilac borer, *Podosesia syringae*
5   Hornet moth, *Sesia apiformis*

PESTS OF DYING TREES AND LOGS

Representatives of insects that attack only dying trees or freshly cut logs are important because of the loss that they cause in both logs and pulpwood, in the former by reducing the grade and in the latter the volume. Some of them also cause indirect injury aiding infection by wood-rotting fungi. The most important of these insects belong to the coleopterous families Cerambycidae and Buprestidae.

The few species that will be discussed here should not be taken as an indication that there are not many more species that damage dying trees and logs. Gardiner (1957) made an intensive survey of the wood-boring beetles causing deterioration of fire-killed pine in Ontario. Following a single fire, he observed 17 species of Cerambycidae that were attacking the fire-damaged trees. In addition, numerous examples of other families were present. Gardiner (1975) and Wickman (1965) reported on deterioration following wind damage. Gardiner estimated a loss in value as a result of reduction in grade of 14.1 percent one year after blowdown in spruce followed by infestation by *Tetropium* species (Coleoptera: Cerambycidae).

Flatheaded Appletree Borer

One of the many buprestids that feed on the phloem and wood is the flatheaded appletree borer, *Chrysobothris femorata* (Coleoptera: Buprestidae). This insect attacks numerous species of forest trees and also a great variety of orchard trees. Some of its many hosts are walnut, pecan, hickory, poplar, willow, beech, oak, elm, and hackberry. Because of its importance as a pest of apple trees, this insect is called the *flatheaded appletree borer.* Because it is a pest of orchards as well as forests, it has received more attention from entomologists than if its activities had been confined to forest trees. As a result, its habits are well known, and it will serve well as an illustration of the phloem wood buprestids.

The flatheaded appletree borer, *C. femorata,* is in the middle ground between the primary and the secondary insects. It apparently cannot successfully attack a vigorous, healthy tree; instead, it attacks trees that are on the decline. The adult females are attracted for oviposition not only to the subnormal trees but also to healthy trees in sunny locations. The eggs hatch, and the young larvae bore into the bark, but they do not enter the phloem while the tree is in a thrifty condition. If, however, such a tree should decline in health, the larvae will promptly penetrate into the phloem and kill it. If, on the other hand, the tree remains in good health, the young larvae will remain in the bark for at least 1 year before they succumb to starvation. This habit has been observed in other species of the same genus, for example. *C. dentipes,* a flatheaded borer of pine. When bark-infested trees die or are cut, the larvae are then able to attack the phloem and later the sapwood; thus, they are able to complete their normal development in the logs.

The adults of the flatheaded appletree borer, *C. femorata,* are short, broad, flat beetles about 12 mm long. In color, they vary from dark brown to black on the top, with grayish spots and bands; the ventral side and the legs are bronze. They emerge from their pupal cells in early spring and are on the wing for about 1 month. The beetles are active only on warm, sunny days; at other times, they remain quiescent, hidden away beneath bark scales or in other suitable places. In warm weather, they are very active, flying and running about. Experiments have shown that beetles of this genus are very resistant to heat and can endure a temperature as high as 52°C. The larvae also are heat-resistant and are able to live on the hot upper side of logs lying in the sun.

The adult beetles feed on the foliage of hardwood trees and sometimes become so numerous that they cause appreciable defoliation. The female beetles deposit their eggs in crevices in the bark of the host trees. These eggs are pale yellow, flattened, and about 1 mm in diameter. After an incubation period of 15 to 20 days, the young larva bores through the underside of the egg and into the bark. This habit is true of all buprestids.

The thoracic region of these larvae is greatly enlarged transversely and flattened dorsoventrally. The head is invaginated into this enlarged thorax so that only the mouthparts protrude. On casual inspection, the thorax appears to be the head of the grub, thus. the name *flatheaded borer.* The larvae are helpless when removed from their burrow. Unless they have convenient surfaces against which they can expand the enlarged thorax, they are unable to move; therefore, they can move about only in a tunnel that fits their bodies. Thus, if a larva should cut through some portion of the eggshell that was not in contact with the bark, it would be unable to progress.

If, on boring through the bark, the larvae find conditions are favorable, they will construct broad feeding tunnels in the phloem and outer sapwood. They grow rapidly and in late summer are ready to bore into the solid sapwood; there they form their pupal cells, in which they pass the winter. In the South, the pupal cell is sometimes formed between the bark and the wood; but in the North, this condition has never been observed. The winter is passed in the larval stage within the pupal cell, and pupation takes place early in the following spring.

Thrifty trees are not susceptible to injury by this borer; therefore, it is a pest only of overmature, decrepit, or dying forest trees. Like many other members of this genus, it does not penetrate deeply into the heartwood and therefore is not so injurious to logs as some other buprestids and cerambycids are. It does, however, loosen the bark and make openings through which decay-causing organisms may enter, and it may tunnel the outer inch of wood to a considerable extent.

The Australian pine borer, *Chrysobothris tranquebarica,* is also commonly called the *mangrove borer.* It breeds in living mangrove and *Casuarina* trees, and the damage to ornamentals and windbreak trees of southern Florida may be severe.

Some of the other buprestids that feed first in the phloem and later in the wood are *Chalcophora virginiensis* and *C. angulicollis.* The first is an eastern species; and the second, a western species. Both attack pines and firs. Several species of the genus *Dicerca* are common phloem wood borers that may at times cause considerable injury to the wood of dying trees and freshly cut logs.

Firtree Borer

The firtree borer, *Semanotus litigiosus* (Coleoptera: Cerambycidae), is a common pest in the coniferous forests of the western United States and Canada (Wickman, 1968*a*). The insect attacks recently dead or dying trees, especially those that have been wind-thrown or damaged in some other fashion. Wickman estimates that the damage may cause value reductions as great as $80 to $100 per 1000 bd ft of timber.

The host trees include *Abies concolor, A. magnifica, A. grandis, A. lasiocarpa, Pseudotsuga menziesii, Picea sitchensis, P. glauca, P. engelmannii, Tsuga* sp. and *Larix* sp. (Wickman, 1968*b*). Adult beetles generally overwinter in pupal chambers in the wood. Emergence occurs as soon as temperatures permit in the spring. Eggs are deposited from March to June, and larval development continues through the summer period. Pupation is completed in early fall, and adults remain in the pupal chamber until the following spring.

The firtree borer, *S. litigiosus,* is one of several insects (Fig. 18-3) that become established in trees following wind damage or logging operations. The prevention of losses by any of these insects is readily accomplished by prompt utilization during the spring and summer months.

Whitespotted Sawyer

The whitespotted sawyer, *Monochamus scutellatus* (Coleoptera: Cerambycidae), is one of the most common wood-boring insects in the coniferous forests of the Northeast. It infests and reduces the value of freshly cut logs and of trees that have been severely injured or killed by fire or by other insects. This insect is so abundant in most localities that practically every log or newly killed tree left in the woods over the summer is almost certain to be infested.

The habits and life history of this species are very similar to those of other closely related species of *Monochamus* (Fig. 18-4). The adult insects are somewhat elongate, cylindrical black beetles about 1 in (2.5 cm) long, with very long

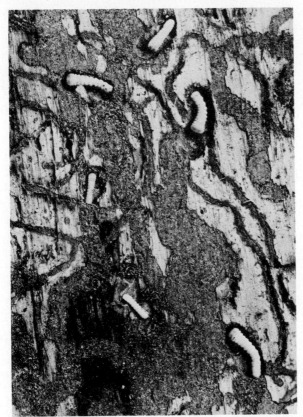

Figure 18-3 Larvae of *Acanthocinus spectabilis* in cambium area of ponderosa pine. *(U.S. Forest Service photo.)*

filiform antennae. The antennae of the male are twice as long as the body; whereas those of the female are only slightly longer than the body. The adult beetles emerge in spring and early summer. During the flight period, they feed on the green bark of pine twigs. Sometimes, when the beetles are abundant, numerous pine twigs may be girdled and killed as a result of this feeding activity. After feeding for a brief period, the beetles deposit their eggs in slits that they cut with their mandibles in the bark of logs and recently killed trees. Oviposition continues throughout the summer season but usually reaches its height before midsummer (Wilson, 1975).

The larvae hatch from the eggs and tunnel in the inner phloem. They are legless, cylindrical white grubs with powerful mandibles. At first, they excavate broad, shallow galleries in the phloem; but as they increase in size, their phloem galleries include not only the green tissues of the inner bark but also the surface of the wood. Later, the larvae penetrate into the sapwood and

Figure 18-4 Adult of the Oregon fir sawyer, *Monochamus oregonensis,* emerging from wood of Pacific Coast silver fir. This species and the whitespotted sawyer, *M. scutellatus,* are two of several species of this genus in North America. *(Weyerhaeuser Company photo.)*

sometimes into the heartwood. When growth is completed, they cut a pupal cell in the sapwood and transform to the pupal stage. The length of the development varies greatly. Under favorable conditions of temperature and moisture, a single year is sufficient; but under less favorable conditions, the life cycle may require 2 years or possibly longer. The larvae of this species, like those of the poplar borer, *S. calcarata,* and the carpenterworm, *P. robiniae,* always keep at least a part of their tunnels clear of frass and chips. Thus, pine sawyers maintain ideal conditions for the infection of the log with wood rots. These wood-destroying organisms are usually associated with *Monochamus* work.

This insect tunnels primarily in the peripheral layer of a tree or log. It is in this layer that the most valuable high-grade wood occurs. Thus, in clear logs, the insect may reduce the grade of lumber sawed from infested parts from a select grade to a no. 3 grade; this means a reduction in the retail value of from 50 to 75 percent.

This and several other species of *Monochamus* have at times caused considerable damage in lumber that has been freshly sawn but not edged. The insects lay their eggs in the bark-covered edges of the boards, and the larvae work first in the phloem and then into the wood, sometimes to a depth of several inches. The southern pine sawyer, *M. titillator,* causes similar problems in the southern states. Still other cerambycids may cause similar injury. An example is the introduced species *Arhopalus ferus* (Coleoptera: Cerambycidae) in New Zealand (Hosking and Bain, 1977). There the insect acts very much like the whitespotted sawyer, *M. scutellatus,* and the southern pine sawyer, *M. titillator,* in the United States. The forester should be aware that such problems exist around the world.

Control of the whitespotted sawyer, *M. scutellatus,* may be accomplished in a variety of ways. Fire-killed trees should be utilized promptly or cut and kept wet with water. Prompt utilization is the best way to handle the borer problem. Water storage of freshly cut material, either by emponding or by sprinkling, is the next-best solution. If neither of these methods is feasible, barking will prevent infestation by destroying the necessary green food for the young larvae.

Not all the species of insects found in dying trees are from the families Cerambycidae or Buprestidae, nor do all of them infest the phloem for a long time before entering the wood.

Horntails

The horntails, another important group of borers that attack recently killed trees or fresh logs, belong to the family Siricidae of the order Hymenoptera. They are very common insects and are responsible for more damage to logs than is usually ascribed to them. Although many species are known to science (Cameron, 1965), little detailed information is known concerning their life histories and habits. The horntails are wood-eating insects (xylophagous); they work in solid wood, maintain no opening to the outside, and apparently are not necessarily associated with fungi. One of the largest and best-known members of this group is said to be an exception to this rule in that it bores only in decaying wood. This is the species known as the pigeon tremex, *Tremex columba* (Stillwell, 1967).

The relationship of horntails to a fungus is developed as a symbiotic relationship in some very damaging species that may attack relatively healthy but stressed trees. *S. noctillio,* the serious pest of *Pinus radiata* in Australia and New Zealand, is an outstanding example (Gilmour, 1965; Zondag, 1968). This is an interesting case of an introduced insect species on an introduced plant species. Furniss and Carolin (1977) provide information on the species of Siricidae in the western United States and the symbiotic fungi *Amylostereum,* which is associated with the genera *Sirex* and *Urocerus.*

The adult horntails are thick-waisted, wasplike insects. They vary greatly in size. Some are less than 12 mm in length; others are more than 5 cm long.

Black is the predominant color, but nearly all species have markings of yellow or orange. Projecting beyond the end of the abdomen, females have a long ovipositor, and both sexes have a short, hornlike process called the *cornus,* from which the group name is derived.

The larvae are white, legless grubs, cylindrical in form and armed with a dark-colored spine on the end of the abdomen (Fig. 18-5).

The adults of most species fly in spring and early summer. The females insert their eggs into solid wood by means of their long ovipositors. This is no mean feat when we consider the flexibility of the ovipositor and the solidity of the wood. Sometimes, a female may get her ovipositor wedged into the wood so that she is unable to extricate herself. Horntails are particularly abundant and especially injurious in fire-killed forests. The scorched trees seem to be especially attractive to the adults. Trees killed by spring fires are much more subject to attack by these insects than are trees killed at other seasons of the year. The blue horntail, *S. cyaneus,* attacks spruce and pine in southern Canada and the northern tier of eastern states and from California to British Columbia in the West.

The horntail larvae tunnel through the wood and cut burrows just large enough for their bodies. The wood passes through their digestive tracts and is packed behind them as they advance. The length of the larval stage appears to be variable, even within the same species, and depends on conditions of temperature and moisture. With most species, the larval stage requires one or two seasons. When growth is completed, the larvae cut their pupal chambers in the peripheral layers of the sapwood. In logs, the pupal cells are usually constructed on the upper side so that, although the larval tunnels may be distributed throughout the log, the majority of exit holes are on top.

Water treatment of logs and prompt utilization are the best ways to prevent injury by these insects. These insects will continue to develop in sawn lumber and may emerge a year or so after construction, leaving holes in walls or floors. Kiln-drying would prevent this type of damage.

AMBROSIA BEETLES

Ambrosia beetles are chiefly pests of green wood, although occasionally they may attack other wood, such as wine casks, in which conditions are somewhat similar to those found in freshly cut wood. With the exception of the genus

Figure 18-5 Horntail larva. The name *horntail* is derived from the hornlike process on the abdomen of both adult sexes, and it is equally appropriate for the larvae, with their hornlike appendage. *(U.S. Forest Service drawing.)*

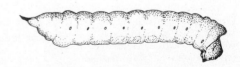

Platypus (Fig. 18-6), which belongs to the family Platypodidae, these insects are all members of the family Scolytidae. Although they belong to the same family as the bark beetles, their habits are very different. Nevertheless, they have many points in common. They are small, dark-colored insects and are more or less cylindrical in form. Instead of working in the phloem, they cut their tunnels into green wood. The tunnel entrances resemble those of bark beetles, but because the ambrosia beetles tunnel into the wood, the frass that they cast out is light-colored instead of red or brown. Although they bore in wood, they are not wood eaters; instead, they feed on fungi growing on their tunnel walls. This type of fungus is always associated with ambrosia beetles, each group of beetles having its own specific fungus. Whenever the wood ceases to be suitable for the fungus, the beetles must leave. When departing, they carry the sticky spores or the mycelium either adhering to their bodies or in special structures, called *mycangia,* in the head or thorax (Finnegan, 1963). Having sought out a suitable freshly cut tree or stump, they proceed to bore directly into the wood. For a short time, they are without food, but soon the spores that they brought with them germinate and grow on the walls of the new tunnels.

This apparent cultivation of food by an insect has caused some observers to ascribe a high degree of intelligence to these beetles. However, the facts do not justify such a conclusion. It is, nevertheless, an interesting example of an effective symbiotic (mutualistic) relationship between a fungus and an insect:

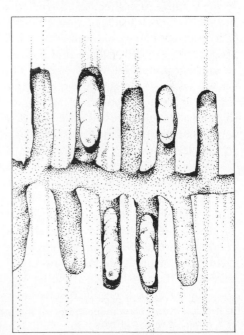

Figure 18-6 Drawing of a section of the tunnel of an ambrosia beetle in wood, with larval cradles containing developing larvae. *(Drawing by Toshiaki Ide.)*

the beetle depending on the fungus for food and the fungus depending on the insect for transportation from an old host to a new one.

Genera of Ambrosia Beetles

There are numerous species of ambrosia beetles, most of them belonging to one of the following genera: (1) *Platypus*, (2) *Anisandrus*, (3) *Xyleborus*, (4) *Gnathotrychus*, (5) *Pterocyclon*, (6) *Trypodendron*, (7) *Xyloterinus*, and (8) *Corthylus*. Some of them are confined to a single species of tree; others may be found in a number of species. Inasmuch as the beetles do not eat the wood but simply use it as a place in which to construct their nests, the specificity of the beetles for certain trees may be the direct result of the specific requirements of their particular fungus.

The arrangement of tunnels varies greatly with different species of ambrosia beetles. *Xyleborus xylographus* constructs simple branching galleries in which the eggs, larvae, and pupae are all found together. Species of *Pterocyclon* excavate several secondary tunnels branching in a horizontal plane from the main entrance tunnel, which is widened to form a nuptial chamber. Along the secondary tunnels, chambers called *larval cradles* are excavated both above and below and parallel with the grain of the wood. Certain others, the striped ambrosia beetle *Trypodendron lineatum,* for example, construct compound tunnels, the main gallery being cut directly into the wood in a generally radial direction and the secondary galleries being cut in the same horizontal plane but in a tangential direction. In the case of *Gnathotrychus materiarius,* the secondary tunnels usually follow an annual ring, and the larval cradles are cut in a series both above and below the main and secondary galleries. With such species, the entire developmental period of a beetle is passed in a cradle. In species where no cradles are formed, the larvae are in the main tunnels, feeding on the ambrosia fungus growing on the walls. In some species, the larvae, immediately prior to pupation, cut pupal cells similar to the cradles. The adults of at least some ambrosia beetle species hibernate in the litter and duff of the forest floor. Beetles of the genus *Trypodendron* are among those that hibernate in such locations (Dyer and Kinghorn, 1961).

The ambrosia beetle colonies are constantly faced with two problems: (1) the selection of trees in which conditions are suitable for the growth of the fungus on which they feed and (2) protection for themselves against the growth of the fungus itself. When conditions for the growth of the fungus are not right, the larvae will starve and the adults will be forced to seek other trees. If conditions are favorable for the fungus but the beetles do not multiply rapidly enough to consume their food as fast as it grows, the rank growth of the fungus may block the tunnels, thus smothering the beetles in their food.

The injury to logs and recently killed trees by ambrosia beetles results in a decided loss of quality of the lumber and of other products cut from it (Chapman and Dyer, 1969; Dyer and Chapman, 1965). This type of injury in lumber is serious because of the resultant reduction in grade, but it is of even greater consequence in other classes of material, for example, stave bolts and wood for furniture. In such materials, pinhole injury may render the product valueless.

Pheromones have been isolated for several of these insects. McLean and Borden (1977) described their tests of ethanol and sulcatol, a primary attractant and population aggregation pheromone, as baits for *G. sulcatus.* Hosts baited with sulcatol were consistently attacked in large numbers. The authors recommend a practice of baiting stumps with sulcatol followed by treatment of the stumps with ethanolic solutions of systemic insecticide for killing field populations and preventing heavy infestation of valuable logs.

In certain cases, ambrosia beetle tunnels cannot be considered serious defects. In oak flooring, for instance, style may demand a few knots and a few wormholes. These defects are said to lend character to a floor, and this onetime low-grade flooring is now being used in some expensive homes. In most cases, however, beetle tunnels constitute a defect that decidedly reduces the value of any forest product so injured.

Columbian Timber Beetle

The Columbian timber beetle, *Corthylus columbianus,* inhabits the trunks of hardwood trees in the eastern United States (Fig. 18-7). Trees appear healthy and are not noticeably affected by the beetles, even after repeated attacks

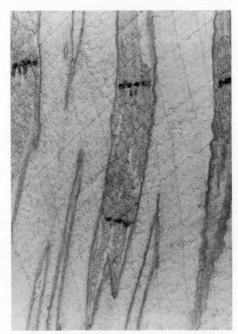

Figure 18-7 The Columbian timber beetle, *Corthylus columbianus,* galleries made in wood and associated stains. *(U.S. Forest Service photo.)*

(Nord and McManus, 1972). The galleries and associated stain cause serious degrade to lumber and veneer.

Kabir and Giese (1966) described the biology of the beetle in some detail. They reported a 26- to 30-day development period under natural conditions. The species overwinters in its galleries as either pupae or adults. In Georgia, three generations per year have been reported; in Indiana, there are at least two.

A second species of *Corthylus*, the pitted ambrosia beetle, *C. punctatissimus*, also attacks living trees (Finnegan, 1967). This pest infests a variety of hardwood species in the regeneration stage. It may cause considerable damage to maple regeneration when it is abundant.

BIBLIOGRAPHY

Baker, W. L. 1972. Eastern forest insects. *USDA For. Serv., Misc. Publ.* no. 1175.

Burke, H. E. 1928. The western cedar pole borer or powder worm. *USDA Tech. Bul.* no. 48.

Cameron, E. A. 1965. The siricinae (Hymenoptera: Siricidae) and their parasites. *Commonwealth Inst. Biol. Control Bull.* 5:1–31.

Chapman, J. A., and Dyer, E. D. A. 1969. Characteristics of Douglas fir logs in relation to ambrosia beetle attack. *Forest Science* 15:95–101.

Donley, D. E., and Terry, J. R. 1977. How to identify damage by major oak borers in the eastern United States. *USDA For. Serv.*

Dyer, E. D. A., and Kinghorn, J. M. 1961. Factors influencing the distribution of overwintering ambrosia beetles, *Trypodendron lineatum* (Oliv.). *Can. Entomol.* 93:746–759.

———, and Chapman, J. A. 1965. Flight and attack of the ambrosia beetle, *Trypodendron lineatum* (Oliv.) in relation to felling date of logs. *Can. Entomol.* 97:42–57.

Finnegan, R. J. 1967. Notes on the biology of the pitted ambrosia beetle, *Corthylus punctatissimus* (Coleoptera: Scolytidae), in Ontario and Quebec. *Can. Entomol.* 99:49–54.

———. 1963. The storage of ambrosia fungus spores by the pitted ambrosia beetle, *Corthylus punctatissimus* (Zimm.) (Coleoptera: Scolytidae). *Can. Entomol.* 95:137–139.

Furniss, R. L., and Carolin, V. M. 1977. Western forest insects. *USDA For. Serv., Misc. Publ.* no. 1339.

Gardiner, L. M. 1957. Deterioration of fire-killed pine in Ontario and the causal wood-boring beetles. *Can. Entomol.* 89:241–264.

———. 1975. Insect attack and value loss in wind-damaged spruce and jack pine stands in northern Ontario. *Canad. J. For. Research* 5:387–398.

Gilmour, J. W. 1965. The life cycle of the fungal symbiont of *Sirex noctilio*. *New Zealand J. Forestry* 10:80–89.

Graham, S. A., and Harrison, R. P. 1954. Insect attacks and hypoxylon infections in aspen. *J. Forestry* 52:741–743.

Hay, C. J. 1962. Reduce red oak borer damage silviculturally. *USDA For. Serv., Central States Expt. Sta. Note* no. 154.

————, and Morris, R. C. 1970. Carpenterworm. *USDA For. Serv., For. Pest Leaf.* no. 64.

Hosking. G. P., and Bain, J. 1977. *Arhopalus ferus* (Coleoptera: Cerambycidae): Its biology in New Zealand. *New Zealand J. For. Sci.* 7:3–15.

Johnson, W. T., and Lyon, H. H. 1976. *Insects that feed on trees and shrubs.* Ithaca: Cornell University Press.

Kabir, A. K. M. F., and Giese, R. L. 1966. The columbian timber beetle, *Corthylus columbianus* (Coleoptera: Scolytidae): I. Biology of the beetle. *Ann. Entom. Soc. Amer.* 59:883–894.

McLean, J. A., and Borden, J. H. 1977. Attack by *Gnathotrichus sulcatus* (Coleoptera: Scolytidae) on stumps and felled trees baited with sulcatol and ethanol. *Can. Entomol.* 109:675–686.

Morris, R. C. 1964. Value losses in southern hardwood lumber from degrade by insects. *USDA For. Serv., Res. Paper* So-8.

Nord, J. C., and McManus, M. L. 1972. The columbian timber beetle. *USDA For. Serv., For. Pest Leaf.* no. 132.

Ross, D. A. 1967. The western larch borer, *Tetropium velutinum* LeConte, in interior British Columbia. *Journ. Ent. Soc. Brit. Columbia* 64:25–28.

Schread, J. C. 1971. Control of borers in trees and woody ornamentals. *Conn. Agr. Expt. Sta., New Haven, Circ.* no. 241.

Solomon, J. D., and Morris, R. C. 1965. White oak borer in Mississippi. *Miss. State Agr. Expt. Sta. Info. Sheet* no. 908.

————, and ————. 1971. Woodpeckers in the ecology of southern hardwood borers. *Proceedings, 2nd Tall Timbers Conf.,* Tallahassee, Fla.: Tall Timbers Research Station, pp. 309–315.

————, and Hay, C. J. 1974. Annotated bibliography of the carpenterworm, *Prionoxystus robiniae. USDA For. Serv., Gen'l. Tech. Rept.* So-4.

Stillwell, M. A. 1967. The pigeon tremex, *Tremex columba* (Hymenoptera: Siricidae), in New Brunswick. *Canad. Entomol.* 99:685–689.

Wickman, B. E. 1965. Insect-caused deterioration of windthrown timber in northern California, 1963–64. *USDA For. Serv., Research Paper* PSW-20.

————. 1968a. Fir tree borer. *USDA For. Serv., For. Pest Leaf.* no. 115.

————. 1968b. The biology of the fir tree borer, *Semanotus litigiosus* (Coleoptera: Cerambycidae) in California. *Canad. Entomol.* 100:208–220.

Wilson, L. F. 1975. White-spotted sawyer, *USDA For. Serv., For. Leaf.* no. 74.

Wollerman, E. H. 1970. The locust borer. *USDA For. Serv., For. Pest Leaf.* no. 71.

Zondag, R. 1968. Entomological problems in New Zealand forests. *Proc. New Zealand Ecol. Soc.* 15:10–14.

Chapter 19

Wood Destroyers (Wood in Use)

In this chapter, we shall discuss the insects and marine borers attacking wood that has been manufactured and is in use for some purpose. These destroyers of wood are extremely important. Their impact is proportionate to the economic value of wood in use, either from the national standpoint or from the individual's concern for his or her tools, home, and furniture.

The insects and marine borers that destroy wood in use are separated into three arbitrary groupings according to the moisture content of the wood and the various ways in which the insects use the wood for food and shelter:

1 *Pests of moist wood.* The insects that attack moist wood, usually in association with the wood-destroying fungi.

2 *Pests of dry wood.* The insects that infest dry wood, not in association with wood-destroying fungi, and often capable of reinfesting the same piece of wood.

3 *Marine borers.* Animals that are mollusks and crustaceans, not insects. They infest wood in salt water, destroying pilings, ships, and boats.

PESTS OF MOIST WOOD

Insects that are pests of moist wood attack living trees or wood that has been cut or dead and is partially moist as a result of contact with the ground or becomes wet from faulty construction, maintenance, or insect activity. These insects are usually associated with protozoa, fungi, and bacteria that aid them by converting the cellulose and lignin in the wood into materials that are more digestible. Because of these microorganisms, moist and decaying wood provides suitable food and shelter for insects. Indirect proof of this is the fact that moist and decaying wood is attacked directly by many insects, even more than those living in green wood. Some of the moist wood insects may be found in scars in the sapwood and heartwood of living trees, but they are usually pests of forest products in contact with the ground.

Carpenter Ants (Hymenoptera: Formicidae)

The carpenter ants are widely distributed in the United States and Canada. The most common species in the East is the black carpenter ant, *Camponotus pennsylvanicus;* in the West, several species are commonly encountered. All carpenter ants belong to the genus *Camponotus.* They are among the largest ants, sometimes 12 mm or more in length. They are social insects, living together in large colonies. Three castes of adults are recognized: the kings and the queens (which are the true males and females) and the workers (which are sexually imperfect females). The term *colony* does not imply many families of one species; actually, a colony is composed of the offspring from one pair of insects. The castes, or forms, are siblings with either arrested or normal sexual development.

In early summer, the young winged males and females leave the nest; sometimes, the air is filled with ants emerging simultaneously from many nests in a mating flight. The female's flight is induced by a mandibular gland secretion of the males so as to coincide with the flight of the males (Brand et al., 1972).

Shortly after mating, the males die. A young female may either be taken into an established colony to replace a queen or may establish a new colony, usually in a small cavity in a tree, in a piece of timber, or in the ground. The female makes up her brood cell by enclosing the cavity, leaving no exit or entrance, and deposits only a few eggs. The emerging larvae are fed a material secreted from her salivary glands, complete their development, and emerge as adult workers without having any other food.

Ants of the first brood are very small workers. They cut parallel galleries longitudinally through the wood to accommodate the enlarging colony (Fig. 19-1). These workers also bring food into the nest through openings cut to the outside, feed the queen, and care for eggs and the larvae that hatch from these

Figure 19-1 A balsam fir split open to show the laminated galleries of a colony of the black carpenter ant, *Camponotus pennsylvanicus.* The winged adults (*a*) have a constricted waist, elbowed antennae, and fore wings longer than hind wings. The workers (*b*) construct the galleries. *(Connecticut Agriculture Experiment Station., New Haven; Canadian Forestry Service.)*

eggs. After the first brood, the young are fed with secretions from the mouths of workers.

The carpenter ants are omnivorous feeders, consuming other insects, meat, refuse, and the sugars in sap, fruit, and root-feeding aphids (Sanders, 1970). They do not feed on wood; instead, they use the wood for shelter of the nest. As the colony grows, more galleries are cut in the wood in order to enlarge the nest. The colony will last until the queen's supply of fertile eggs is exhausted and the colony does not acquire a replacement queen. The winged forms are not produced until the colony's population reaches several thousand, usually after 3 years.

The carpenter ants build nests in a great variety of places. They may select the dead, moist heartwood of living trees or fallen trees, logs, timbers, lumber, or telephone poles. In the spruce-fir forests of New Brunswick, Sanders (1964) found that the ants initially established a brood tree, then tunnelled in the soil about 12 mm underground to numerous satellite trees. Tunneling occupied 90 percent of the ground area in young stands of spruce-fir but declined to 20 percent in overmature stands (Sanders, 1970). The number of colonies per acre apparently is limited by the number of permanent nesting sites.

Intercolonial wars occur when a colony's territory is invaded; hundreds of workers have been wounded or killed in a single major battle. The workers attack with open mandibles, severing opponents' appendages and paralyzing wounded enemies with sprays of formic acid.

The damage to trees in the forest varies with the tree species, site, and product. Sanders estimates as much as 12 percent of the commercial-size balsam fir [6 in (15 cm) dbh and larger] may be infested. If the basal 5 ft (1.5 m) are culled, the corresponding cubic volume loss would be 12.5 percent. Although the cull can be used for pulp, the loss is nearly total if the trees were to have been used for lumber or poles. The carpenter ants are serious pests of northern white cedar pole timber; infestations have ranged from 11 percent in New Brunswick to 70 percent in Minnesota (Graham, 1918).

In the South, the carpenter ants readily infest tulip poplar and black locust attacked by the locust borer, *Megacyllene robiniae*. The tunnels are often in the grass (*Zoysia* spp.) above ground, disfiguring lawns in residential areas.[1] The presence of carpenter ants in wooded residential areas is serious. The trees with butt scars may be weakened by the ants' galleries and, with time, may be subject to windthrow.

The prevention of damage by carpenter ants in the forest begins with the avoidance of butt scars from fire, thinning, and ax blazes. The next logical step is to harvest stands of soft-wooded trees such as aspen, balsam fir, cedar, and tulip poplar when they reach commercial size. When cruising overmature stands, the forester should sound for hollow butts and confirm infested trees

[1]Personal communication from Dr. K. L. Hayes, Auburn University, Alabama.

by kicking the litter to expose the tunnels. The cull should be deducted from the estimated volume.

In residential and recreational areas, butt-scarred trees should be examined annually. These insects are a good reason to keep construction sites cleaned of stumps and waste wood and to avoid scarring trees. Wounds should be treated with an insecticide and pore sealer.

Subterranean Termite (Isoptera: Rhinotermitidae)

The termites are in the order Isoptera, one of the oldest orders of the primitive insects. There are approximately 40 species in the United States, 25 in the West and 15 in the East. There is also an introduced species of serious economic importance, the Formosan subterranean termite, *Coptotermes formosanus.*

Among the wood destroyers, the termites constitute the most important group. These insects may be separated into three divisions on the basis of their habits: (1) the damp wood termites, (2) the subterranean termites, and (3) the drywood termites. The damp wood termites are important ecologically yet relatively unimportant economically because they are usually confined to damp and decaying wood that is largely buried in the ground. These termites seldom attack materials of value. The subterranean termites are by far the most important and widely distributed and will be discussed in this section. The drywood termites are more restricted in distribution but may be extremely destructive in localities in which they occur. They will be discussed in the section "Pests of Dry Wood."

Subterranean termites are social insects and live in colonies composed of a number of castes. Each caste has a more or less specific function to perform. The primary reproductives, usually called *kings* and *queens,* are the only members of the colony that at any time have functional wings (Fig. 19-2). These dark-colored, long-winged males and females swarm out of the nest at certain seasons of the year for the purpose of mating and establishing new colonies. The most common species in the East, the eastern subterranean termite, *Reticulitermes flavipes,* swarms primarily in the early spring. After a brief flight period, adults break off their wings, and the pairs that survive seek suitable locations for new nests in wood buried in the soil. Eggs are deposited, and the first broods of young are cared for by the young pair. Later, the pair devotes its efforts entirely to egg production.

The termite undergoes incomplete metamorphosis but is unique in that the polymorphism is partially reversible. The main forms, or castes, are the workers, soldiers, supplementary reproductives, and the alates. Depending on the needs of the colony, the nymphs may develop into any of the forms, such as workers or soldiers. Similarly, immature workers may develop into soldiers or nymphs. These forms may also degenerate backward; for example, a nymph may regress to a worker. However, development ends when the soldier, alate,

Figure 19-2 The eastern subterranean termite, *Reticulitermes flavipes,* is the most economically important destroyer of wood in use. The damage (*a*) is caused by the workers (*b*), usually feeding with the grain in the springwood. The colony is guarded by soldiers (*c*). The primary reproductives are the only forms with wings (*d*). *(U.S. Forest Service photo.)*

or reproductive form is reached. The termites regulate the numbers of each form by chemicals passed by trophallaxis (mutual exchange of nutrients), either by stomodaeal (mouth) or proctodaeal (hind gut) feeding.

The subterranean termites make their nests in wood in contact with the ground where the required moisture conditions prevail. They are able to work out from the nest and invade wooden structures above ground. They must, however, always maintain contact with moisture, even when they feed in dry wood above ground. To do this, their aboveground activities must be connected with the subterranean nest. The nest may be in a tree stump, in pieces of waste wood buried near the foundation or under the building, or in other similar wood materials. From these nests, the termites construct tunnels through the ground and covered passages over foundation walls and other objects too hard for them to tunnel through and even across treated lumber. Thus, their injurious work may be some distance from the nest.

When new sources of food are found, subcolonies may be formed. If, for any reason, a subcolony is cut off, supplementary reproductive forms soon develop in the group, and it becomes a new colony. Division of colonies is believed to be the principal way subterranean termites spread in the northern part of the United States.

Formosan Subterranean Termite (Isoptera: Rhinotermitidae)   The Formosan subterranean termite, *Coptotermes formosanus,* is an introduced species first discovered in Houston, Texas, in 1965. This species has also been discovered in port cities in Louisiana and South Carolina. First identified in Formosa, the insect is widespread in the tropical and subtropical lands of the Pacific and has also become established in South Africa.

This pest is especially destructive because of its high fecundity; a queen may lay 1000 eggs per day. Single colonies may contain more than 350,000 workers. A nest may be several cubic feet in size, with tunnels extending more than 10 ft (3.0 m) underground and outward for more than 200 ft (61 m). In Hawaii, the walls of new homes have been destroyed within 3 months (Baker, 1972). The biology and ecology of the Formosan subterranean termite, *C. formosanus,* are similar to those of other subterranean termites.

Wood, Fungi, and Termites   The insects inhabiting moist wood in use have one common denominator: decayed wood. Our knowledge of the possible symbiotic relations between insects and the wood-destroying fungi is perhaps best understood from studies of the termites. Termites ingest wood, and the wood is converted into food by a cellulose-digesting protozoa and bacteria complex within the termite's gut. These microorganisms are shed along with the cuticle when an insect molts but reestablished by proctodaeal feeding from other members of the colony that have not recently molted. An average colony of subterranean termites could consume 1 bd ft of pine within 236 days, and the Formosan subterranean termite, *C. formosanus,* can do this within 38 days (Haverty, 1976).

The subterranean termites are attracted to wood decayed by the brown rot fungus, *Gloeophyllum trabeum* (= *Lenzites trabeum*). The discovery of the olfactory feeding and aggregation stimuli has a fascinating, involved history (Sands, 1969). Pioneering studies were done by Hendee (1935); she demonstrated that the presence of wood-destroying fungi in Monterey pine and Douglas-fir was essential in the termites' diet. Hendee hypothesized that the fungi provided a source of proteins and vitamins and also rendered the toxic substances in the wood harmless.

Perhaps the next major contribution was made by Esenther et al. (1961). They noticed tubes of the subterranean termites leading up to decayed branch stubs of shade trees in Sheboygan, Wisconsin, and hypothesized that the termites were following a concentration gradient emitted from decaying wood. Esenther et al. discovered an olfactory attractant produced by the brown rot fungus, *G. trabeum.*

Hendee's hypothesis regarding fungal deterioration of toxic substances was proven by Williams (1965), who found that when the toxic and repellent resins of the resistant heartwood of *Pinus caribaea* were broken down by the cubical brown rots, the heartwood became a preferred feeding and nesting site for termites.

Our knowledge of the relationship of wood-destroying fungi and termites continues to expand. Recently, Amburgey and Beal (1977) found that southern yellow pine colonized by white rot decay fungi is avoided by subterranean termites. There are differences in tube formation and feeding associated with the various types of wood-destroying fungi (Amburgey and Smythe, 1977; Sands, 1969).

Although this discussion of fungi and termites is brief, the essential point is the close interrelations between wood, water, fungi, and termites. Your awareness of these interactions will provide a better understanding of prevention and control techniques.

Prevention and Control

The prevention and control of subterranean termites destroying moist wood in use is based predominantly on the same factors that affect the growth of wood-destroying fungi. The conditions necessary for the growth of these fungi are the subject of forest pathology; a review of the subject is available in the standard textbooks (Baxter, 1952; Boyce, 1961). The optimum temperatures for most wood-destroying fungi range from 76°F (24°C) to 86°F (30°C), with extremes of 40° and 94°F (4° and 34°C). The optimum moisture content is above the fiber saturation point, with the upper limit being 150 percent. Fungal growth is greatly retarded in wood of 25 to 30 percent average moisture content and stops below 20 percent. The inhibiting limits are more than exceeded when lumber is kiln-dried; temperatures far exceed 94°F (34°C), and the moisture content is reduced below 20 percent. Lumber within the home normally has a moisture content between 10 and 20 percent, depending on the climatic conditions of the area. However, as stated previously, subterranean termites can attack even this dry wood if they have access to a source of moisture.

Nevertheless, the best way to avoid attacks of subterranean termites is to build wooden structures in such a manner as to keep the wood dry. Detailed specifications for the termite-proofing of buildings will be found in references cited in the Bibliography. In general, protection against both termites and decay requires that the following points be observed:

1 Remove all tree stumps before grading, and clean up *all* wood (forms, grade stakes, and waste) under and around buildings before backfilling.
2 Treat soil with insecticides before pouring foundation.
3 Set buildings on concrete foundations that have an 8-in (20-cm) clearance between wood and ground, using a good cement mix either for a poured wall or floor or for laying a stone or block wall. The spaces between and within blocks should be capped by filling to a depth of 4 in (10 cm) with dense concrete.[2]
4 Buildings with crawl spaces should have an 18-in (45-cm) clearance between joists and ground.
5 Grade land to slope away from the building; on flat sites, install drain tiles.
6 Wood closest to the ground or foundation, especially soleplates, should be pressure treated. All untreated wood should be kept away from contact with the ground.
7 Prevent lumber in the home from becoming wet from rain, leaks, and condensation. Adequate ventilation must be maintained in crawl spaces.
8 Inspect building annually.

Structural and sanitary measures do not give absolute protection against termites; therefore, a chemical barrier is needed. The best time to install the barrier under a building is at the time of construction. However, if a building becomes infested, steps must be taken to construct the chemical barrier between the nest and the infested wood. Also, nests and potential nesting places near the building should be eliminated if they can be found.

It is advisable to use sills that have been pressure treated to prevent the entrance of fungi and termites. A number of materials may be used in treating wood. The most satisfactory and effective method is pressure treatment with oil-borne or water-borne preservatives such as pentachlorophenol or copper chromarsenate. Brush treatments give some protection, but they do not have the deep penetration of wood that is possible with the pressure method. However, it must be remembered that termites will readily tunnel over such treated wood to wood that has not been treated.

With the advent of modern insecticides, especially the chlorinated hydrocarbons, the use of chemical barriers for preventing access of termites to buildings received a great impetus. Chemical barriers are usually applied to the soil before the foundation is laid. If the foundation is not solid, the treatment

---

[2]The capping of foundations with metal shields is no longer recommended because of failure to install or maintain them properly.

should be made to the depth of the footings. For many years, tests have been conducted at the Gulfport Laboratory of the U.S. Forest Service, Southern Forest Experiment Station, to determine the effectiveness and longevity of various chemicals. These continuing tests have now been in progress for 30 years, and several insecticides are still effective. In addition, progress is being made on the development of insecticidal baits to control termites.

Termites seem to be increasing in abundance in the northern and central parts of the United States and in eastern Canada. Modern styles in architecture that demand central heat, heated basements, and low foundations may be contributing factors to the higher incidence of these insects. Another factor of importance in causing these insects to increase may be indirectly associated with the use of bulldozers for basement excavations. This method of excavation usually leaves a large space in the ground outside the foundation wall into which wood waste inevitably falls during construction. There is a tendency on the part of the builders not only to leave this wood in the trench but also to throw in other waste materials, knowing that they will be effectively hidden when backfilled. In this way, ideal conditions for termites are created. The major contributing factor to the termite attacks is the lack of soil treatment prior to laying the foundation and the soil that is used for the backfill.

Several states and many individual communities have incorporated into their building codes the minimum protective practices that should be employed in localities in which termites are serious pests. Such regulations provide for approval by building inspectors and appear to be the only way to ensure satisfactory construction and the use of antitermite measures where needed.

PESTS OF DRY WOOD

There are comparatively few genera of insects that are able to live throughout their entire development period in dry wood. Entomologists have been attempting to learn how insects digest sterile wood. They have discovered that termites accomplish this by utilizing wood that contains wood-destroying fungi and through the action of their own intestinal fauna, capable of digesting cellulose. It has not been demonstrated for most other wood eaters. It seems likely that this symbiotic relationship may be characteristic of most (if not all) of the drywood insects.

The coleopteron families causing the most commonly encountered damage in dry wood are the Lyctidae, Anobiidae, and Cerambycidae. In the past, the name *powderpost beetles* has been loosely applied to these families. However, they differ greatly in their biologies, the age and type of wood within which they feed, and the rapidity and extent of the damage. These beetles should be referred to by their generic names.

Lyctid Beetles (Coleoptera: Lyctidae)

The most common species of beetles infesting dry hardwoods is the southern lyctus beetle, *Lyctus planicollis,* which occurs throughout the United States (Fig. 19-3). One full year is usually required to complete the life cycle. The adults emerge primarily in early spring. However, in heated buildings, emergence may occur at any time. After the brief period of flight, mating takes place. Then the eggs are laid in pores (vessels) of the sapwood. When the young larvae emerge, their first meal consists of the remains of the eggs. Then the larvae cut irregular winding galleries in the wood, packing behind them the finely pulverized frass. The winter is passed in the larval stage. In early spring, they pupate within the wood. The adult beetles then chew their way to the outside. These insects will repeatedly oviposit in infested wood.

The lyctid beetles are pests of hardwood species, especially ash, hickory, oak, and walnut that has usually been in place less than 10 years. Because the heartwood is usually immune, the injury is confined mostly to the sapwood. The preference of lyctid beetles for certain hardwoods is based on the starch content and pore size of the wood. The female seldom oviposits in wood with a starch content of less than 3 percent and with pores as small or smaller than her ovipositor. The fine pores of woods such as apple, beech, and magnolia provide a hindrance to oviposition, hence a degree of resistance (Hickin, 1972).

It is not at all uncommon during the spring months for dealers in hardwood products suddenly to discover that their products have been infested with lyctid beetles. When an infestation is severe, an entire supply of hardwood may sometimes be destroyed before the insects are discovered. Occasionally,

(a)   *(b)*

Figure 19-3 The *Lyctus* beetles infest the sapwood of hardwoods. The most common species is the southern lyctus beetle, *Lyctus planicollis* (*a*). Emergence tunnels may perforate the heartwood (*b*). *(U.S. Forest Service photo.)*

infested flooring may be laid and finished without anyone realizing the wood is not sound, and it is not until the beetles emerge in the spring that their presence is recognized. Even furniture may be infested. When this occurs, it usually happens before the wood has been finished. After the insects emerge, the wood is peppered with the small, round exit holes.

The lyctid beetles were formerly thought to feed only in dry, well-seasoned hardwood. However, the lyctid beetles have been known to infest and complete their development in freshly sawn hickory and oaks. This is caused by the rapid drying at the surface of the boards. As a rule, a full year of air drying is required before the wood becomes suitable for these beetles, although small pieces may dry out more rapidly and become susceptible to attack in 8 months after cutting. Kiln-dried material is susceptible to attack as soon as it cools after leaving the kiln.

Anobiid Beetles (Coleoptera: Anobiidae)

The anobiid beetles feed in both the sapwood and the heartwood of conifers and hardwood, usually in the forest (Fig. 19-4). Only a few species are of consequence in wood products. The anobiid beetles seem to exhibit a preference for old wood, wood that has been in place for 10 or more years. The most serious pest in the eastern United States is *Xyletinus peltatus.* The eggs are laid in the surface cracks of wood or in the adult-emergence holes, and the larvae feed and pupate within the wood. Their long life cycle varies from 1 to 5 years or more and is largely responsible for the seeming preference for old wood. Infestations usually occur in exposed, untreated pine joists and floorboards over crawl spaces. The majority of infestations develop rather slowly and are usually not detected until after several generations of adults have emerged.

Other anobiid species include the furniture, death watch, and drugstore beetles. Although these insects are occasionally encountered, they are not

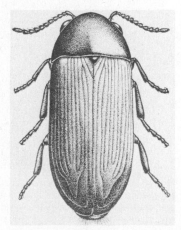

Figure 19-4 The anobiid beetles infest both the sapwood and heartwood of both conifers and hardwoods, exhibiting a preference for old wood. The most common eastern species is *Xyletinus peltatus. (U.S. Forest Service photo.)*

usually serious pests of wood in use. The furniture beetle, *Anobium punctatum,* is named for its common occurrence in antique furniture. The death watch beetle's name, *Xestobium rufovillosum,* was derived perhaps from the mating signals produced by both sexes; they bang their heads against the walls of their tunnels. This faint sound is usually heard during the still of the night. The drugstore beetle, *Stegobium paniceum,* is found in association with books and stored food products. Concentrations of these insects may be due to confinement within cupboards (Mallis, 1969). The drugstore beetle does not feed on wood, but it occasionally chews holes into the surface, in which it pupates.

Bostrichid Beetles (Coleoptera: Bostrichidae)

The bostrichid beetles attack partially seasoned hardwoods and conifers while the bark is on the wood. The females may lay eggs in wood cracks or bore through bark to the sapwood and oviposit along galleries cut across the grain. The larvae feed parallel with the grain throughout the sapwood until the wood becomes dry (Fig. 19-5). The life cycle is usually 1 year. Thus, the emerging beetles are often associated with recently made bark-covered furniture and timbers; firewood that has been brought indoors and recently processed wood may also be infested. The most common species in the eastern United States is *Xylobiops basilaris,* which infests hickory and persimmon.

Figure 19-5 Bostrychid tunnels in eucalyptus. The larvae of the leadcable borer, *Scobicia declivis,* pack the frass in the tunnels as they bore through the wood. This habit is characteristic of most bostrychids. *(U.S. Forest Service photo.)*

Old House Borer (Coleoptera: Creambycidae)

The old house borer, *Hylotrupes bajulus,* is a cerambycid beetle that has spread from Europe throughout the world. The pest was introduced into the United States more than 100 years ago. In all probability, this borer will be found in recently dead conifers in the forest and may continue to disperse into the West. The common name, *old house borer,* is inappropriate in this country because here it is more common in partly dried lumber of newer homes (Cannon, 1978).

The adult beetle is brownish black in color, with gray hairs on the head and thorax; the length ranges from ½ to ¾ in (16 to 20 mm) (Fig. 19-6). There are two dark, eyelike spots on the thorax and two whitish bands on the elytra. The larvae are typical roundheaded borers, light in color. The female oviposits in the checks and cracks of wood and also between closely stacked lumber. The larvae mine within the wood without breaking through the surface. They mine in the sapwood; their galleries are packed with powdery borings and small pellets of frass. Fortunately, some of the dust falls out and can be observed. The length of the life cycle varies greatly, from 3 to 5 years in the South and from 6 to 8 years in the North.

The old house borer, *H. bajulus,* exhibits a preference for new softwood lumber. However, its presence is usually not detected for several years because of the long life cycle. The symptoms of the initial infestation include the

(a)

(b)

Figure 19-6   The old house borer, *Hylotrupes bajulus,* introduced from Europe, is actually a serious wood destroyer in new homes. The female (*a*) is attracted to previously infested wood. The larvae will feed across the wood grain (*b*). *(U.S. Forest Service photo.)*

distinct, irritating sound of larval feeding, an occasional small amount of boring dust, and the blistering of the wood surface. Their presence can be readily verified by probing. The adults are often found against window screens within the house, especially in the attic.

These are serious, frequently encountered pests. The females are attracted to, and oviposit on, previously infested wood. Individual timbers and pieces of lumber may be structurally destroyed within 4 years.

Carpenter Bees (Hymenoptera: Apidae (= Xylocopidae)

The carpenter bees are listed as a separate family Xylocopidae by some authorities, but others place them in the hymenopterous family Apidae, which includes the bumblebees and honeybees. Carpenter bees resemble bumblebees except that they have smooth and shiny (rather than hairy) abdomens and no pollen baskets on the hind legs. The carpenter bee, *Xylocopia virginica*, is the most common species in the United States.

Carpenter bees are univoltine. In the spring, a female bee will construct a nest in sound but exposed wood by cutting an entrance hole straight into the wood for a short distance, then turning at a right angle and tunneling with the grain for 12 in (30 cm) or more. Within the tunnel, the female places an egg on a pollen-nectar mass; a series of eggs are individually chambered off by sealing the tunnel with macerated wood cemented in place by the insect.

A larva feeds on the pollen nectar, matures, and overwinters as a young adult within its chamber. The adults of this next generation emerge through the entrance holes. The carpenter bees feed on pollen during the day; they do not eat the wood cut from the tunnel. The female spends the night within the nest, and the male spends the night in a protected place outside of the nest.

Although the carpenter bees are usually not serious pests, their tunnels can be damaging if the wood is used annually for nesting. Although all species of wood may be infested, these bees exhibit a preference for cypress, cedars, pines, firs, and redwoods. The female bees will sting if provoked or handled.

Drywood Termites (Isoptera: Kalotermitidae)

The drywood termites are extremely serious pests in the tropics but are limited in their distribution in the United States to a belt along the Atlantic and Gulf coasts from Virginia to Texas, westward along the Mexican border into California, and as far north as San Francisco. The drywood termites have been widely distributed throughout the world by the shipment of infested wood. Fortunately, colonies in northern areas do not become permanently established; yet, an introduced colony can destroy the wood they are in before the colony dies out. They attack buildings, posts, poles, fences, and other structures of seasoned wood (Snyder, 1966).

The drywood termites require no contact with the ground. The attack may be at any point where the newly mated adults can find unpainted wood in a protected situation. However, winged sexuals can chew through many types of protective materials to become established. The colony is composed chiefly of nymphs; however, some eggs hatch into soldiers with long mandibles designed for purposes of defense. Some of the nymphs develop into primary winged sexuals; others may become secondary sexuals. There is no true worker caste in the drywood termites.

The winged primary sexuals attack exposed wood in pairs. Checks or cracks provide favorable locations for these primary attacks. At first, the new colony grows very slowly. Not until a lapse of several years would a single primary attack develop to appreciable proportions. Nevertheless, buildings can become heavily infested in a season or two. This condition can be explained only as multiple colonies. Many pairs may start primary colonies close together. Soon their tunnels coalesce and appear to be a single large colony.

The presence of drywood termites is indicated by the blistering of the wood surface, the faint sound of moving termites, and the piles of frass pellets ejected from infested wood. An infestation can be verified by probing and examining the galleries, which, unlike those of subterranean termites, often cut across the grain. These termites feed on both spring and summer wood, constructing broad, rather open chambers. When the termites swarm, they are attracted to light, and large numbers may be found on windowsills.

Prevention and Control

Although numerous species of insects infest dry, seasoned wood, prevention and control should be based on the insect genera that have rather similar characteristics of behavior and food requirements. The correct identification of the genus is important because the rate and amount of damage vary considerably. The presence of boring dust is a matter of concern, but the rate of damage is sufficiently slow to allow time for correct identification and application of appropriate control measures if needed.

The damage from lyctid beetles can be prevented by denying the insects the time and place to oviposit. These beetles are primarily pests of new hardwood lumber; infestations are rare in wood in use for more than 5 years. Prevention is the keynote of lyctid beetle control. Wood infested by these insects is fit for nothing except firewood. In lumberyards and manufacturing plants, where large quantities of hardwood stock must be kept on hand, thorough inspection of wood should be made frequently. An infestation is usually indicated by small quantities of fine borings that sift out of infested pieces. Rapid utilization will greatly reduce the chance of an expanding infestation in storage sheds. Woodworking industries should separate heartwood from sapwood if feasible and utilize the oldest stock first. All waste wood should be burned or hauled away; under no circumstances should it be allowed to accumulate as a breeding place for the beetles.

One of the most effective means of protecting susceptible wood from the attacks of lyctid beetles is by treatment with a sealer such as linseed oil, varnish, paraffin, or paint. These sealers will close the pores and prevent oviposition. Millwork, which must be paintable, should not be treated with a wax-type sealer.

Ordinarily, the safer procedure in dealing with infested material is to burn it. Sometimes, however, a part of the stock that has not been too heavily injured may be saved if the insects working in it can be killed. To accomplish this, the operator of a dry kiln can subject the insects to lethal temperatures.

Within a home, lyctid damage is initially confined to a few pieces of wood. Early replacement of these infested pieces is usually sufficient to eliminate the problem.

The damage from anobiid beetles can best be prevented by using treated lumber and avoiding humid conditions by maintaining adequate ventilation, especially within crawl spaces.

Infestation of bostrichid beetles can be prevented by the prompt utilization of hardwood logs and the debarking of wood used for sheathing, rustic furniture, and house timbers. Preventing emergence of bostrichids from firewood in the home can be readily achieved by burning immediately or by stacking out of doors for 1 year before burning.

The old house borer, *H. bajulus,* can be prevented from damaging wood by using kiln-dried lumber that has been properly stored. Such lumber can be readily recognized by its fresh, clean appearance; avoid weathered and stained lumber. Once an infestation is detected, the extent of damage must be determined. Fortunately, the old house borer tends to reinfest the same piece of wood. Major damage will require the replacement of the infested pieces. Occasionally, an entire house will be infested, indicating that the wood was infested prior to construction. When such infestation occurs, only fumigation is feasible.

The carpenter bees are easily controlled by placing an insecticide in the tunnel.

The drywood termites are serious pests. Good construction, with tight joints and with all pieces nailed firmly into place, is the first step in protection. This should be followed by thorough painting of all exposed surfaces. Moreover, all painted surfaces should be repainted often enough to maintain a protective covering over the wood. Back-painting (i.e., painting the interior surface) of siding and wood trim with a priming coat of paint or a wood preservative is an excellent practice. Finally, watchfulness is of prime importance if infestations are to be detected at an early stage, when they can be most effectively treated.

When an infestation by the drywood termites occurs, treatments can be made using chemicals as discussed in Chapter 11. The only treatment registered for control of drywood termites is fumigation. Badly damaged timbers will, of course, have to be replaced. Because control work is always expensive and not permanent, preventive practices are far more desirable.

MARINE WOOD BORERS

The wood-boring insects are so numerous and their work so conspicuous and familiar that we are sometimes prone to think only of insects when we consider wood-boring organisms. As a matter of fact, wood provides food or shelter for many other forms of life. Among the most important of these from the economic standpoint are the marine borers. In the strict interpretation of the scope of forest entomology, these organisms would be excluded. However, inasmuch as they are responsible for tremendous losses wherever wood is used in salt water, a brief discussion of them is included. The most destructive of the marine wood borers are the bivalve mollusks called *shipworms* and several genera of marine crustaceans. The annual damage in the United States alone exceeds $50 million (Hochman, 1973).

The marine borers exert a direct influence on forestry, especially in the South and Pacific Northwest. The stumpage value for quality poles and piling may be two to three times above regular stumpage. The value of a tree sold as a pole or pile may often be three times that of its use for lumber or plywood. The demand for treated lumber is substantial; approximately 10 percent of the annual lumber cut is treated by more than 400 wood treatment plants in the United States. The rapid destruction of unprotected wood and the high cost of replacement explain the major industry to produce pressure-treated wood for use in the coastal waters of North America.

Shipworms (Mollusca: Pelecypoda)

The shipworms are mollusks belonging to the genera *Teredo* and *Bankia* in the family Teredinidae. They have been known to maritime people for centuries and have been studied periodically from earliest times. Many species exist, and each one is limited geographically, just as terrestrial organisms are. Some of the larger species of shipworms may attain a length of 4 ft (over 1 m) under favorable conditions; other species may never exceed 5 in (12.5 cm) in length.

The most common shipworm in the coastal waters of North America is *Teredo navalis*. The teredo in early life are free-swimming larvae and are provided with a bivalve shell. After certain stages have been passed in the water, the larvae attach themselves to submerged wood. They then bore into the wood, leaving very small openings to the outside. Once inside the wood, they grow rapidly and enlarge the tunnel to suit their increased size; the body of the teredo is thus as long as its tunnel. In its burrow, the animal is elongate and soft-bodied, with the small paired shell at the anterior end of the body. The shell is used as a rasping tool and is no longer needed for protection of the body. At the posterior end, there are two tubes that are the exhalant and inhalant siphons (Fig. 19-7). As the shipworm develops within its tunnel, it secretes a calcareous, shell-like material to line its burrow. This lining is thicker in soft, porous woods and thinner in harder woods. Although a piece of wood may be riddled with teredo tunnels, this mollusk never drills into a neighboring tunnel. The life cycle is usually completed in 1 year.

The food of the teredo is obtained from the seawater that passes through the inhalant siphon and from the wood that passes through the digestive tract.

Figure 19-7 The shipworm, *Toredo navalis* (*a*), is a bivalve mollusk that destroys wood in salt water. This untreated piling (*b*) was destroyed within 90 days. *(Batelle's W. F. Clapp Laboratories.)*

Considerable controversy exists concerning the means by which the wood may be digested and the amount that may be utilized. Early researchers believed that the shipworm breaks down the cellulose within its gut by secreting the enzyme cellulase (Lane, 1959). Later researchers favor the cellulolytic bacteria theory (Turner and Johnson, 1971).

Shipworms will attack any untreated wood submerged in salt water. The greatest damage is done to piling. Untreated piles may last less than 1 month in the warm waters of the Gulf States and less than 2 years in the cooler Alaskan waters. The replacement cost of piling destroyed annually by these animals is tremendous.

Temperature, however, is not the only limiting factor in the life of these animals; the salinity and purity of the water have an influence on them. Fresh water is decidedly detrimental; this probably explains the comparatively long life of some untreated pilings in the extreme inner harbor at Norfolk, Virginia. In New York harbor, the damage by marine borers began to increase during the 1970s as efforts to reduce water pollution succeeded.

Owing to the fact that the entrance hole into the wood is very small and inconspicuous, there is usually little external indication of the presence of the borers. More than one-half of the volume of a pile might easily be destroyed without any evidence of injury being apparent on the pile's surface. Only by cutting into the pile can its condition be ascertained. The greatest amount of attack in piling usually occurs just above the mud line, although entrance holes may be found throughout the submerged area.

Limnoria (Crustacea: Isopoda)

The *Limnoria,* commonly known as the *gribbles,* are marine borers that do considerable damage along coastal waters. They are confined to clear salt water and cannot endure fresh or turbid water. Their temperature tolerance is much greater than that of the teredos, and as a result, they are found much farther north. The *Limnoria* are very common along the North American coasts, occurring as far north as Alaska and Newfoundland.

The *Limnoria* are crustaceans, only about ⅛ in (3 mm) long, and suggest in form a very small pill bug (Fig. 19-8). They are gregarious and attack pilings and other structures in large numbers, usually near low-water mark. As a general rule, the greater the difference between high and low tide, the greater the vertical distribution of these organisms will be. One species, *L. tripunctata,* predominantly a warm-water species, is creosote-tolerant and causes millions of dollars damage to pilings. Severe attack produces the familiar hourglass shape in the intertidal zone.

The individual gallery is short, about ½ in (12 mm) in length, and penetrates directly into the wood. When large numbers attack a pile, their tunnels almost touch, so that the thin walls between them are quickly worn away by wave action, leaving a new surface of wood ready for reinfestation. The

(b)

(a)

Figure 19-8 The gribbles *(Limnoria* spp.) are small crustaceans (*a*) that gnaw away the surface of wood in salt water. Pilings are reduced in diameter (*b*) in the water depths inhabited by gribbles. *(Batelle's W. F. Clapp Laboratories.)*

progress of this injury is slower but more easily seen than the work of the teredos. Heavily infested piling may lose 1 in (2.5 cm) a year in diameter as a result of *Limnoria* attack.

Limnoria find both food and shelter in the wood they attack. The wood is digested by cellulase enzymes present in the intestines (Ray, 1959). Marine bacteria in the *Limnoria* galleries convert both cellulose and lignum into bacterial protoplasm, which is ingested by the *Limnoria* (Kalnins, 1976).

Prevention and Control

Theoretically, control of marine borers is a simple matter; but practically, it is anything but simple. As with a given insect problem, the first steps are to identify the species of marine borer in the waters of concern and to obtain an expert opinion regarding the potential problem. Protecting the susceptible parts of piling with any sort of covering will prevent the attack of these pests. Metal sheathing is effective but expensive and not very durable owing to the corroding action of salt water. Cement casings have not proved satisfactory because of cracks that develop as the result of expansion and contraction, wave action, or battering of driftwood. Similarly, coatings of tar, asphalt, pitch, and other similar materials give protection only while they remain intact. In recent years, both plastic and fiber glass have been utilized with some success.

The most successful method of protecting piling against the shipworms that has been developed so far is impregnation with marine-grade creosote. This technique has been used for more than a century and is still the most common method. Surface treatment is not sufficient. Only by pressure treatments (20 lb/ft^3) can the wood be sufficiently impregnated to resist attack. Satisfactory impregnation is impossible to obtain with most woods, especially dense hardwoods. Generally, sapwood is more treatable than heartwood; therefore, only pine, Douglas-fir, or other softwoods with a generous width of treatable sapwood should be used.

The protection of wood from *Limnoria* attacks is a far more difficult task than preventing shipworm damage. Some species of *Limnoria* in warmer waters are resistant to creosote. This resistance may be due to the more rapid deterioration and leaching of creosote in warmer waters. The use of certain copper salts (i.e., chromated copper arsenate and ammoniacal copper arsenate) provides very effective protection from *Limnoria* attack. Numerous other copper or tin salts, fungicides, and insecticides, especially the chlorinated hydrocarbons added to creosote, have shown various degrees of effectiveness in preventing *Limnoria* attack, but their use is limited by environmental regulations. In areas where severe attack by both types of marine borers is present, a dual treatment is used. This method calls for a pressure treatment with the copper salt followed by a pressure treatment with marine-grade creosote.

Movable wooden structures and boats can be protected by an unbroken covering of paint. When borers have gained entrance to wooden vessels, they can be killed by running the boats into fresh water or dry dock for at least 30 days.

Foresters responsible for moving log rafts or storing logs in coastal waters should minimize the time in salt water and move the logs upriver to where the water is fresh at all depths. Foresters should be alert for advances in basic research (Kalnins, 1976) for discoveries in the symbiotic relationship between marine bacteria and the *Limnoria*. This research suggests the possibility that bactericidal preparations or antibodies may be useful control agents.

BIBLIOGRAPHY

General Texts

Baker, W. L. 1972. Eastern forest insects. *USDA For. Serv. Misc. Pub.* no. 1175.
Furniss, R. L., and Carolin, V. M. 1977. Western forest insects. *USDA For. Serv. Misc. Pub.* no. 1339.
Hickin, N. E. 1968. *The insect factor in wood decay.* London: Hutchinson & Co. (Publishers).
Mallis, A. 1969. *Handbook of pest control.* 5th ed. New York: MacNair-Dorland Co.

Pests of Moist Wood

Carpenter Ants

Brand, J. M., et al. 1972. Caste-specific compounds in male carpenter ants. *Science* 179:388–389.
Graham, S. A. 1918. The carpenter ant as a destroyer of sound wood. *Minn. St. Entomol. Rept.* 17:32–40.
Sanders, C. J. 1964. The biology of carpenter ants in New Brunswick. *Can. Entomol.* 96:894–909.
————. 1970. The distribution of carpenter ants in colonies in the spruce-fir forests of Northwestern Ontario. *Ecology* 51:865–873.
Smith, M. R. 1965. House infesting ants of the eastern United States. *USDA Agric. Res. Serv. Tech. Bull.* no. 1326.
Townsend, L. H. 1945. Literature of the black carpenter ant. *Kent. Agric. Exp. Sta., Univ. Kent., Lexington, Circ.* no. 59.

Subterranean Termites

Amburgey, T. L. 1979. Interactions between wood-inhabiting fungi and subterranean termites: A presentation of the literature. *Sociobiology* (in press).
————, and Beal, R. H. 1977. White rot inhibits termite attack. *Sociobiology* 3:35–38.
————, and Smythe, R. V. 1977. Shelter tube construction and orientation by *Reticulitermes flavipes* in response to stimuli produced by brown-rotted wood. *Sociobiology* 3:1–34.

Esenther, G. B., et al. 1961. Termite attractant from fungus-infected wood. *Science* 134:50.

Haverty, M. I. 1976. Termites. *Pest Control* 44:12–17, 46–49.

Hendee, E. C. 1935. The role of fungi in the diet of the common damp-wood termite, *Zootermopsis angusticollis. Hilgardia* 9:499–525.

Krishna, K., and Weesner, F. M. eds. 1969. *Biology of termites.* vols. 1 and 2. New York: Academic Press.

Sands, W. A. 1969. The association of termites and fungi. In *Biology of termites,* ed. K. Krishna and F. M. Weesner, vol. 1, pp. 495–524. New York: Academic Press.

Williams, R. M. C. 1965. Termite infestation of pines in British Honduras. Overseas Res. Publ. no 11. London: H. M. Stationary Office.

Formosan Subterranean Termite

Haverty, M. I. 1976. Termites. *Pest Control* 44:12–17, 46–49.

King, E. G., Jr., and Spink, W. T. 1969. Foraging galleries of the Formosan subterranean termite, *Coptoterines formosanus,* in Louisiana. *Ann. Entomol. Soc. Am.* 62:536–542.

Smyth, R. V., and Carter, F. L. 1970. Feeding responses to sound wood by *Coptotermes formosanus, Reticulitermes flavipes* and *R. virginicus. Ann. Entomol. Soc. Am.* 63:841–850.

Wood Decay

Amburgey, T. L. 1971. Annotated bibliography of prevention and control of decay in wooden structures (including boats). *USDA For. Serv. South. For. Exp. Sta., Gulfport* (mimeographed).

Baxter, D. V. 1952. *Pathology in forest practice.* 2d ed. New York: John Wiley & Sons.

Boyce, J. S. 1961. *Forest pathology,* 3d ed. New York: McGraw-Hill Book Company.

Gilbert, R. J., and Lovelock, D. W. eds. 1975. *Microbial aspects of the deterioration of materials.* London: Academic Press.

Hunt, G. M., and Garett, G. A. 1967. *Wood preservation.* 3d ed. New York: McGraw-Hill Book Company.

Nicholas, D. D., ed. 1972. *Wood deterioration and its prevention by preservative treatment.* Degradation and protection of wood, vol. 1. Syracuse: Syracuse University Press.

Rowell, R. M., et al. 1977. Protecting log cabins from decay. *USDA For. Serv., For. Prod. Lab., Gen. Tech. Rep.* no. 11.

Scheffer, T. C., and Verrall, A. F. Principles of protecting wood buildings from decay. *USDA For. Serv. Res. Pap.* FPL-190.

Prevention and Control

Bisterfeldt, R. C., et al. 1973. Finding and keeping a healthy home. *USDA For. Serv. Misc. Publ.* no. 1284.

Johnston, H. R., et al. 1972. Subterranean termites, their prevention and control in buildings. *USDA Home and Gard. Bull.* no. 64.

St. George, R. A. 1973. Protecting log cabins, rustic work, and unseasoned wood from injurious insects in the eastern United States. *USDA Farmers' Bull.* no. 2104.

Spear, P. J. 1970. Principles of termite control. In *Biology of termites,* ed. K. Krishna and F. M. Weesner, vol. 1. New York: Academic Press.

Pests of Dry Wood

Powderpost Beetles

Christian, M. B. 1940–41. Biology of the powderpost beetle. *La. Conserv. Rev.* 9:56–59; 10:40–42.

Gergerg, E. J. 1957. A revision of the New World species of the powderpost beetles belonging to the family *Lyctidae. USDA Tech. Bull.* no. 1157.

Hickin, N. E. 1972. *The woodworm problem.* 2d ed. London: Hutchinson and Co.

Mallis, A. 1969. *Handbook of pest control.* 5th ed. New York: MacNair-Dorland Co.

St. George, R. A., and McIntyre, T. 1959. Powderpost beetles in buildings: What to do about them. *USDA Leaf.* no. 358.

Snyder, T. E. 1950. Preventing damage by Lyctus powderpost beetles. *USDA Farm. Bull.* no. 1477.

Williams, L. H., and Mauldin, J. K. 1974. Anobiid beetle, *Xyletinus peltatus,* oviposition on various woods. *Can. Ent.* 106:949–955.

————, and Johnson, H. R. 1972. Controlling wood destroying beetles in buildings and furniture. *USDA Leaf.* no. 558.

Wright, C. G. 1960. Biology of the southern Lyctus beetle, *Lyctus planicollis. Ann. Entomol. Soc. Am.* 53:285–292.

Drywood Termites

Kofoid, C. A., ed. 1934. *Termites and termite control.* 2d ed. Berkeley: University of California Press.

Snyder, T. E. 1966. Control of nonsubterranean termites. *USDA Farm. Bull.* no. 2018.

Weesner, F. M. 1970. Termites of the Nearctic region. In *Biology of termites,* ed. K. Krishna and F. M. Weesner, vol. 2. New York: Academic Press.

Old House Borer

Cannon, K. F. 1978. Aspects of oldhouse borer biology and distribution in Virginia. M.S. thesis, Virginia Polytechnic Institute and State University.

Durr, H. J. R. 1954. The European house borer, *Hylotrupes bajulus* (L.) and its control in the Western Cape Province. *Un. S. Afr. Dept. Agr. Bull.* no. 337.

————. 1957. The morphology and bionomics of the European houseborer, *Hylotrupes bajulus.* Union of South Africa Dept. Agr. Ent. Memoirs vol. 4.

Hicken, N. E. 1972. *The woodworm problem.* London: Hutchinson & Co. (Publishers).

Higgs, M. D., and Evans, D. A. 1978. Chemical mediators in the oviposition behavior of the house longhorn borer, *Hylotrupes bajulus. Specialie* 15:46–47.

McIntyre, T., and St. George, R. A. 1961. The oldhouse borer. *USDA Leaf.* no. 501.

Carpenter Bee

Balduf, W. V. 1962. *Ann. Entomol. Soc. Am.* 55:263–271.

Marine Borers

Bramhall, G. 1960. Protection of wooden structures in British Columbia waters. *Can. For. Br. For. Prod. Lab. Can. Bull.* no. 126.

Clapp, W. F., and Kenk, R. 1963. Marine borers: An annotated bibliography. ACR-74, Off. Naval Res., Dept. Navy.

Hickin, N. E. 1972. *The woodworm problem.* London: Hutchinson & Co. (Publishers).

Hochman, H. 1973. Degradation and protection of wood from Marine organisms. In *Wood deterioration and its prevention by preservative treatments,* ed. D. D. Nicholas, vol. 1. Syracuse: Syracuse University Press.

Jones, E. B. G., and Eitringham, S. K., eds. 1971. Marine borers, fungi and feeding organisms of wood. Paris: Organization for Economic Cooperation and Development Publication no. 2.

Kalnins, M. A. 1976. Characterization of the attack on wood by the marine borer, *Limnoria tripunctata* (Menzies). *Proc. Am. Wood-Pres. Asso.* 72:250–262.

Lane, C. E. 1959. Some aspects of the general biology of Teredo. In *Marine boring and fouling organisms,* ed. D. L. Ray, pp. 137–156. Seattle: University of Washington Press.

———. 1961. The teredo. *Sci. Am.* 204:132–142.

Ray, D. L., ed. 1959. *Marine boring and fouling organisms.* Seattle: University of Washington Press.

Richards, B. R. 1969. Marine borers. In *The encyclopedia of marine resources,* ed. F. Firth, pp. 377–380. New York: Van Nostrand Reinhold Company.

Turner, R. D., and Johnson, A. C. 1971. Biology of wood boring mollusks. In *Marine borers, fungi and feeding organisms of wood,* ed. E. B. G. Jones and S. K. Eltringham, pp. 259–301. Paris: Organization for Economic Cooperation and Development Publication no. 2.

Chapter 20

Forest Arthropods of Human Importance

The importance of insects and other arthropods to our use and enjoyment of the forest and also to our health and well-being is the reason for this chapter. The concepts and suggestions were originally prepared by Dr. Samuel A. Graham, with the cooperation of Dr. Earl C. O'Roke, for the benefit of University of Michigan foresters. During World War II, Graham and O'Roke published their knowledge in a small manual, *On Your Own* (Graham and O'Roke, 1943).

Graham, O'Roke, and many other early foresters and wildlife biologists did much of their fieldwork out of contact with civilization. For weeks and even months, they were on their own. They accumulated a vast personal knowledge of how to get along in the wilderness of mountains, forests, and plains. Few young foresters have had the opportunity to profit from such experiences. When faced with the prospect of working in and enjoying the North American forests and possibly the forests throughout the world, it is well to be prepared to meet the arthropod contingencies that may arise.

Of all the animals that torment and irritate humans in their forest activities, by far the greater number are arthropods, and most of these are insects.

Mosquitoes, lice, fleas, ticks, chiggers, wasps, and ants are samples of arthropodous pests. These animals usually cause little direct injury; however, some are vectors of disease-causing organisms. It is of value to know the carriers of some of the more common diseases and where they are to be found. The suggestions in this chapter are designed to assist the individual in caring for himself or herself. Prevention of infection is the keynote of this chapter.

FLYING AND BITING INSECTS

Mosquitoes (Diptera: Culicidae)

The mosquitoes are most numerous in the subarctic and temperate regions and most dangerous in the subtropics and tropics. In the North, they may constitute only a source of irritation; whereas in more southern lands, some species are carriers of malaria, yellow fever, encephalitis, and other diseases.

Each mosquito species has definite requirements that determine where and when it will breed. Some species are almost always restricted to wild lands; others are frequently associated with human habitations. Some species remain close to their breeding places; others may migrate for miles.

The larvae of these insects always live in quiet water, and very small quantities of water may meet their needs. For instance, mosquito larvea may develop in fresh water lakes and ponds, saltwater swamps and estuaries, flood waters, and pools of spring meltwater. Other possible breeding spots for mosquitoes are water standing in the footprints of animals, road ruts, abandoned cars and tires, discarded cans, cavities in trees, cisterns, and sewer catch basins.

After completing their development, the larvae pupate; adults then emerge and live in the terrestrial environment for a short time, usually varying from a few days to a few weeks. During the adult stage, mosquitoes live on liquid food, the sap of plants or the blood of animals. The female feeds on blood if she is to produce eggs, but there is evidence that some autogenous species may produce generation after generation without having had a blood meal.

The adults cannot endure a combination of high temperature and low humidity. In the daytime, they usually retire to the shelter of vegetation and emerge at dusk or during the night. An exception to this generalization is the saltmarsh mosquito, *Aedes sollicitans*, which is found along the Atlantic Coast and is often quite bothersome by day. In the northern forests, the nights are likely to be too cold to permit activity, and mosquitoes are active during the relatively cool days.

The time of year at which mosquitoes are most abundant varies with climatic conditions but is always associated with an abundance of surface water combined with moderate temperatures and high humidity.

The mosquitoes are vectors of at least three types of disease-causing organisms: protozoa, viruses, and filarial worms.

Common Malaria Mosquito Perhaps the most important worldwide disease of humans is malaria. Although malaria is thought of at present as a tropical disease, outbreaks have occurred as far north as Finland, and the disease did affect the colonization of the eastern United States.

In eastern North America, the principal vector for human malaria is the common malaria mosquito, *Anopheles quadrimaculatus.* This species prefers to breed in fresh, clean, quiet waters, partially sunlit and covered with floating organic debris. This multivoltine species may produce as many as 10 broods annually in the South. The nocturnal adults feed on plant nectar and fluids. However, the female must obtain blood to produce eggs; she may lay as many as 12 batches totaling more than 3000 eggs.

The causal organism of the disease, human malaria, is one or more species of the protozoa genus *Plasmodium,* which reproduces sexually within the mosquitoes and asexually within humans. Once within a person, the *Plasmodium* enter the liver cells, transform, escape into the circulatory system, then infest red blood cells. When this stage of development is completed, the *Plasmodium* burst the cells, releasing associated toxins. Their increased numbers can impede the circulatory system, especially the capillaries. The *Plasmodium* are siphoned into the female mosquito's body when she obtains human blood. Within her body, sexual reproduction of *Plasmodium* occurs. The *Plasmodium* move to the salivary glands and are injected into another human during blood feeding, thus completing the cycle.

Yellow-Fever Mosquito Outbreaks of yellow fever have occurred frequently in the United States since colonial times, especially in coastal cities. During outbreaks, many people died, and others fled the cities for higher elevations, perhaps a contributing factor to the establishment of health spas throughout the Appalachian Mountains. The last major outbreak occurred in New Orleans in 1905.

Yellow fever is a tropical viral disease that is now known to have two forms. The urban form is transmitted by several species of mosquitoes, principally the yellow-fever mosquito, *Aedes aegypti.* The sylvatic or jungle form is transmitted by several species of forest mosquitoes, especially the *Haemagogus* spp. in the Americas.

The causal agent of yellow fever is a group B viscerotropic virus that invades and multiplies within the body cells and organs of humans. The destruction of the liver and circulatory system provides the classic terminal symptoms: jaundice (a yellowing of the skin) and the vomiting of darkened blood. Human mortality may approach 50 percent in populations that have not been previously exposed to the disease.

The transmission of yellow fever by the yellow-fever mosquito, *A. aegypti,* was reported by Walter Reed and associates in 1900; this was the first human disease proved to be caused by a virus and transmitted by a mosquito. The yellow-fever mosquito is believed to be native to Africa; however, its distribution is now almost worldwide. It was perhaps originally transported in the bilge water of sailing ships.

The yellow-fever mosquito, *A. aegypti,* is diurnal, breeding in artificial containers in or near houses. This species seldom flies more than a few hundred feet; thus, infections are localized.

The successful control of yellow fever quickly followed Reed's proof of transmission by mosquitoes. But in the 1930s, outbreaks of the disease in forested areas of Central and South America led to the discovery of the sylvatic form of yellow fever. The vectors were not *Aedes* spp. but principally *Haemagogus* spp. that inhabit the forest canopy. These species breed in tree holes and leaf axils, usually feed on monkeys during the day, and are known to disperse more than 7 mi (11.2 km), which may account for the rapid, widespread outbreaks of sylvatic yellow fever. Fortunately, intrahuman transmission is low, humans being dead-end hosts. The incidence of the disease is directly related to the amount of exposure in the forests, being highest among loggers.

House Mosquitoes At present, the most important disease in the United States transmitted by mosquitoes is St. Louis encephalitis, an inflammation of the brain first discovered in St. Louis in 1933. Sporadic outbreaks have occurred rather frequently throughout the United States. During the 1975 outbreak, more than 2000 confirmed cases were reported from 30 states, with 95 known deaths.

The vectors of St. Louis encephalitis are the nocturnal house mosquitoes, *Culex pipiens* and *C. quinquefasciatus.* These species usually breed in small collections of stagnant water that is high in organic content, especially around buildings.

The causal organism of this disease is a group A neurotropic virus that invades, multiplies in, and destroys nerve cells. The virus is primarily a parasite of birds, including such common species as sparrows, robins, and pigeons. The human mortality rate in localized outbreaks has reached 20 percent. The initial symptoms range from headaches and fever to stiff neck, tremors, and drowsiness.

Blackflies (Diptera: Simuliidae)

Blackflies of the family Simuliidae are extremely bothersome biting flies, especially in temperate regions. As larvae and pupae, they live attached to rocks or vegetation in the sunny riffles and waterfalls of swift-flowing, clear streams (Fig. 20-1). The adults are terrestrial but oviposit in the water.

Blackflies are diurnal feeders, especially numerous within 1 ft (30 cm) or so of the ground. They are likely to creep up loose trouser legs and bite the individual who is unaware of their presence. F. C. Craighead, in correspondence

Figure 20-1 The habitat of the blackfly larvae (*Simuliidae*). Swarms of this biting fly seriously detract from the enjoyment of outdoor recreation. *(Science Service, New York State Museum.)*

with S. A. Graham,[1] reported that he had known men in the northern woods to lose considerable weight (as much as 10 to 15 lb/4.5 to 6.8 kg in a month) as a result of continued irritation from blackflies. In bad years, they can be so numerous that animals and children die from a loss of blood or suffocation from a blockage of the bronchial tubes caused by inhaling the flies.

Some people react severely to the bites and may become ill. Most individuals, after exposure to blackflies for several years, develop a degree of immunity, although at the first of each season, each bite raises an uncomfortable itching, reddish wheal. All too frequently, blackfly bites become infected as a result of scratching.

The blackflies and mosquitoes can transmit several serious filarial diseases. The filarial worms are actually nematodes that develop to adults in an animal's lymphatic or circulatory system. Among the more important tropical diseases are onchocerciasis, or river blindness, and elephantiasis. Filariae of dogs, commonly known as *heartworms,* are now common in parts of the eastern and central United States and are becoming more apparent in humans.

[1]Personal communication.

Biting Midges (Diptera: Ceratopogonidae)

The biting midges, commonly called *punkies, no-see-ums, sand flies, or gnats,* are extremely bothersome in the deciduous northeastern forests and along the salt marshes of the Atlantic Coast. They are extremely small and difficult to see. They breed in a variety of wet locations: in wet leaves, along the margins of pools and streams, in fresh and salt marshes, and in the accumulations of water in stumps and hollow trees. The adults hide in the litter on the forest floor, emerging and biting viciously in the late afternoon and early evening and again in the early morning. In the subtropics and tropics, some species are vectors of protozoa, viruses, and filarial worms.

Often, these midges congregate inside houses and tents, resting quietly until the moisture and temperature combination is right for evening feeding. They are able to pass through standard screening.

Horseflies and Deerflies (Diptera: Tabanidae)

The bloodsucking flies of the family Tabanidae (Fig. 20-2) are often pests of animals in forested areas. The most commonly encountered are the horseflies and deerflies.

Horseflies are large, diurnal flies. The male feeds on plant material and does not bite. However, the female feeds on large warm-blooded animals such as horses, moose, deer, and humans. Her bladelike mouthparts deeply lance the skin, causing a flow of blood that she laps up. She is easily disturbed and may cut and feed several times in different places before completing her blood meal.

The female oviposits on objects near or over water. The emerging larvae may either burrow into the damp ground or drop into the water. The larvae

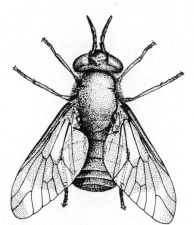

Figure 20-2 The horseflies and deer-flies, Tabanidae, are vectors of numerous diseases, especially anthrax and tularemia. *(Daly et al., An Introduction to Insect Biology and Diversity, 1978.)*

feed on organic matter and are predaceous on other animals in the soil. The adult males usually live for 1 week or so, but the females live and blood feed repeatedly throughout the growing season.

The tabanids can be especially bothersome to swimmers and loggers and in recreation areas where horses are used. Blood loss in livestock, wild animals, and humans can be considerable. These pests are mechanical vectors of numerous mammalian diseases, especially the dreaded disease anthrax, which is caused by *Bacillus anthracis.*

Deerflies are smaller than the horseflies yet similar in habit and habitat except that they attack humans more often. The deerflies, *Chrysops* spp., are also mechanical vectors of many diseases, especially tularemia.

Tularemia, rabbit fever, or deerfly fever, as it is variously known, is a fatal bacterial disease of wild rodents. The causal organism is *Francisella tularensis.* The usual source of human infection is through a scratch or wound coming into contact with a diseased animal, for example, when skinning an infected rabbit. Transmission may also occur through the feeding of an infected tick, flea, mosquito, or biting fly. A serious outbreak in a Civilian Conservation Corps camp in Utah was traced to men who, stripped to their waists, were bitten by numerous deerflies that apparently had previously fed on diseased rodents.

Prevention and Control

In the forest, control of flying and biting insects is out of the question except in the immediate vicinity of permanent camps. The forester who must work in places infested with these pests must resort to methods of individual protection. In northern regions, where these insects seldom transmit serious diseases, comfort is the main consideration. A few bites under such conditions are of minor consequence, and so protection is seldom carried to extremes. One expects to be bitten to some extent in forestry.

Reasonably good protection can be obtained by using the various kinds of insect repellents, especially those containing N, N-diethylmeta-toluamide, on exposed skin. Most repellents are irritating to the eyes and sometimes cause severe conjunctivitis (inflammation of the inner lining of the eyelid and eyeball); consequently, they should be used according to the instructions on the label.

Suitable clothing the best protection in the field, especially when combined with the use of repellents. Clothing of closely woven fabric through which the insects cannot bite, trousers tucked into boots or under socks to prevent insects from crawling up the legs, and a handkerchief tucked under the hard hat with the loose edge hanging about the neck and ears will prevent many insect bites.

Head nets and gloves give better protection than the repellents but can be very uncomfortable in warm, moist weather. Nets should be so constructed that they will be held away from the head.

Camps and recreational areas should be located where there is little underbrush and ample opportunity for circulation of air. In such a location, flying and biting insects will seldom become objectionable during the day, although they may swarm in at night. When practicable, living and sleeping quarters, whether they be tents or buildings, should be tightly screened, and beds should be covered with canopies of netting with a 24 mesh. Insect repellents sprayed on the screens may help to keep out the biting insects, especially the midges.

The local control of the flying and biting insects is a complex problem involving a detailed consideration of the habits of numerous different species. Nevertheless, it should be mentioned that local control is usually possible and can be accomplished by measures directed against the centers of breeding. For example, control of mosquitoes can be accomplished by draining standing water, removing receptacles containing water, suffocating the immature stages with oil, poisoning the insecticides in standing water that cannot be drained, or planting minnows, *Gambusia* spp., in suitable pools.

Protection from arthropod-borne diseases is largely a matter of avoiding bites by infested vectors. Very few of the potential vectors are infected because only a few have had opportunity to feed on a diseased host. The wounds caused by the flying and biting insects should be treated with an antiseptic such as tincture of Methiolate or alcohol. Above all, to prevent infection the wounds should not be scratched.

Diagnosis of human protozoa and viral infections is a highly technical matter based on the preparation of material by the use of special techniques and examination by competent, skilled personnel. Facilities for such diagnosis have been greatly increased in the parts of the world in which these diseases are common, and methods of treatment have been much improved in recent years.

In localities where malaria, yellow fever, or encephalitis prevail, any individual of a carrier species may be infected. Under such conditions, protection against these insects is a question not only of comfort but also of public health. The control of malaria through mosquito eradication in centers of populations and the treatment of the disease through the administration of quinine and the modern synthetic antimalarial drugs have become classic. An effective vaccine has been developed for yellow fever, and foresters who will work in areas in which yellow fever occurs should be immunized. At present, the only treatment for encephalitis is supportive.

The treatment for anthrax includes antiserum and antibiotic therapy. Although the number of cases of anthrax in humans is low, the mortality can be high unless prompt and thorough medical treatment is provided. It is best to wear rubber or plastic gloves while dressing an animal; at least, wash your hands thoroughly afterward with soap and water and rinse with a disinfectant.

WINGLESS, BLOODSUCKING ARTHROPODS

Wingless, bloodsucking arthropods are sometimes as bothersome as the winged pests and are also dangerous disease carriers. Fleas, lice, bedbugs, and ticks are common in almost all climates. All except perhaps the bedbug can transmit diseases, but fortunately, only a small percentage of them are infected.

Fleas (Siphonaptera: Pulicidae)

The more common fleas of the family Pulicidae in North America include the human flea, *Pulex irritans,* the dog and cat fleas, *Ctenocephalides canis* and *C. felis,* and the oriental rat fleas, *Xenopsylla cheopis* (Fig. 20-3). These fleas infest domestic and wild animals as well as humans, depending to some extent on the availability of the host. As adults, they are active, running through the hair of mammals or on the clothing of humans. Their hind legs are modified for jumping, and when disturbed, they may hop several feet. Thus, they can easily move on and off the host animal many times.

In the adult stage, they feed only on blood. The eggs are laid on the host. The emerging larvae fall to the ground, usually in the host's nest, where they feed on frass and organic detritus. The life cycle may range from 2 weeks to 2 years, and longevity of adults may reach almost 3 years. The fleas are transmitters of several serious diseases of humans, especially plague.

Plague Perhaps the most feared disease of humans that has an association with insects is plague. Bubonic plague, or black death, is a bacterial disease caused by the bacillus *Yersinia* (=*Pasteurella*) *pestis.* Once the bacilli are within the body, they spread out and multiply throughout the lymph system.

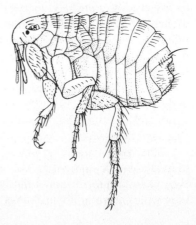

Figure 20-3 The oriental rat flea, *Xenopsylla cheopis,* is a vector of the bacterial disease plague. This flea is usually associated with rats. *(Daly et al., An Introduction to Insect Biology and Diversity, 1978.)*

The increased numbers of bacilli may create an enlargement and rupture of the lymph nodes. The bacilli then escape through the skin or even the circulatory system and destroy such organs as the lungs, liver, spleen, and brain.

When the bacilli are confined to the lymph system, the disease is in its bubonic form, which can be spread by fleas. In the septicemic or pneumonic form, the bacilli are within the blood and organs. This form is spread, not by fleas, but by infected airborne droplets from the lungs or by direct contact with blood, pus, or other bodily discharges.

Plague is a disease for which rodents and fleas are primarily responsible. The persistence of plague is due to the fact that the causal organism is never entirely eliminated from populations of wild rodents. Numerous endemic reservoirs are widely distributed among wild rodents throughout the world. The disease is established in the rodent population of western North America as far east as Nebraska. Therefore, foresters and recreationists in western North America should avoid contact with sick or dying rodents that may be infested with fleas.

Sucking Lice (Anoplura: Pediculidae)

Human lice are now so uncommon in America that one forgets that only a few generations ago, infestations were general even among the most fastidious. It was then no disgrace to be moderately lousy. Today, in many parts of the world, people are still habitually infested; and elsewhere, others become infested periodically.

Different species of lice infest the various species of mammals. Humans are plagued by three types: (1) the head louse, *Pediculus humanus capitis,* (2) the body louse, *P. humanus humanus,* and (3) the pubic or crab louse, *Pthirus pubis.*

The head louse, *P. humanus capitis,* is usually found on the head; the presence of numerous eggs on the hair betrays an infestation often before the lice themselves are seen. The lice move rapidly about among the hairs and spread readily from person to person through contacts among individuals, the exchange of hats or other clothing, and the sharing of sleeping quarters with infested individuals. The eggs usually hatch in about 1 week from the time of oviposition. About 10 days are required for development of the young lice to the adult stage. The adults are somewhat larger than the body louse (Fig. 20-4).

The body louse, *P. humanus humanus,* or cootie, infests the parts of the body covered with clothing, especially about the chest and under the arms. From time to time, the lice crawl onto the skin and feed by sucking blood through their piercing mouthparts.

The adult females oviposit their eggs (nits) on the clothing of infested individuals. The eggs hatch in about 4 days to 2 weeks; the young lice resemble their parents in form. Under the favorable temperature conditions on the body, they will mature in about 8 days after hatching. The newly hatched lice can

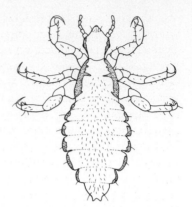

Figure 20-4 The body louse, *Pediculus humanus humanus,* or cootie, is the principal vector of the rickettsial disease typhus fever. *(Daly et al., An Introduction to Insect Biology and Diversity, 1978.)*

live only 1 day or 2 days away from the human body, but the adults can live for 1 week or more without eating.

The most favorable temperature for body lice, *P. humanus humanus,* is between 80° and 90°F (27° and 32° C). At higher temperatures, they become restless and may leave the body. This tendency is especially significant in connection with the spread of epidemic typhus; the insects leave as body temperatures rise, greatly increasing the danger of spreading the disease from person to person.

Because these insects may wander away from the body, living quarters become infested, and eggs may be laid on blankets and other objects. However, the relatively short life of the insects when separated from a source of blood prevents such infestations from persisting long in the absence of humans. In warm weather, they will seldom persist for more than 2 weeks and never for more than 1 month. It is necessary, however, to clean up infested quarters and clothing as a part of the delousing process in order to prevent reinfestation.

The crab louse, *P. pubis,* infests the pubic region but may also occur on other body hairs, even eyelashes and eyebrows. These lice move about very little and therefore are less rapidly disseminated from person to person. They cling tightly to the hair on which they rest, feeding almost continually. They are crablike in form and are much shorter and broader than the body louse, *P. humanus humanus,* or head louse, *P. humanus capitis.* The eggs are cemented to the hairs and hatch in about 1 week after oviposition; development to the adult stage is completed within 2 to 3 weeks.

The crab louse, *P. pubis,* spreads from individual to individual through sexual contact and occasionally on shed hairs to which they are clinging. Public toilet seats, the benches in barracks or recreational buildings where people may sit while dressing, and other similar places can afford an opportunity for infestation. However, these sources of infestation are not so common as is generally believed.

Typhus Fever   Infestation of body lice, *P. humanus humanus,* becomes especially important when louse-borne diseases such as typhus are prevalent. Typhus fever, or epidemic typhus, is a rickettsial disease caused by *Rickettsia prowazeki.* The *Rickettsia* are small parasites (less than 1 micron) that inhabit arthropod tissues and feed on the host cells. Once *Rickettsia* spp. enter a vertebrate host, they multiply and destroy the endothelial (i.e., the lining) cells of the lymph and circulatory systems and body cavities.

Humans are the natural reservoir for typhus, and the vector is usually the body louse, *P. humanus humanus.* The *Rickettsia* spp. are excreted by the louse while defecating. The transmission occurs when the louse bite is scratched and the fecal matter enters either the bite or the wound. The body louse will die within 2 weeks of being infested with *Rickettsia* from a diseased human.

The human mortality rate may vary from 10 to 100 percent, being higher in populations that have not been previously exposed to the disease, are in a reduced state of health, and exist in crowded conditions.

Bed bugs (Hemiptera: Cimicidae)

The bed bug, *Cimex lectularius,* is not pleasant to think about and even less pleasant to sleep with. This flattened, brown bug (Fig. 20-5) possesses a peculiarly pungent odor that may often betray its presence to one who has contacted them before. The bed bug spends its resting hours hidden away in cracks or other hiding places, only venturing out at night in search of blood, preferably human. These habits lead it to concentrate around beds, where its food is most likely to be found sometime during the night.

Ticks (Acari: Ixodidae)

The ticks are arthropods of the class Acari, the mites and ticks. They are widely distributed throughout the world, and many species attack humans. However, humans are not the chief host of any of them. Any tick that will

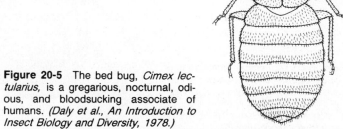

Figure 20-5 The bed bug, *Cimex lectularius,* is a gregarious, nocturnal, odious, and bloodsucking associate of humans. *(Daly et al., An Introduction to Insect Biology and Diversity, 1978.)*

attack a human is a potential disease carrier, but fortunately, tick-borne diseases are relatively uncommon in most localities; and even where ticks occur, only a small percentage of them may be infested.

The American dog tick, *Dermacentor variabilis,* is the principal vector of the disease Rocky Mountain spotted fever in central and eastern North America (Fig. 20-6). The female, after engorging on blood, drops to the ground, where she lays her eggs, often more more than 6000. The emerging larvae climb upward on low vegetation, then attach themselves to passing animals such as small rodents. The larvae feed on blood for about 1 week, then drop to the ground and molt into nymphs, which climb upward, wait, attach themselves to another host, feed, drop, and molt to adults. The adults repeat the process, mating on the host while the females are feeding. The life cycle may be completed within 3 months; however, unfed adults may live for several years.

The Rocky Mountain wood tick, *D. andersonii,* which is widely distributed throughout western North America, has a similar biology. However, the adults exhibit a feeding preference for large mammals such as horses, deer, and bears, and the nymphs feed on small mammals.

There is no reason to worry if you become infested with ticks. Remove your clothes, and pick them off. They will usually crawl around for several hours before biting. The bite of most ticks may cause temporary local irritation and sometimes persistent small lesions, but unless they carry disease-causing organisms, they are not really dangerous. However, the ticks are vectors of numerous diseases, perhaps the best known being Rocky Mountain spotted fever.

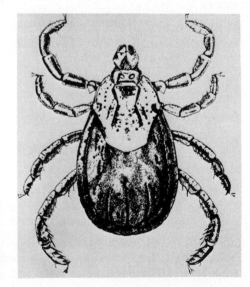

Figure 20-6 The female American dog tick, *Dermacentor variabilis,* when unengorged. This species is the principal vector of Rocky Mountain spotted fever in the eastern United States. *(USDA Bureau of Entomology and Plant Quarantine.)*

Rocky Mountain Spotted Fever   Originally, Rocky Mountain spotted fever was known to occur only in the northern Rocky Mountain states; but today, the disease has spread generally over North America and some parts of South America. During the 1970s, Virginia and North Carolina reported the most confirmed cases. This rickettsial disease is transmitted by ticks, the reservoir of infection being several species of wild rodents. Foresters frequently acquire this disease because their work brings them into contact with ticks in the field. In the eastern United States, people other than foresters acquire the disease because of their contact with pets infested with dog ticks, *D. variabilis*.

Chiggers (Acari: Trombiculidae)

Chiggers are tiny mites whose ability to irritate humans is entirely out of proportion to their size. They are very small, barely visible to the naked eye. In attacking humans, the chigger attaches itself to the surface of the skin and, if not disturbed, will become firmly attached. The chiggers do not suck blood; rather, they inject a fluid that penetrates the skin and disintegrates the cells. The cell parts are siphoned for food. This requires from 4 to 6 hours. The irritated spot swells and continues to itch severely for a few days to 2 weeks. The first indication of the chigger's presence is the development of a whitish wheal with a tiny central red dot. This combination is followed by the development of a small central blister that is usually scratched open and may become infected with secondary organisms.

Chiggers are distributed throughout North and South America, especially in the southern and central United States. They are also found in many isolated spots in the northern states. Chiggers are vectors of the rickettsial disease scrub typhus in the Orient and epidemic hemorrhagic fever, which was epidemic during the Korean conflict.

Although there is considerable variation in the effects of chigger bites on different people, almost everyone is affected. Although some individuals develop considerable immunity to their effects, many continue to react to them after years of exposure. People who are especially susceptible may be so seriously affected that they become ill, and the efficiency of most individuals is materially reduced when they have been subjected to chigger attack.

These mites are especially common in brier patches, grasses, and other heavy vegetation, but in some localities, they seem to be almost everywhere.

Prevention and Control

Control of wingless, bloodsucking arthropods can be accomplished in a variety of ways when the habits and requirements of the pests are understood.

Fleas   The prevention of flea infestation is, of course, aided by good housekeeping. Regular removal of dust accumulations in residences, frequent fresh bedding for domestic animals, and procedures that eliminate breeding places are helpful. But the best housekeeper cannot remove all possible flea-breeding places in human residences, and an infestation, once started, is almost certain to increase unless it is controlled by special means.

Chlorinated hydrocarbons have been the standard flea remedy for many years and even under best conditions kill only a part of the fleas in the places treated. However, continual use keeps down the number of fleas. To control fleas, a diluted powder should be dusted in the beds of domestic animals and in other places about human habitations where fleas may breed.

Fleas and other ectoparasites from freshly killed animals may infest clothing and be troublesome if the dead animals are placed immediately into a hunting coat or pocket. This trouble may be avoided by waiting; the ectoparasites leave the body of an animal as soon as it becomes cool.

Bubonic plague   Control of bubonic plague is dependent on flea control and the elimination of rodents, especially the common rat, from the neighborhood of human living quarters. The treatment for plague victims consists of isolation and the use of antibiotics. With early diagnosis and immediate chemotherapy, the mortality rate has been reduced to 5 percent.

The mortality rate from plague in the bubonic form when not treated is about 50 percent; the mortality rate from the septicemic form is extremely high, often 90 percent. Three pandemics are known to have swept the world, directly changing the course of human history. The last pandemic began in China in 1894, spread through all the continents, and began to decline in the 1930s. The possibility of a fourth pandemic does exist.

Lice   The prevention of lice infestations is difficult for people who are moving from place to place where good hotels are not available. Lice will inevitably be contacted. This is true even in the United States, where the incidence of lousiness is much lower than in most parts of the world. It is difficult to believe, but nevertheless true, that in many places lice are an endemic condition.

The bedding and clothing of infested individuals will become contaminated with these pests, and the lice creep from person to person when they have the opportunity. Thus, people living in close contact, as is the case on a forest survey or expedition or in a military camp, are especially subject to general infestation. One lousy individual, if allowed to remain in that condition, can infest the entire group.

Under ordinary circumstances, dusting with recommended powders will prevent infestations. The removal of lice from clothing can be accomplished by washing garments at a temperature of 150°F (65°C). Also, the heat normally encountered within a sauna will kill lice. Materials placed in storage for 30 days or longer will become free of living lice provided no infested equipment is introduced into the storeroom during the period.

Bed Bugs   The curse of bed bugs demands extermination as the only satisfactory degree of control, but this is something difficult to accomplish.

Tight buildings may best be treated by fumigation. If fumigation is impracticable, spraying bedding, mattresses, and all cracks and crevices with recommended sprays is effective if done thoroughly. Often, such treatments must be repeated several times before complete control is accomplished.

Ticks   It is advisable to protect yourself by wearing proper clothing during the growing season. Ticks tend to crawl upward and can usually be detected and picked off as they creep onto bare skin. A band of tape, sticky on both sides, wrapped around the trouser legs will trap ticks. Insect repellent containing N, N-diethylmeta-toluamide should be used as directed.

Examination of the body is essential. It usually takes several hours after a tick gets on a person before it becomes attached. Consequently, examination twice a day will usually prevent attachment. If one is not traveling with close friends, a mirror is helpful in examination.

Ticks are usually removed by picking them off with the fingers or with a pair of forceps. The wound should be disinfected and not scratched. The use of fingers is somewhat dangerous because of the chance of infection from disease-causing organisms in the tick's fecal matter. You may induce them to loosen their hold by touching them with a drop of kerosene, gasoline, alcohol, or clear nail polish. The old belief that ticks should be unscrewed from their attachment is wrong, and attempts to do this sometimes result in leaving the mouthparts in the skin. This may be followed by serious infection.

Protection from some tick-borne diseases such as Rocky Mountain spotted fever can be obtained by the use of immunizing serums or vaccines. Anyone working in areas where the disease is common should be immunized.

Chiggers   The prevention of feeding by chiggers is best accomplished by the same procedures suggested for ticks. If possible, avoid brush, tall grasses, green brier patches, and sitting on the litter of the forest floor. Because chiggers are not noticed when they first attach themselves, it is best to bathe or shower every evening and to scrub *hard* with soap and a washcloth. Once a person is infested with chiggers, wounds should be disinfected and not scratched. Severe infestations of chiggers are serious, and a physician should be consulted.

MECHANICAL VECTORS OF DISEASE

The mechanical vectors are those arthropods that transmit disease-causing organisms that accidently adhere to the mouthparts, appendages, or body. The pathogens are then transmitted to humans, usually when deposited on food but also by blood feeding with contaminated mouthparts. The vectors usually feed and breed in excrement or diseased animals and are associated with such diseases as typhoid fever, cholera, and dysentery. The most common and best-known mechanical vector is the house fly.

House Fly (Diptera: Muscidae)

The house fly, *Musca domestica,* may complete its life cycle within 10 days in warm weather and may produce more than 10 generations during the growing season. The female exhibits an oviposition preference for material high in organic matter, usually manure, garbage, or feed troughs. The larvae, called maggots, feed on microorganisms, develop, and pupate within the filth. The emerging adults push their way upward and out.

The adults feed by emitting fluids to dissolve the food surface. The liquid food is then sponged and siphoned into the body. The fly usually defecates while feeding.

The flies acquire disease-causing organisms by walking over infected material that adheres to the hairs of their legs and body, by their feeding habits, and when the adults emerge through filth. Flies are thus associated with a multitude of organisms that cause various kinds of dysentery, typhoid fever, and other human diseases.

Amoebic dysentery is an extremely severe, debilitating diarrhea. It is caused by *Entamoeba histolytica,* a microscopic protozoan of the human intestine that is now widely distributed in temperate and tropic regions. Infection results from ingesting the cysts of this protozoan, which are passed in the feces of infected individuals. Eating contaminated food and putting fingers or other objects that may be contaminated into the mouth are the usual ways in which the disease is contracted. Flies and other food-soiling insects are also involved. Water that has been contaminated with human excrement is a very dangerous source of infection. Human carriers of *E. histolytica* are numerous throughout the world and do not necessarily exhibit the symptoms of the disease.

Contamination of food with disease-causing organisms by flies is a mechanical process; thus, these insects are only one cause of the problem. In some parts of the world, human excrement is used for fertilizer. This results in contamination of the soil and of all surface water; food can be infested before it is purchased. The eating of raw vegetables and fruits is especially dangerous in some localities. If contamination is likely to occur, vegetables should always be thoroughly cooked and served hot. All drinking water should be boiled for 20 minutes. Doubtful water that cannot be boiled should be avoided; if it must be used, such water should be chemically treated.

Control of flies is usually a community matter and need not be considered here in detail. Nevertheless, much can be done to reduce the number of flies in camps and recreational areas. The problem of protection by screening and insecticides can be greatly simplified if the number of these insects is not excessive. Garbage should be either burned or buried under at least 1 ft (0.3 m) of earth. If it is buried less deeply, any eggs or larvae already present may continue their development, and adults may eventually emerge.

Latrines should be covered with fly-tight boxes, and seats should be equipped with lids hinged in such a way that they cannot be left open. If this is impossible, all feces should be well covered with earth immediately. The use of insecticides inside latrines will help to control flies and prevent them from ovipositing.

VENOMOUS ANIMALS

Many animals are more or less venomous, but relatively few are really dangerous beyond the temporary discomfort that may follow their bite or sting. However, the absolute degree of toxicity of the venom of these animals is not necessarily a measure of their danger. For instance, the venom of the honeybee is said to be more toxic, volume for volume, than the venom of a rattlesnake, but the quantity injected when the bee stings is so small that the effect is usually local. Nevertheless, children or even adults may be seriously affected by numerous stings.

Bees, Wasps, and Hornets (Hymenoptera: Apidae, Vespidae)

These insects, because of the hemolytic and histaminic character of the materials injected with the sting, must be classed with the poisonous animals. Keep a sharp watch for the nest of these insects, and avoid them. Wasps and hornets often nest in the ground; before making camp, it is best to stomp the ground to establish their nonpresence. They attack only in self-defense or in defense of their nests (Fig. 20-7).

Bees have barbed stings, and when a stinging bee is brushed off, she usually leaves the stinging organ and venom sac attached to the wound. To avoid forcing the poison into the wound, scrape off the stinging organ and venom sac with a fingernail or knife blade. Do not pick them off with the fingers or even with forceps. Wasps and hornets have smooth stingers and can therefore withdraw them for another attack.

Ammonia often relieves the pain of stings if it is applied promptly after attack. Swelling and pain resulting from severe stings can be reduced by applying ice and wet Epsom salts or baking soda packs. Do not use mudpacks; they only increase the probability of infection.

Bees and wasps can be controlled around a camp by the use of quick-knockdown insecticides.

Ants (Hymenoptera: Formicidae)

Several species of ants are vicious and are equipped with stinging organs. They may attack humans. Some of these are much feared in tropical America because they attack in force with painful or even more serious results when their nest is disturbed. Fire ants, especially the red imported fire ant, *Solenopsis invicea,* and the black imported fire ant, *S. richteri,* which now infest the South, are able to attack and kill small animals and children. Ants can bite

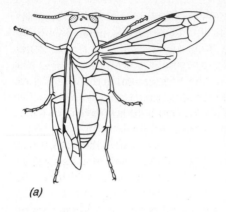

(a)

(b)

Figure 20-7 The vespid wasps (*a*) are serious venomous animals with smooth stingers and are capable of repeated attacks. These pests often nest underground. This nest (*b*) of the European hornet, *Vespa crabro,* was found in a loblolly pine stump in Virginia. *(Daly et al., An Introduction to Insect Biology and Diversity, 1978, and Virginia Polytechnic Institute and State University.)*

fiercely, and the larger species are able to puncture the human skin. Formic acid is injected with the bite, causing acute pain.

Spiders (Araneae: Theridiidae, Loxoscelidae)

The spiders are also venomous, but in practically no case are they dangerous to humans because their fangs are unable to penetrate the skin except in very thin spots. Even the notorious tarantulas seldom attempt to bite people and are usually quiet animals living in burrows in the ground. However, their bite is painful and possibly dangerous, especially those species found in the subtropics and tropics.

The black widow spider, *Latrodectus mactans,* is a medium-sized, dark-colored spider with a red hourglass-shaped marking on the underside of the abdomen of the adult female. The bite of this spider can be dangerous and might possibly result in death because it injects a neurotoxin. The bite may not be felt immediately, but a small swelling and two red spots will occur. Like other spiders, this one is not aggressive toward humans, and bites that do occur are accidental. These spiders frequent sheltered, semidark places, where they wait for insects to become entangled in their very sticky webs. A favorite place for spinning their webs is beneath the seats of outdoor privies, in telephone and electric control boxes, around piles of stones and bricks, and under flat stones. There they wait for flies. A light touch on the web will cause the spider to rush out to attack the fly that has presumably been caught in the web. Before using outdoor facilities, it is well to break away any spider webs that may be present. A small stick to be used for this purpose should be standard equipment for all outdoor toilets from the Canadian border southward.

The brown recluse spider, *Loxosceles reclusa,* is found in the central and western United States and is most commonly found in unfrequented places within the home. The bite injects a neurotoxin that destroys the skin cells. The wound may be quite large, difficult to heal, and leave a large scar.

Spider bites are serious, especially in young children and older people. The victim should be kept quiet and taken to a physician immediately.

OTHER INSECTS

Many insect families, especially the hemipteron, cause serious wounds in humans. For example, the assassin bugs (Reduviidae) can pierce human skin and inflict a painful wound when handled. Also, the giant water bugs (Belostomatidae) and the back swimmers (Notonectidae) can pierce human skin when handled.

Among the forest insects causing human dermal problems are many lepidopterous species such as the gypsy moth, *Lymantria dispar,* and the tussock moths, *Orgyia* spp. The larvae have urticating (stinging) hairs that contain toxins. The toxins are extremely irritating to the skin and eyes of many people. During and after outbreaks, these urticating hairs are a serious problem in forest areas, especially to loggers.

GENERAL RECOMMENDATIONS

It is well to remember that people have lived in almost any forest for generations, often without the information we have concerning the arthropod-borne diseases and personal care. You should understand, preferably in advance, the entomological conditions in the forest in which you expect to be, should use reasonable precautions against avoidable risks, and then should get the most from your use of the forest without worrying about the things you cannot alter.

By way of summary, the following list of recommendations is presented:

1 Become familiar with the animals and the diseases common to a new locality.
2 Obtain medical information and emergency equipment and have the equipment always on hand when in the forest.

3 Wear suitable clothing and boots constructed to provide maximum protection from arthropods and venomous animals. Avoid going barefoot, especially on ground that may be contaminated by human feces.

4 Sleep behind screens or covered with a net in any locality in which diseases transmitted by nocturnal insects are prevalent.

5 Water that may be contaminated should be boiled or, if this is impracticable, disinfected before drinking. Foods, especially vegetables and meat, should be well cooked.

6 Avoid exposure to unnecessary risks; when you are on your own there is only one opportunity to make a mistake.

With these recommendations, we close this chapter and the text. Our best wishes and good luck to you in your forestry career.

BIBLIOGRAPHY

General References

Frankie, G. W., and Ehler, L. E. 1978. Ecology of insects in urban environment. *Ann. Rev. Entomol.* 23:367–387.

Gerhardt, R. R. 1973. Public opinion on insect pest management in coastal North Carolina. *N.C. Agric. Ext. Serv. Misc. Publ.* no. 97.

Graham, S. A., and O'Roke, E. C. 1943. *On your own.* Minneapolis: University of Minnesota Press.

Gray, B. 1972. Economic tropical forest entomology. *Ann. Rev. Entomol.* 17:313–354.

Hansens, E. J. 1951. The stable fly and its effect on seashore recreational areas in New Jersey. *J. Econ. Entomol.* 44:482–487.

Linley, J. R., and Davis, J. B. 1971 Sandflies and tourism in Florida and the Bahamas and Caribbean area. *J. Econ. Entomol.* 64:264–278.

Newson, H. D. 1977. Arthropod problems in recreational areas. *Ann. Rev. Entomol.* 22:333–353.

Schrock, C. E. 1977. Techniques for the evaluation of insect repellents: A critical review. *Ann. Rev. Entomol.* 22:101–119.

Medical Entomology

Gillis, M. T. 1974. Methods for assessing the density and survival of blood-sucking diptera. *Ann. Rev. Entomol.* 19:345–362.

James, H. T., and Harwood, R. H. 1969. *Herms's medical entomology.* New York: Macmillan Publishing Company.

Jamnback, H. 1969. *Bloodsucking flies and other outdoor nuisance arthropods of New York State.* New York State Museum Memoir no. 19.

Smith, K. G. V., ed. 1973. *Insects and other arthropods of medical importance.* Somerset, N.J.: John Wiley and Sons.

U.S. Department of Agriculture. Index-catalogue of medical and veterinary zoology.

Animal Diseases

Benenson, A. S., ed. 1975. *Control of communicable diseases in man.* 12 ed. Washington, D.C.: American Public Health Association.

Davis, J. W., and Anderson R. C. 1971. *Parasitic diseases of wild animals.* Ames: Iowa State University Press.

Hopla, C. E. 1974. The ecology of tularemia. *Adv. Vet. Sci. Comp. Med.* 18:25–53.

Hubbert, W. T., et al., eds. 1975 *Diseases transmitted from animals to man.* Springfield, Ill.: Charles C Thomas, Publisher.

Schwabe, C. W. 1969. *Veterinary medicine and human health.* Baltimore, Md.: Williams & Wilkins Company.

Mosquitoes

Bates, M. 1949. *The natural history of mosquitoes.* New York: Macmillan Publishing Company.

Carpenter, S. J., and LaCasse, W. J. 1955. *Mosquitoes of North America.* Berkeley: University of California Press.

Gillett, J. D. 1972. *The mosquito: Its life, activities, and impact on human affairs.* New York: Doubleday Publishing Company.

Malaria

Harrison, G. 1978. *Mosquitoes, malaria and man.* New York: E. P. Dutton.

Wright, J. W., et al. 1972. Changing concepts of vector control in malaria eradication. *Ann Rev. Entomol.* 17:75–102.

Yellow Fever

Duffy, J. 1968. Yellow fever in the continental United States during the nineteenth century. *New York Acad. Med.* 44:687–701.

Strode, G. K., ed. 1951 *Yellow fever.* New York: McGraw-Hill Book Company.

Encephalites

McLintock, J. 1978. Mosquito-virus relationships of American encephalitides. *Ann. Rev. Entmol.* 23:39–68.

Reeves, W. C. 1965. Ecology of mosquitoes in relation to arboviruses. *Ann. Rev. Entomol.* 10:25–46.

U.S. Department of Health, Education, and Welfare. 1976. *Control of St. Louis encephalitis.* USDHEW, Public Health Serv., Center Disease Control, Bur. Trop. Dis. Vector, Biol. and Control Div. Vector Topics I.

Blackflies (Diptera: Simuliidae)

Anderson, J. 1961. Feeding behavior and host preferences of some black flies in Wisconsin. *Ann. Entomol. Soc. Amer.* 54:716–729.

Davis, D. M., and Petersen B. V. 1956. Observations on the mating, feeding, ovarian development, and oviposition by adult black flies. *Cand. J. Zool.* 34:615–655.

Jamnback, H. 1973. Recent developments in control of blackflies. *Ann. Rev. Entomol.* 18:281–304.

Stone, A., and Jamnback, H. 1955. *The black flies of New York State.* New York State Museum Bulletin no. 349.

Biting Midges (Diptera: Ceratopogonidae)

Kettle, D. 1962. Biting ceratopogonids as vectors of human and animal diseases. *Acta Tropica* 22:356–362.

———. 1977. Biology and bionomics of bloodsucking Ceratopogonids. *Ann. Rev. Entomol.* 22:35–51.

Linley, J. R. 1979. Biting midges of mangrove swamps and salt marshes. In *Marine insects,* ed. L. Cheng, pp. 335–376. Amsterdam: North Holland Publishing Company.

Horseflies and Deerflies (Diptera: Tabanidae)

Axtell, R. C. 1979. Coastal horse flies and deer flies. In *Marine insects,* ed. L. Cheng, pp. 415–445. Amsterdam: North Holland Publishing Company.

Jollison, W. L. 1950. Tularemia, geographical distribution of "deer-fly fever" and the biting fly, *Chrysops discalis. U. S. Pub. Health Repts.* 65:1321–1329.

Pechuman, L. L. 1961. The Tabanidae (Diptera) of Ontario. *Proc. Entomol. Soc. Ont.* 91:77–121.

Wingless, Bloodsucking Arthropods

Friend, W. G., and Smith, J. J. B. 1977. Factors affecting feeding by bloodsucking insects. *Ann. Rev. Entomol.* 22:309–331.

Hocking, B. 1971. Blood-sucking behavior of terrestrial arthropods. *Ann. Rev. Entomol.* 16:1–26.

Fleas (Siphonaptera: Pulicidae)

Bibikova, V. A. 1977. Contemporary views on the interrelationships between fleas and the pathogens of human and animal diseases. *Ann. Rev. Entomol.* 22:23–32.

Hirst, L. F. 1953. *The conquest of plague.* Oxford: Oxford University Press.

Langler, W. L. 1964. The black death. *Sci. Amer.* 210:114–121.

Rothschild, M. 1975. Recent advances in our knowledge of the order Siphonaptera. *Ann. Rev. Entomol.* 20:241–260.

Sucking Lice (Anoplura: Pediculidae)

Cushing, E. C. 1957. *History of entomology in World War II.* Washington, D.C.: Smithsonian Institution Press Publication no. 4294.

Ziegler, P. 1969. *The black death.* New York: Harper & Row Publishers.

Zinsser, H. 1935. *Rats, lice and history.* Boston: Little, Brown & Company.

Chiggers (Acarina: Trombiculidae)

Jones, B. M. 1950. The penetration of the host tissue by the harvest mite. *Parasitology* 40:247–260.

Sasa, M. 1961. Biology of the chiggers. *Ann. Rev. Entomol.* 6:221–244.

Wharton, G. W., and Fuller, H. S. 1952. A manual of the chiggers. Entomol. Soc. Wash., Memoir no. 4.

Williams, R. W. 1946. A contribution to our knowledge of the bionomics of the common North American chigger. *Amer. J. Trop. Med.* 26:243–250.

Bed Bugs (Hemiptera: Cimicidae)

Usinger, R. L. 1965. *Monograph of Cimicidae.* Thom. Say Found. vol. 7.

Ticks (Arcari: Ixodidae)

Gregson, J. D. 1956. The Ixodoidea of Canada. *Canada Dept. Agric. Sci. Serv. Entomol. Div. Publ.* no. 930.

Hoogstraal, H. 1967. Ticks in relation to human diseases caused by *Rickettsia* species. *Ann. Rev. Entomol.* 12:377–420.

———. 1972. Ticks in relation to human diseases caused by viruses. *Ann. Rev. Entomol.* 11:261–308.

Pratt, H. D., and Barnes, R. 1962. Ticks of public health importance and their control. *U.S. Pub. Health Serv. Publ.* 772:1–42.

Sonenshine, D. E. et al. 1966. The ecology of ticks transmitting Rocky Mountain spotted fever in a study area in Virginia. *Ann. Entomol. Soc. Amer.* 59:1234–1262.

Mechanical Vectors

House Fly (Diptera: Muscidae)

Greenberg, B. 1971. *Flies and disease, I and II.* Princeton, N. J.: Princeton University Press.

Sacca, G. 1964. Comparative bionomics in the genus *Musca. Ann. Rev. Entomol.* 9:341–358.

West, L. S. 1951. *The house fly.* Ithaca: Comstock Publishing Associates.

Venomous Animals

Human Reactions

Arnold, R. E. 1973. *What to do about bites and stings of venomous animals.* New York: Collier Books.

Feingold, B. F., et al. 1968. The allergic responses to insect bites. *Ann. Rev. Entomol.* 13:137–158.

Minton, S. A., Jr. 1974. *Venon diseases.* Springfield, Ill.: Charles C Thomas, Publisher.

Shulman, S. 1967. Allergic responses to insects. *Ann. Rev. Entomol.* 12:323–346.

Parrish, H. M. 1959. Deaths from bites and stings of venomous animals and insects in the United States. *Archives Internal Med.* 104:198–207.

Bees, Wasps, and Hornets

Akre, R. D., et al. 1974. Yellowjacket literature (Hymenoptera: Vespidae). *Melandria.* 18:67–93.
————, and Davis, H. G. 1978. Biology and pest status of venomous wasps. *Ann. Rev. Entomol.* 23:215–238.
MacDonald, J. R. 1976. Evaluation of yellow-jacket abatement. *Bull. Entomol. Soc. Am.* 22:397–401.
Michener, D. C. 1975. The Brazilian bee problem. *Ann. Rev. Entomol.* 20:399–416.
Spradbery, J. P. 1973. *Wasps: An account of the biology and natural history of social and solitary wasps.* Seattle: Universtiy of Washington Press.

Ants

Lofgren, C. S., et al. 1975. Biology and control of imported fire ants. *Ann. Rev. Entomol.* 20:1–30.

Spiders

Baerg, W. J. 1958. *The tarantula.* Lawrence: Universtiy of Kansas Press.
————. 1959. The black widow spider and five other venomous spiders in the United States. *Univ. Ark. Exp. Sta. Bull.* no. 608.
Hite, J. M., et al. 1966. Biology of the brown recluse spider. *Univ. Ark. Agric. Exp. Sta. Bull.* no. 711.
Turnbull, A. L. 1973. Ecology of the true spiders (Aranemorphae). *Ann. Rev. Entomol.* 18:305–345.

Other Insects

Goldman, L. 1960. Investigative studies of skin irritations from caterpillars. *J. Invest. Dermatol.* 34:67–79.
Perlman, F., et al. 1977. Health hazards to timber and forestry workers from the Douglas-fir tussock moth. *Archives Environ. Health* 32:206–210.

Appendix One

Common and Scientific Names of Insects[1]

I.   Order Orthoptera: grasshoppers, walkingsticks

Family Phasmatidae
  walkingstick, *Diapheromera femorata* (Say)

Family Acrididae
  post-oak locust, *Dendrotettix quercus* Packard
  migratory grasshopper, *Melanoplus sanguinipes* (Fabricius)

II.   Order Isoptera: termites

Family Kalotermitidae
  western drywood termite, *Incisitermes minor* (Hagen)

Family Rhinotermitidae
  Formosan subterranean termite, *Coptotermes formosanus* Shiraki
  eastern subterranean termite, *Reticulitermes flavipes* (Kollar)

[1]The majority of these names are in *Common Names of Insects and Related Organisms* (1978 revision), Entomological Society of America, College Park, Md. The remainder are from *Eastern Forest Insects* (Baxter, 1972), *USDA Forest Service Misc. Publ.* 1175; and *Western Forest Insects* (Furniss and Carolin, 1977), *USDA Forest Service Misc. Publ.* 1339.

III.   Order Anoplura: sucking lice

Family Pediculidae
  head louse, *Pediculus humanus capitis* DeGeer
  body louse, *Pediculus humanus humanus* Linnaeus
  crab louse, *Pthirus pubis* (Linnaeus)

IV.   Order Thysanoptera: thrips

Family Phlaeothripidae
  slash pine flower thrips, *Gnophothrips fuscus* (Morgan)

V.   Order Hemiptera: true bugs

Family Cimicidae
  bed bug, *Cimex lectularius* Linnaeus

Family Coreidae
  leaffooted pine seed bug, *Leptoglossus corculus* (Say)
  western conifer seed bug, *Leptoglossus occidentalis* Heidemann

Family Miridae
  tarnished plant bug, *Lygus lineolaris* (Palisot de Beauvois)

Family Pentatomidae
  spined soldier bug, *Podisus maculiventris* (Say)
  shieldbacked pine seedbug, *Tetrya bipunctata* (Herrich-Schaeffer)

Family Reduviidae
  wheel bug, *Arilus cristatus* (Linnaeus)

Family Rhopalidae
  boxelder bug, *Leptocoris trivittatus* (Say)

Family Tingidae
  sycamore lace bug, *Corythucha ciliata* (Say)
  elm lace bug, *Corythucha ulmi* Osborn and Drake

VI.   Order Homoptera: aphids, scales, spittlebugs

Family Aphididae
  balsam twig aphid, *Mindarus abietinus* Koch

Family Cercopidae
  pine spittlebug, *Aphrophora parallela* (Say)
  Saratoga spittlebug, *Aphrophora saratogensis* (Fitch)

Family Cicadidae
  periodical cicada, *Magicicada septendecim* (Linnaeus)

Family Coccidae
  cottony maple scale, *Pulvinaria innumerabilis* (Rathvon)
  pine tortoise scale, *Toumeyella parvicornis* (Cockerell)

Family Diaspididae
  pine needle scale, *Chionaspis pinifoliae* (Fitch)
  oystershell scale, *Lepidosaphes ulmi* (Linnaeus)
  San Jose scale, *Quadraspidiotus perniciosus* (Comstock)

Family Eriococcidae
  beech scale, *Cryptococcus fagisuga* Lindinger

European elm scale, *Gossyparia spuria* (Modeer)

Family Margarodidae
cottonycushion scale, *Icerya purchasi* (Maskell)
pine twig gall scale, *Matsucoccus gallicola* Morrison
red pine scale, *Matsucoccus resinosae* Bean and Goodwin
Prescott scale, *Matsucoccus vexillorum* Morrison

Family Phylloxeridae
eastern spruce gall adelgid, *Adelges abietis* (Linnaeus)
Cooley spruce gall adelgid, *Adelges cooleyi* (Gillette)
balsam woolly adelgid, *Adelges piceae* (Ratzeburg)

Family Pseudococcidae
spruce mealybug, *Puto sandini* Washburn

VII. Order Lepidoptera: moths and butterflies

Family Arctiidae
fall webworm, *Hyphantria cunea* (Drury)
silverspotted tussock moth, *Halisidota argentata* Packard
spotted tussock moth, *Halisidota maculata* (Harris)

Family Coleophoridae
larch casebearer, *Coleophora laricella* (Hübner)

Family Cossidae
carpenterworm, *Prionoxystus robiniae* (Peck)

Family Danaidae
monarch butterfly, *Danaus plexippus* (Linnaeus)

Family Dioptidae
California oakworm, *Phryganida californica* Packard

Family Gelechiidae
lodgepole needleminer, *Coleotechnites milleri* (Busck)

Family Geometridae
fall cankerworm, *Alsophila pometaria* (Harris)
elm spanworm, *Ennomos subsignarius* (Hübner)
hemlock looper, *Lambdina fiscellaria fiscellaria* (Guénee)
pine looper, *Phaeoura mexicanaria* (Groté)
western hemlock looper. *Lambdina fiscellaria lugubrosa* (Hulst)
winter moth, *Operophtera brumata* (Linnaeus)
spring cankerworm, *Paleacrita vernata* (Peck)
spearmarked black moth, *Rheumaptera hastata* (Linnaeus)

Family Gracillariidae
aspen blotchminer, *Phyllonorycter tremuloidiella* (Braun)

Family Lasiocampidae
eastern tent caterpillar, *Malacosoma americanum* (Fabricius)
western tent caterpillar, *Malacosoma californicum* (Packard)
Pacific tent caterpillar, *Malacosoma constrictum* (Henry Edwards)
forest tent caterpillar, *Malacosoma disstria* Hübner
southwestern tent caterpillar, *Malacosoma incurvum* (Henry Edwards)

Family Lymantriidae
pine tussock moth, *Dasychira pinicola* (Dyar)
browntail moth, *Euproctis chrysorrhoea* (Linnaeus)
satin moth, *Leucoma salicis* (Linnaeus)
gypsy moth, *Lymantria dispar* (Linnaeus)
whitemarked tussock moth, *Orgyia leucostigma* (J. E. Smith)
Douglas-fir tussock moth, *Orgyia pseudotsugata* (McDonnough)

Family Noctuidae
darksided cutworm, *Euxoa messoria* (Harris)

Family Notodontidae
saddled prominent, *Heterocampa guttivitta* (Walker)
variable oakleaf caterpillar, *Heterocampa manteo* (Doubleday)
redhumped oakworm, *Symerista canicosta* Franclemont
orangehumped mapleworm, *Symerista leucitys* Franclemont

Family Nymphalidae
viceroy butterfly, *Limenitis archippus* (Cramer)
mourningcloak butterfly, *Nymphalis antiopa* (Linnaeus)

Family Olethreutidae
white fir needleminer, *Epinotia meritana* Heinrich
shortleaf pine cone borer, *Eucosma cocana* Kearfott
eastern pine shoot borer, *Eucosma gloriola* Heinrich
western pine shoot borer, *Eucosma sonomana* Kearfott
western pine cone borer, *Eucosma tocullionana* Heinrich
cottonwood twig borer, *Gypsonoma haimbachiana* (Kearfott)
slash pine seedworm, *Laspeyresia anaranjada* Miller
hickory shuckworm, *Laspeyresia caryana* (Fitch)
longleaf pine seedworm, *Laspeyresia ingens* Heinrich
codling moth, *Laspeyresia pomonella* (Linnaeus)
eastern pine seedworm, *Laspeyresia toreuta* (Groté)
northern pitch twig moth, *Petrova albicapitana* (Busck)
pitch twig moth, *Petrova comstockiana* (Fernald)
European pine shoot moth, *Rhyacionia buoliana* (Schiffermüller)
western pine tip moth, *Rhyacionia bushnelli* (Busck)
Nantucket pine tip moth, *Rhyacionia frustrana* (Comstock)
southwestern pine tip moth, *Rhyacionia neomexicana* (Dyar)
pitch pine tip moth, *Rhyacionia rigidana* (Fernald)
subtropical pine tip moth, *Rhyacionia subtropica* Miller

Family Pieridae
pine butterfly, *Neophasia menapia* (Felder and Felder)

Family Psychidae
bagworm, *Thyridopteryx ephemeraeformis* (Haworth)

Family Pyralidae
southern pine coneworm, *Dioryctria amatella* (Hulst)
Zimmerman pine moth, *Dioryctria zimmermani* (Groté)

Family Saturniidae
    orangestriped oakworm, *Anisota senatoria* (J. E. Smith)
    pandora moth, *Coloradia pandora* Blake
    greenstriped mapleworm, *Dryocampa rubicunda* (Fabricius)

Family Sesiidae
    ash borer/lilac borer, *Podosesia syringae* (Harris)
    hornet moth, *Sesia apiformis* (Clerck)
    Douglas-fir pitch moth, *Synanthedon novaroensis* (Hy. Edwards)
    lesser peachtree borer, *Synanthedon pictipes* (Groté and Robinson)
    pitch mass borer, *Synanthedon pini* (Kellicott)
    rhododendron borer, *Synanthedon rhododendri* Beutenmüller
    dogwood borer, *Synanthedon scitula* (Harris)
    sequoia pitch moth, *Synanthedon sequoiae* (Hy. Edwards)

Family Sphingidae
    catalpa sphinx, *Ceratomia catalpae* (Boisduval)

Family Tortricidae
    western blackheaded budworm, *Acleris gloverana* (Walsingham)
    fruittree leafroller, *Archips argyrospilus* (Walker)
    oak leafroller, *Archips semiferanus* (Walker)
    large aspen tortrix, *Choristoneura conflictana* (Walker)
    spruce budworm, *Choristoneura fumiferana* (Clemens)
    western spruce budworm, *Choristoneura occidentalis* Freeman
    jack pine budworm, *Choristoneura pinus* Freeman
    obliquebanded leafroller, *Choristoneura rosaceana* (Harris)

Family Yponomeutidae
    arborvitae leafminer, *Argyresthia thuiella* (Packard)

VIII. Order Coleoptera: beetles

Family Anobiidae
    furniture beetle, *Anobium punctatum* (DeGeer)
    Pacific powderpost beetle, *Hemicoelus gibbicollis* (LeConte)
    death watch beetle, *Xestobium rufovillosum* (DeGeer)
    drugstore beetle, *Stegobium paniceum* (Linnaeus)

Family Bostrichidae
    leadcable borer, *Scobicia declivis* (LeConte)

Family Buprestidae
    Pacific oak twig girdler, *Agrilus angelicus* Horn
    bronze birch borer, *Agrilus anxius* Gory
    hickory spiral borer, *Agrilus arcuatus* var. *torquatus* LeConte
    twolined chestnut borer, *Agrilus bilineatus* (Weber)
    bronze poplar borer, *Agrilus liragus* Barter and Brown
    turpentine borer, *Buprestis apricans* Herbst
    golden buprestid, *Buprestis aurulenta* Linnaeus
    sculptured pine borer, *Chalcophora angulicollis* (LeConte)
    flatheaded appletree borer, *Chrysobothris femorata* (Olivier)

Pacific flatheaded borer, *Chrysobothris mali* Horn
Australianpine borer, *Chrysobothris tranquebarica* (Gmelin)
California flatheaded borer, *Melanophila californica* Van Dyke
flatheaded fir borer, *Melanophila drummondi* (Kirby)
hemlock borer, *Melanophila fulvoguttata* (Harris)
western cedar borer, *Trachykele blondeli* Marseul

Family Carabidae
fiery hunter, *Calosoma calidum* (Fabricius)

Family Cerambycidae
twig pruner, *Elaphidionoides villosus* (Fabricius)
red oak borer, *Enaphalodes rufulus* (Haldeman)
sugar maple borer, *Glycobius speciosus* (Say)
white oak borer, *Goes tigrinus* (DeGeer)
old house borer, *Hylotrupes bajulus* (Linnaeus)
locust borer, *Megacyllene robiniae* (Forster)
Oregon fir sawyer, *Monochamus oregonensis* (LeConte)
whitespotted sawyer, *Monochamus scutellatus* (Say)
southern pine sawyer, *Monochamus titillator* (Fabricius)
pole borer, *Parandra brunnea* (Fabricius)
poplar borer, *Saperda calcarata* Say
linden borer, *Saperda vestita* Say
firtree borer, *Semanotus litigiosus* (Casey)
western larch borer, *Tetropium velutinum* LeConte

Family Chrysomelidae
alder flea beetle, *Altica ambiens* LeConte
cottonwood leaf beetle, *Chrysomela scripta* Fabricius
Colorado potato beetle, *Leptinotarsa decemlineata* (Say)
locust leafminer, *Odontota dorsalis* (Thunberg)
elm leaf beetle, *Pyrrhalta luteola* (Müller)

Family Cleridae
blackbellied clerid, *Enoclerus lecontei* (Wolcott)
redbellied clerid, *Enoclerus sphegeus* (Fabricius)

Family Coccinellidae
vedalia, *Rodolia cardinalis* (Mulsant)

Family Curculionidae
black walnut curculio, *Conotrachelus retentus* Say
poplar-and-willow borer, *Cryptorhynchus lapathi* (Linnaeus)
pecan weevil, *Curculio caryae* (Horn)
filbert weevil, *Curculio occidentis* (Casey)
pine reproduction weevil, *Cylindrocopturus eatoni* Buchanan
Douglis-fir twig weevil, *Cylindrocopturus furnissi* Buchanan
pales weevil, *Hylobius pales* (Herbst)
pine root collar weevil, *Hylobius radicis* Buchanan
strawberry root weevil, *Otiorhynchus ovatus* (Linnaeus)
black vine weevil, *Otiorhynchus sulcatus* (Fabricius)
pitch-eating weevil, *Pachylobius picivorus* (Germar)

northern pine weevil, *Pissodes approximatus* Hopkins
deodar weevil, *Pissodes nemorensis* Germar
Monterey pine weevil, *Pissodes radiatae* Hopkins
Yosemite bark weevil, *Pissodes schwarzi* Hopkins
white pine weevil (Sitka spruce weevil, Engelmann spruce weevil), *Pissodes strobi* (Peck)
lodgepole terminal weevil, *Pissodes terminalis* Hopping
willow flea weevil, *Rhynchaenus rufipes* (LeConte)

Family Lyctidae
southern lyctus beetle, *Lyctus planicollis* LeConte

Family Scarabaeidae
pine chafer, *Anomala oblivia* Horn
Japanese beetle, *Popillia japonica* Newman

Family Scolytidae
red pine cone beetle, *Conophthorus resinosae* Hopkins
sugarpine cone beetle, *Conopthorus lambertianae* Hopkins
white pine cone beetle, *Conopthorus coniperda* (Schwarz)
Columbian timber beetle, *Corthylus columbianus* Hopkins
pitted ambrosia beetle, *Corthylus punctatissimus* (Zimmerman)
roundheaded pine beetle, *Dendroctonus adjunctus* Blandford
Mexican pine beetle, *Dendroctonus approximatus* Dietz
western pine beetle, *Dendroctonus brevicomis* LeConte
southern pine beetle, *Dendroctonus frontalis* Zimmerman
Jeffrey pine beetle, *Dendroctonus jeffreyi* Hopkins
European spruce beetle, *Dendroctonus micans* (Kugelann)
lodgepole pine beetle, *Dendroctonus murrayanae* Hopkins
mountain pine beetle, *Dendroctonus ponderosae* Hopkins
Douglas-fir beetle, *Dendroctonus pseudotsugae* Hopkins
Allegheny spruce beetle, *Dendroctonus punctatus* LeConte
spruce beetle, *Dendroctonus rufipennis* (Kirby)
eastern larch beetle, *Dendroctonus simplex* LeConte
black turpentine beetle, *Dendroctonus terebrans* (Olivier)
red turpentine beetle, *Dendroctonus valens* LeConte
native elm bark beetle, *Hylurgopinus rufipes* (Eichhoff)
small southern pine engraver, *Ips avulsus* (Eichhoff)
California fivespined ips, *Ips paraconfusus* Lanier
pine engraver, *Ips pini* (Say)
balsam fir bark beetle, *Pityokteines sparsus* (LeConte)
silver fir beetle, *Pseudohylesinus sericeus* (Mannerheim)
hickory bark beetle, *Scolytus quadrispinosus* Say
smaller European elm bark beetle, *Scolytus multistriatus* (Marshaw)
fir engraver, *Scolytus ventralis* LeConte
striped ambrosia beetle, *Trypodendron lineatum* (Olivier)

IX. Order Hymenoptera: sawflies, wood wasps, bees, ants

Family Cynipidae
mossyrose gallwasp (woody oak gall), *Diplolepis rosae* (Linnaeus)

Family Diprionidae
  nursery pine sawfly, *Diprion frutetorum* (Fabricius)
  introduced pine sawfly, *Diprion similis* (Hartig)
  European spruce sawfly, *Gilpinia hercyniae* (Hartig)
  arborvitae sawfly, *Monoctenus juniperinus* MacGillvray
  eastern cedar sawfly, *Monoctenus melliceps* (Cresson)
  balsam fir sawfly, *Neodiprion abietis* (Harris)
  lodgepole sawfly, *Neodiprion burkei* Middleton
  brownheaded jack pine sawfly, *Neodiprion dubiosus* Schedl
  piñon sawfly, *Neodiprion edulicolus* Ross
  redheaded pine sawfly, *Neodiprion lecontei* (Fitch)
  red pine sawfly, *Neodiprion nanulus nanulus* Schedl
  white pine sawfly, *Neodiprion pinetum* (Norton)
  pitch pine sawfly, *Neodiprion pinirigidae* (Norton)
  jack pine sawfly, *Neodiprion pratti banksianae* Rohwer
  redheaded jack pine sawfly, *Neodiprion rugifrons* Middleton
  European pine sawfly, *Neodiprion sertifer* (Geoffroy)
  Swaine jack pine sawfly, *Neodiprion swainei* Middleton
  loblolly pine sawfly, *Neodiprion taedae linearis* Ross
  hemlock sawfly, *Neodiprion tsugae* Middleton

Family Formicidae
  Texas leafcutting ant, *Atta texana* (Buckley)
  black carpenter ant, *Camponotus pennsylvanicus* (DeGeer)
  red imported fire ant, *Solenopsis invicta* Buren
  black imported fire ant, *Solenopsis richteri* Forel

Family Pamphiliidae
  Monterey pine sawfly, *Acantholyda burkei* Middlekauff
  pine false webworm, *Acantholyda erythrocephala* (Linnaeus)
  nesting-pine sawfly, *Acantholyda zappei* (Rohwer)

Family Siricidae
  blue horntail, *Sirex cyaneus* Fabricius
  pigeon tremex, *Tremex columba* (Linnaeus)

Family Tenthredinidae
  twolined larch sawfly, *Anoplonyx laricivorus* (Rohwer and Middleton)
  western larch sawfly, *Anoplonyx occidens* Ross
  pear sawfly, *Caliroa cerasi* (Linnaeus)
  maple petiole borer, *Caulocampus acericaulis* (MacGillivray)
  European alder leafminer, *Fenusa dohrnii* (Tischbein)
  birch leafminer, *Fenusa pusilla* (Lepeletier)
  elm leafminer, *Fenusa ulmi* Sundevall
  birch leaf-mining sawfly, *Heterarthus nemoratus* (Fall.)
  yellowheaded spruce sawfly, *Pikonema alaskensis* (Rohwer)
  greenheaded spruce sawfly, *Pikonema dimmockii* (Cresson)
  larch sawfly, *Pristiphora erichsonii* (Hartig)
  mountain-ash sawfly, *Pristiphora geniculata* (Hartig)

Family Torymidae
  balsam fir seed chalcid, *Megastigmus specularis* Walley
  Douglas-fir seed chalcid, *Megastigmus spermotrophus* Wachtl

Family Xylocopidae (= Apidae)
  carpenter bee, *Xylocopia virginica* (Linnaeus)

X.   Order Diptera: flies

Family Agromyzidae
  native holly leafminer, *Phytomyza ilicicola* Loew
  holly leafminer, *Phytomyza ilicis* Curtis

Family Calliphoridae
  screwworm, *Cochliomyia hominivorax* (Coquerel)

Family Cecidomyiidae
  Douglas-fir cone midge, *Contarinia oregonensis* Foote
  boxwood leafminer, *Monarthropalpus buxi* (Laboulbène)
  juniper tip midge, *Oligotrophus betheli* Felt
  balsam gall midge, *Paradiplosis tumifex* Gagné
  Monterey pine midge, *Thecodiplosis piniradiatae* (Snow and Mills)
  red pine needle midge, *Thecodiplosis piniresinosae* Kearby

Family Culicidae
  yellowfever mosquito, *Aedes aegypti* (Linnaeus)
  saltmarsh mosquito, *Aedes sollicitans* (Walker)
  common malaria mosquito, *Anopheles quadrimaculatus* Say
  northern house mosquito, *Culex pipiens* Linnaeus
  southern house mosquito, *Culex quinquefasciatus* Say

Family Cuterebridae
  torsalo, *Dermatobia hominis* (Linnaeus, Jr.)

Family Muscidae
  house fly, *Musca domestica* (Linnaeus)
  stable fly, *Stomoxys calcitrans* (Linnaeus)

Family Tephritidae
  walnut husk fly, *Rhagoletis completa* Cresson

XI.   Order Siphonaptera: fleas

Family Pulicidae
  human flea, *Pulex irritans* Linnaeus
  dog flea, *Ctenocephalides canis* (Curtis)
  cat flea, *Ctenocephalides felis* (Bouché)
  oriental rat flea, *Xenopsylla cheopis* (Rothchild)

Glossary

The terms and definitions in this glossary are the basic entomological terms used in the text. The forestry terminology is based on the Society of American Foresters' *Terminology of Forest Science, Technology, Practice and Products,* edited by F. C. Ford-Robertson, 1971 (SAF Multilingual Forest Terminology Series no. 1). Students needing more entomological definitions should consult J. R. DeLa Torre-Bueno, *A Glossary of Entomology,* New York Entomological Society, New York, 1973.

Abdomen Posterior of the three main body divisions of an insect; *see also* Head; Thorax.

Adult, callow Newly molted adult insect; exoskeleton has not hardened; often pale in color.

Adult, parent Sexually mature insect that has produced young.

Aedeagus Male insect's intromittent organ; penis.

Aestivation Dormancy resulting from increased temperatures; *cf.* Hibernation.

Allopatric Inhabiting geographically separate areas; *cf.* Sympatric.

Ametabolous Without metamorphosis; *cf.* Hemimetabolous.

Anal Posterior basal part; last abdominal segment.

Antenna (*pl.,* antennae) Appendages located on an insect's head; usually two, located above the eyes.

Anterior Frontward; in front of; *cf.* Posterior.

Anus Posterior opening of alimentary tract.

Apical Toward the outside; outermost part; tip end.

Apodous Without legs.

Apterous Wingless.

Aquatic Living in water.

Arachnida Arthropod class having four pairs of legs—such as spiders and mites.

Arthropoda Animal phylum including such classes as crustaceans, spiders, and insects.

Asexual Reproduction by a single individual (parthenogenesis); *cf.* Bisexual.

Asymmetrical Not alike on opposite sides.

Atrophy Reduce in size; waste away.

Autogenous Producing eggs without a blood meal.

Balance of nature When the reproductive potential of plant and animal populations is equal to the environmental resistance; the population densities of the species are neither infinitely increasing nor decreasing.

Basal Toward the base; point of attachment.

Biological control Living component of environmental resistance, especially microorganisms and animals such as disease-causing organisms, parasites, and predators.

Biological pest suppression Use by humans of biological-control organisms to reduce populations of plant or animal pests.

Biotype Biological strain of a species which is morphologically indistinguishable but has distinct physiological differences.

Bisexual Reproduction involving two separate sexes, male and female; *cf.* Asexual.

Bivoltine Producing two generations per year; *cf.* Multivoltine, Univoltine.

Brood Individuals produced by one female which hatch and mature synchronously.

Carnivorous Feeding on animals.

Caste Specialized form of a species, usually in social insects.

Caudal Pertaining to posterior bodily locations.

Cephalothorax Combined head and thorax regions of arthropods, such as spiders and crustaceans.

Chelicera Mouthparts in such arachnids as spiders.

Class Group of related orders; subdivision of a phylum, such as phylum Arthropoda, class Insecta.

Cocoon Silken case spun by a larva, in which it pupates.

Compound eye Eye composed of numerous individual elements, each with an external facet; *see also* Ocellus.

Crawler The mobile first instar of scale insects.

Crepuscular Active at dusk or twilight; *cf.* Diurnal, Nocturnal.

Cryptic Secret; concealed.

Diapause State of arrested development or movement.

Diurnal Active during the day; *cf.* Crepuscular, Nocturnal.

Dormancy Inactive, quiescent condition.

Dorsum The top, back, or upper side.

Ecdysis Molting; shedding of exoskeleton.

Eclosion   Emergence from the egg.

Economic threshold   Density of an insect species that will cause economic damage unless action is taken.

Ectoparasite   Parasite that lives outside its host; *cf.* Endoparasite.

Encephalitis   Disease of animals; inflamation of the brain.

Endemic   Low level of incidence; normal condition; species native to a region; *cf.* Epidemic.

Endoparasite   Parasite that lives inside its host; *cf.* Ectoparasite.

Endoskeleton   Supporting structure within the body; *cf.* Exoskeleton.

Entomophagous   Insect-eating; insectivorous.

Enzootic   Having a constantly low incidence of a disease (or pest) condition.

Epidemic   High level of incidence; abnormal condition; *cf.* Endemic.

Epizootic   Having an unusually high density and incidence of a disease (or pest) condition.

Exoskeleton   Supporting structure on the outside of the body.

Exotic   Species introduced into an area; not native; *cf.* Indigenous.

Exuviae   Cast skins of nymphs and larvae.

Facultative parasite   Species that can develop without its host; *cf.* Obligate parasite.

Family   Subdivision of an order having related genera.

Frass   Insect excrement.

Gall   Abnormal growth of plant tissue caused by an external stimulus.

Generation   Time span from a given stage of the life cycle to the same stage in the offspring, usually n adult to $n+1$ adult.

Genitalia   Sexual organs.

Genus   Group of closely related species; the name is Latinized and the first letter is capitalized. When written, both the genus and the species are italicized or underlined; *see also* Scientific name.

Glabrous   Smooth; not hairy.

Gregarious   Together; in groups.

Grub   Larva of the Coleoptera, usually Scarabaeidae.

Haltere   Reduced hindwings of Diptera.

Head   Anterior body region of an insect; *see also* Abdomen, Thorax.

Hemimetabolous   With incomplete or gradual metamorphosis; *cf.* Ametabolous, Holometabolous.

Hermaphrodite   Individual having both male and female reproductive organs.

Hexapoda   Arthropoda class having three pairs of legs: Insecta.

Hibernation   Dormancy resulting from reduced temperatures; *cf.* Aestivation.

Holometabolous   With complete metamorphosis; *cf.* Ametabolous, Hemimetabolous.

Honeydew   Liquid anal discharge of some homopteran insects.

Host   Organism on (or in) which an animal feeds.

Host, alternate   Second organism suitable for the development of a species, either plant or animal.

Hyperparasite   Parasite that parasitizes a parasite.

Imago   Adult insect.

Indigenous   Species native to an area; *cf.* Exotic.

Infestation   Insect outbreak widely and evenly distributed over a given area.

Insect Air-breathing Arthropod, usually having, when an adult, three body regions, three pairs of legs, and two pairs of wings.

Insect, primary Organism capable of completing development in an apparently healthy, living tree.

Insect, secondary Organism found generally in an unhealthy or dead tree.

Insecticide Material that kills insects.

Insecticide, contact Material effective when penetrating or coating body wall.

Insecticide, soil Material effective when applied to the soil, either to kill insects or to prevent insect habitation.

Insecticide, stomach Material effective when taken into the alimentary tract.

Insecticide, systemic Material absorbed into the internal system of a plant or animal.

Instar Developmental stage of the insect between successive molts; *see also* Stadium.

Integrated pest management Compatible, systematic, and intelligent use of all available forms of control (silvicultural, biological, and chemical) to achieve safe reduction of a pest problem.

Intolerance Lack of capacity to develop, such as failure to grow in shaded areas, polluted air, or droughty soils; *cf.* Tolerance.

Introduction Transfer of a plant or animal from its indigenous area by either design or accident; *see also* Exotic.

Iso Same; a line drawn; used as a prefix.

Isopleth Line drawn through equal places or points.

Key factor Environmental resistance factor most responsible for change in the density of an insect population from one generation to the next.

Kinesis Nondirected movement of an organism resulting from a stimulus; *cf.* Taxis.

Larva (*pl.*, larvae; *adj.*, larval) Immature stage of a holometobolous insect, between the egg and the pupal stage.

Lateral On or pertaining to the side, right or left.

Life table Age-specific distribution and effect of specific mortality causes, expressing the changing density of the affected population in time for a given place.

Maggot Larva of the Diptera.

Mandible Jaw.

Meson Theoretical midline dividing an insect's body longitudinally, left and right.

Metamorphosis Change in an insect's form during development from egg to adult.

Mimicry Resemblance of one organism to another.

Molt Process by which an insect sheds its exoskeleton; also termed *ecdysis.*

Monophagous Feeding or using only one species; *cf.* Oligophagous, Polyphagous.

Morphology Science of form and structure.

Multivoltine Producing two or more generations annually; *cf.* Bivoltine, Univoltine.

Myiasis Disease of humans caused by invasion of dipterous larvae (maggots).

Naiad Aquatic nymph.

Neoteny Sexual maturity during an insect's larval stage.

Niche Ecological position of an organism in relation to its habitat.

Nocturnal Active during the night; *cf.* Crepuscular, Diurnal.

Nuptial chamber Place where mating occurs, especially with Coleoptera: Scolytidae.

Nuptial flight Dispersal and mating flight, especially in Hymenoptera and Isopoda.

Nymph Immature hemimetobolous insect.

Obligate parasite Species that cannot develop away from its host; *cf.* Facultative parasite.

Ocellus (*pl.,* ocelli) Simple eye of adult insect, usually three; *see also* Compound eye.

Oligophagous Using a limited number of hosts; *cf.* Monophagous, Polyphagous.

Omnivorous Having variable feeding habits.

Oral Pertaining to the mouth.

Order Subdivision of a class, containing a group of related families.

Outbreak, insect Sudden increase in the destructiveness of an insect species (*syn.,* epidemic).

Ovary Egg-producing organ in the female.

Oviparous Egg-laying; *cf.* Viviparous.

Oviposit To lay eggs.

Ovipositor Egg-laying apparatus; external genitalia.

Parasite Animal species that lives on or in a larger animal host, frequently destroying it; *cf.* Predator.

Parthenogenesis Reproduction with fertilization by a single individual.

Phenology Time of appearance of characteristic events in the life cycle of organisms, especially regarding weather, such as when tree buds expand and insect larvae emerge.

Phytophagous Feeding on plants.

Polyembryony Production of more than one individual from a single egg.

Polymorphism More than one form.

Polyphagous Having broad feeding tolerances; *cf.* Monophagous, Oligophagous.

Population Sum of all interacting units or organisms of one kind.

Posterior Hindward or rearward; in hind of; *cf.* Anterior.

Predaceous Being a predator.

Predator Animal species that lives upon smaller animals (prey), frequently eating them rapidly and completely; *cf.* Parasite.

Prehensile Adapted for grasping.

Primary parasite Parasite of an organism that is not itself parasitic.

Pubescent Covered with short, dense hairs.

Pupa (*adj.,* pupal; *pl.,* pupae; *v.,* pupate) In holometabolous insects, the instar during which transformation from larva to adult occurs.

Puparium Pupal case in the Diptera, formed from exoskeleton of last instar.

Queen Primary female reproductive, usually in social insects, such as bees and termites.

Range Geographical area of occurrence of plants or animals.

Repellent Substance used to prevent animal feeding, resting, or oviposition.

Rudimentary Underdeveloped.

Saprophagous Feeding on decaying organic matter.

Saprophyte Organism that obtains food from dead organic matter.

Scavenger Animal—such as a termite—that feeds on dead, decaying, or waste organic matter.

Scientific name Internationally recognized italicized, Latinized name of a species; has generic and specific names and author; example, *Dendroctonus frontalis* (Zimm.).

Semiaquatic Living partially in water or wet places.

Solitary   Singly or in pairs, not social.

Species   Group of individuals similar in structure and physiology and capable of producing fertile offspring.

Spine   Rigid outgrowth of the insect exoskeleton.

Spiracle   External opening of tracheal system.

Spittle   Frothy secretion of immature spittlebugs.

Stadium   Time period between molts; *see also* Instar.

Sting   Modified ovipositor adapted for injecting venom, such as in wasps.

Stridulation   Production of sound by rubbing one body surface against another.

Stylet   Piercing mouthparts of sucking insects.

Swarm   Simultaneous emergence or grouping of large numbers of one insect species.

Symbiosis   Living together of two organisms, usually for mutual benefit.

Sympatric   Inhabiting the same geographic area; *cf.* Allopatric.

Systematics   Taxonomy.

Taxis   Direct response, with movement of an organism toward (positive) or away from (negative) a stimulus; *cf.* Kinesis.

Taxonomy   Classification into ranked categories that are described and named.

Terrestrial   Living on land.

Thorax   Midbody division of an insect; *see also* Head, Abdomen.

Tolerance   (1) General term for the relative ability of an organism to endure an adverse environmental influence; (2) capacity of a tree to develop and grow in the shade of and in competition with other trees.

Trachea   Tube of respiratory system; *see also* Spiracle.

Tropism   Orientation toward (positive) or away from (negative) a stimulus; *cf.* Taxis.

Tympanum   Auditory membrane.

Univoltine   Producing one generation per year; *cf.* Bivoltine, Multivoltine.

Vector   Organism that conveys, usually a disease-causing organism.

Ventral   Pertaining to the underside.

Viviparous   Producing living young—not egg-laying; *cf.* Oviparous.

Wheal   Rise in the skin.

Zoophagous   Feeding on animals.

Index

Note: Common names are used throughout the index for those insects listed in Appendix One.